Recent Progress in Desalination, Environmental and Marine Outfall Systems

Mahad Baawain • B.S. Choudri
Mushtaque Ahmed • Anton Purnama
Editors

Recent Progress in Desalination, Environmental and Marine Outfall Systems

Springer

Editors
Mahad Baawain
Center for Environmental Studies and Research (CESAR)
Sultan Qaboos University
Muscat, Oman

B.S. Choudri
Center for Environmental Studies and Research (CESAR)
Sultan Qaboos University
Muscat, Oman

Mushtaque Ahmed
College of Agricultural and Marine Sciences
Sultan Qaboos University
Muscat, Oman

Anton Purnama
College of Science
Sultan Qaboos University
Muscat, Oman

ISBN 978-3-319-19122-5 ISBN 978-3-319-19123-2 (eBook)
DOI 10.1007/978-3-319-19123-2

Library of Congress Control Number: 2015949387

Springer Cham Heidelberg New York Dordrecht London

Printed on acid-free paper

Springer International Publishing AG Switzerland is part of Springer Science+Business Media (www.springer.com)

Preface

This special issue of edited book series of Springer Proceedings in Environmental Science contains 22 selected papers presented at the International Conference on "Desalination, Environment and Marine Outfall Systems" held during April 13–16, 2014 at Sultan Qaboos University, Muscat, Sultanate of Oman. The conference was organized by the Center for Environmental Studies and Research (CESAR) of Sultan Qaboos University (SQU) in association with the International Water Association (IWA) and International Association for Hydro-Environment Engineering and Research (IAHR) and supported by the Research Council (TRC) of Oman. The conference, focused on desalination industry, was held under the patronage of the Undersecretary of the Ministry of Environment and Climate Affairs, Sultanate of Oman.

The conference included presentations about desalination; brine and industrial discharges; resource recovery and carbon emission from desalination plants; nanotechnology and desalination; environmental costs of desalination; desalination and renewable energy source; harmful algal bloom and threat to seawater desalination plants; advances in desalination technology to mitigate environmental impacts; environmental impact assessment for desalination plants; intake and outfall systems for cooling water and wastewater treatment plants; and experimental and computational techniques. Speakers from Europe, the USA, Australia, South Asia, Middle East including Gulf Cooperation Countries (GCC) presented papers about their work related to desalination industry and relevant issues. Overall, the conference saw participation of over 150 people from 20 countries, in eleven technical sessions, with 58 papers presented. Many of the presentations were focused on the studies conducted in the Middle East including Sultanate of Oman and United Arab Emirates.

This book highlights current scientific information on advanced technologies and management practices associated with the desalination industry in the Middle East and elsewhere around the world. The book opens with an introductory chapter, which briefly gives an overview of the desalination technology, and describes the current state of development in the field. Further, the book focuses on recent developments in desalination technologies, which are specifically aimed at reducing energy consumption and cost, and minimizing environmental impact. We hope that the book will be of benefit to all associated with the desalination industry.

Muscat, Sultanate of Oman

Mahad Baawain
B.S. Choudri
Mushtaque Ahmed
Anton Purnama

Acknowledgements

We are grateful to the authors of the book chapters for sharing their expertise and providing scientific information on advanced technologies and management practices associated with the desalination industry in the Middle East and elsewhere around the world.

The financial support provided by the Sultan Qaboos University and the Research Council, Oman (Conference Grant) to organize the International Conference on Desalination, Environment and Marine Outfall Systems and publish this book in a successful manner is greatly acknowledged. We are also thankful to the Deputy Vice-Chancellor, Post-graduate studies, Deanship (Research) and other staff members of Sultan Qaboos University for extending their full support and encouragement in successfully publishing this book.

Thanks to the Staff of Springer Publications for their patience and assistance at various stages of this book publication and thanks to Mr. Ali Sherazee of English Language Center at Sultan Qaboos University for English language editing of the chapters. The Center for Environmental Studies and Research (CESAR) staff at Sultan Qaboos University deserves special thanks for their untiring efforts in organizing this significant event.

Mahad Baawain
B.S. Choudri
Mushtaque Ahmed
Anton Purnama

Contents

Contributors

L.H. Adnani Department of Natural Science and Public Health, College of Sustainability Sciences & Humanities, Zayed University, Dubai, United Arab Emirates

Mushtaque Ahmed College of Agriculture and Marine Sciences, Sultan Qaboos University, Muscat, Oman

Mohammed Al-Abri Department of Mechanical and Industrial Engineering, College of Engineering, Sultan Qaboos University, Muscat, Sultanate of Oman

U. AlAlami Department of Natural Science and Public Health, College of Sustainability Sciences & Humanities, Zayed University, Dubai, United Arab Emirates

Mansour Al-Haddabi School of Chemical and Petroleum Engineering, Curtin University, Perth, WA, Australia

Department of Soils, Water and Agricultural Engineering, Sultan Qaboos University, Muscat, Sultanate of Oman

Hamed Al-Hasni Public Authority of Electricity and Water, Muscat, Oman

Aliaa Al-Kalbani Department of Soils, Water and Agricultural Engineering, College of Agricultural and Marine Sciences, Sultan Qaboos University, Muscat, Oman

Sara Al-Kiyumi Department of Soils, Water and Agricultural Engineering, College of Agricultural and Marine Sciences, Sultan Qaboos University, Muscat, Oman

Sulaiman Al Obaidani Department of Mechanical and Industrial Engineering, College of Engineering, Sultan Qaboos University, Muscat, Sultanate of Oman

Nabeel Al-Rawahi Department of Mechanical and Industrial Engineering, College of Engineering, Sultan Qaboos University, Muscat, Sultanate of Oman

Salwa Al-Rawahi Department of Soils, Water and Agricultural Engineering, College of Agricultural and Marine Sciences, Sultan Qaboos University, Muscat, Oman

Abdullah M. Al-Shabibi Department of Mechanical and Industrial Engineering, Sultan Qaboos University, Muscat, Oman

Zainab Ambu Saidi Department of Soils, Water and Agricultural Engineering, College of Agricultural and Marine Sciences, Sultan Qaboos University, Muscat, Oman

M.A. Antar Department of Mechanical Engineering, King Fahd University of Petroleum and Minerals, Dhahran, Saudi Arabia

M. Atif Department of Mechanical Engineering, King Fahd University of Petroleum and Minerals, Dhahran, Saudi Arabia

Mahad Baawain Center for Environmental Studies and Research (CESAR), Sultan Qaboos University, Muscat, Oman

Silene Cristina Baptistelli Companhia de Saneamento Básico do Estado de São Paulo (SABESP), São Paulo, Brazil

Aliyu Salisu Barau Research Fellow, Earth System Governance Project, IPO, Lund University, Lund, Sweden

Faculty of Built Environment, Universiti Teknologi Malaysia, Johor Bahru, Johor, Malaysia

A.H. Basioni ADCO, AADNOC, Abu Dhabi, United Arab Emirates

M.C.M. Bruijs Department Process & Cooling Water, DNV GL Energy, Arnhem, The Netherlands

B.S. Choudri Center for Environmental Studies and Research (CESAR), Sultan Qaboos University, Muscat, Oman

Redouane Choukr-Allah Complex of Horticulture, Hassan II Institute of Agronomy and Veterinary Medicine, Agadir, Morocco

Jean-Philippe Croué Water Desalination and Reuse Center, King Abdullah University of Science and Technology (KAUST), Thuwal, Saudi Arabia

Curtin Water Quality Research Centre, Curtin University, Bentley, Perth, Western Australia

D.R. Danish Institute for Ocean Management, Department of Civil Engineering, Anna University, Chennai, Tamil Nadu, India

Clifford Dansoh CKD Consultants Limited, Worthing, West Sussex, United Kingdom

Shivani Dhage National Environmental Engineering Research Institute, Mumbai Zonal Centre, Worli, Mumbai, India

G. Dhinesh National Institute for Ocean Technology, Chennai, Tamil Nadu, India

Mohamed El-Otmani International Center for Biosaline Agriculture, Dubai, United Arab Emirates

W.E. Elshorbagy Department of Civil and Environmental Engineering, United Arab Emirates University, AlMaqam, AlAin, United Arab Emirates

Eman Hasan Misurata University, Bani Waled, Libya

National Water Research Center, Cairo, Egypt

Abdelaziz Hirich International Center for Biosaline Agriculture, Dubai, United Arab Emirates

Complex of Horticulture, Hassan II Institute of Agronomy and Veterinary Medicine, Agadir, Morocco

Prakash Kelkar Formaly, Advisor, Rajiv Gandhi Science and Technology Commission, Government of Maharashtra, Nagpur, India

M.A. Khan Department of Natural Science and Public Health, College of Sustainability Sciences & Humanities, Zayed University, Dubai, United Arab Emirates

B.V. Mudgal Centre for Water Resources, Department of Civil Engineering, Anna University, Chennai, Tamil Nadu, India

Maolida Nihemaiti Water Desalination and Reuse Center, King Abdullah University of Science and Technology (KAUST), Thuwal, Saudi Arabia

Neil T. Palmer National Centre of Excellence in Desalination Australia, Rockingham, WA, Australia

Anton Purnama College of Science, Sultan Qaboos University, Muscat, Oman

K.G.A. Qalandri Emirates Authority for Standardization & Metrology, Dubai, United Arab Emirates

M.V. Ramanamurthy National Institute for Ocean Technology, Chennai, Tamil Nadu, India

Abdellatif Rami Faculty of Science, University IbnZohr, Agadir, Morocco

M.N. Rashed Chemistry Department, Aswan Faculty of Science, Aswan University, Aswan, Egypt

Julien Le Roux Water Desalination and Reuse Center, King Abdullah University of Science and Technology (KAUST), Thuwal, Saudi Arabia

Ahmad Sana Department of Civil and Architectural Engineering, College of Engineering, Sultan Qaboos University, Muscat, Oman

A. Sankaran Department of Natural Science and Public Health, College of Sustainability Sciences & Humanities, Zayed University, Dubai, United Arab Emirates

Peter Schmaeh Herrenknecht AG, Schlehenweg 2, Schwanau, Germany

M. Tahat Department of Mechanical and Industrial Engineering, Sultan Qaboos University, Muscat, Oman

F. Tahir Department of Mechanical Engineering, King Fahd University of Petroleum and Minerals, Dhahran, Saudi Arabia

Ritesh Vijay National Environmental Engineering Research Institute, Nagpur, India

Hari Vuthaluru School of Chemical and Petroleum Engineering, Curtin University, Perth, WA, Australia

Farhad Yazdandoost Department of Civil Engineering, K N Toosi University of Technology, Tehran, Iran

Hussein Znad School of Chemical and Petroleum Engineering, Curtin University, Perth, WA, Australia

About the Editors

Dr. Mahad Baawain is currently the Director of Center for Environmental Studies and Research (CESAR) as well as an Associate Professor in the Department of Civil and Architectural Engineering of Sultan Qaboos University. He obtained his BSc in Civil Engineering from Sultan Qaboos University, Oman in 1998. Then he continued his postgraduate studies in the area of Environmental Engineering. He obtained MSc from Imperial College, UK in 2000 and PhD from University of Alberta, Canada in 2007. His research interest covers several areas among environmental engineering including fundamentals of water and wastewater treatment processes, advanced oxidation processes, water and wastewater reactors modeling, nanotechnology applications for water and wastewater treatment, climate change and impacts on environmental systems, and air pollution modeling and management.

Dr. B.S. Choudri has PhD in Environmental Management and his special interests have been environmental management issues related to developmental activities including global climate change. He is presently working as Senior Researcher with the Center for Environmental Studies and Research (CESAR) at Sultan Qaboos University, Oman. Dr. Choudri has more than 15 years of experience in leading research projects and capacity building activities in the area of environmental management at the national and international level.

Dr. Mushtaque Ahmed obtained his PhD from Iowa State University, USA in 1988. He is currently an Associate Professor of the Department of Soils, Water and Agricultural Engineering of Sultan Qaboos University, Oman. His current research interests are biosaline agriculture, managed aquifer recharge, climate change and adaptability, wastewater treatment and reuse. He has published more than 170 scientific papers in peer reviewed refereed journals, book chapters, conference proceedings, and manuals, as well as serving as Editor of various publications.

Dr. Anton Purnama obtained his PhD from University of Cambridge, UK. He is an Associate Professor of the Department of Mathematics and Statistics of Sultan Qaboos University, Oman. Prior to that he worked in UK at the NERC Water Resource Systems Research Unit, Department of Civil Engineering, University of Newcastle upon Tyne, and Department of Mathematical Sciences, Loughborough University. His current research is focused on mathematical models for brine discharges from seawater desalination plants and the marine outfall with multiport diffusers and sediment transport.

Part I
Desalination Systems

Chapter 1
An Overview: Desalination, Environmental and Marine Outfall Systems

Mahad Baawain, B.S. Choudri, Mushtaque Ahmed, and Anton Purnama

1.1 Introduction

Desalination is the method of removing salts from brackish or sea water in order to produce fresh water. Desalting is a natural and continual process which is an essential part of the water cycle. After rainfall, rain water carries dissolved minerals and other materials along the way to the sea, which makes the water increasingly salty. Through the sun's energy, the evaporation of water leaves the salts behind and the resulting water vapor forms clouds that produce rain thus continuing the water cycle.

Use of desalinated water by people has a history over the centuries and one of the first references to the phenomenon was by Aristotle, who wrote of seawater distillation in 320 BCE (Burke 2013). Different desalination techniques were used throughout the ages. Rome's Pliny the Elder described seawater distillation by condensing it on fleece in 70 CE and the Greek Alexander of Aphrodisias described seawater distillation with condensation on sponges 130 years later. The successful distillation of seawater, during a voyage to Brazil in 1565, was reported by the French explorer Jean De Lery, and James Cook desalinated seawater during his circumnavigation of the world during the late 1760s and the 1770s (Sanz 2012).

M. Baawain (✉) • B.S. Choudri
Center for Environmental Studies and Research (CESAR), Sultan Qaboos University, Muscat, Oman
e-mail: msab@squ.edu.om; bchoudri@squ.edu.om

M. Ahmed
College of Agricultural and Marine Sciences, Sultan Qaboos University, Muscat, Oman
e-mail: ahmedm@squ.edu.om

A. Purnama
Department of Mathematics and Statistics, College of Science, Sultan Qaboos University, P.O. Box 36, Al-Khod PC123, Muscat, Sultanate of Oman
e-mail: antonp@squ.edu.om

M. Baawain et al. (eds.), *Recent Progress in Desalination, Environmental and Marine Outfall Systems*, DOI 10.1007/978-3-319-19123-2_1

Modern desalting technologies are useful in a number of applications and the most prevalent use is for providing potable water from seawater to meet the requirements of domestic and municipal supplies. However, the use of desalination and desalination technologies for industrial applications has been increasing in recent times, particularly within the oil and gas industry (Global Water Intelligence 2012).

1.2 Developments in Desalination Technology

The techniques of desalination, such as evaporation and distillation, were commonly used in the mid-1900s and the development of desalination processes took a major step forward in the 1940s during the Second World War since military establishments operating in arid areas needed a way to supply potable water to their troops (Birkett 1984). During the late 1960s, in different parts of the world, commercial desalting units were installed, with most of them using thermal processes, producing approximately 8,000 m^3/day (Henthorne 2009). Post-World War II, researchers and scientists began to study osmotic processes to desalinate water, and the term "reverse osmosis" (RO), a very popular present-day desalination technology, first appeared in the US Department of Interior's Office of Saline Water Commission's annual report in 1955 (Brandt 1985). In the 1970s, development efforts continued to focus on commercial membrane processes, such as RO and electrodialysis (ED), and these technologies came to be used extensively (Henthorne 2009). The techniques of ED desalted brackish water more economically than other processes such as distillation, and more interest was focused on using desalination as an alternative means to provide water for municipalities which had brackish water available but had limited fresh water resources (IDA 1987). Desalination technology became a fully commercial enterprise by the 1980s, and its use for municipal water supplies was commonplace by the 1990s, with major desalination processes adopting membrane as well as thermal technologies. Presently, RO technology accounts for approximately 60 % of installed capacity (Maalouf 2014).

Desalination in agriculture is being practiced in some countries, which under certain special circumstances is viable in arid regions. Furthermore, desalination and membrane technology is becoming popular in wastewater treatment. The Membrane Bio Reactor (MBR) is now being used in wastewater treatment in many places with better quality effluent at lower costs and other benefits. Opportunities for using ceramic membranes in the treatment of oil production water is being investigated. Challenges remain though, especially in lowering energy use and decreasing production costs. The problem of environmentally-sound brine disposal is still far from being solved. On the other hand, recent research developments show that in the next 10 years, desalination technologies for producing freshwater will be affordable and environmentally- sustainable in many countries.

1.3 Reducing Energy Consumption and Addressing Environmental Concerns

Today, developments in desalination technologies are specifically aimed at reducing energy consumption and cost, as well as minimizing environmental impact. Advancements include such new and emerging technologies as forward osmosis, low temperature distillation, membrane distillation, pressure retarded osmosis as well as biomimetic and graphene membranes. Hybrid plants that apply multiple-effect disinfection and reverse osmosis are gaining wider use in the Middle East, which has traditionally been home to facilities using more energy-intensive thermal technologies such as multi-stage flash distillation. There is also a push to utilize renewable energy to power desalination plants in the Gulf Cooperation Council (GCC) countries, with Saudi Arabia, for example, making a major investment in solar energy for desalination (Sood and Smakhtin 2014).

The desalination industry is also paying a great deal of attention to environmental considerations. Environmental safeguards have become increasingly important in the siting and construction of new plants. Monitoring programs are also being more widely utilized. Lower energy consumption reduces a plant's carbon footprint, and in addition, new technologies are being used successfully to lessen disruptions to marine life during the intake and outfall processes. Studies such as IDA's Blue Paper (IDA 2011) on Desalination and the Gulf have raised awareness of steps to leverage best practices and mitigate potential harmful environmental effects of desalination. While energy consumption has been significantly reduced in the past two decades, IDA's Energy Task Force has specifically challenged the desalination industry to achieve a further reduction of 20 % in energy requirements for seawater desalination by 2015.

1.4 Brine Disposal from Desalination Plants

One common by-product of all types of desalination plants is brine. The amount of brine as a percentage of feed water varies depending on the choice of method, initial salinity of feed water, and factors affecting the choice of the disposal method. It has been reported that RO plants generally produce 25 % of the total feed water flow as reject brine. The reject brine, also known as concentrate, reject water, or wastewater, from these plants cannot be economically discharged into the ocean, as is done with most coastal desalination plants. In such instances, evaporation ponds may be useful. In other cases, alternatives such as waste minimization, discharge to surface water, discharge to wastewater treatment plants, deep well injection, land application, and wastewater evaporators may be appropriate (Ahmed et al. 2004).

Evaporation ponds have been used over the centuries to remove water from saline solutions. Evaporation ponds are relatively easy to construct, while requiring low maintenance and little operator attention compared to mechanical systems

(Ahmed et al. 2001). In many instances, evaporation ponds are frequently the least costly means of disposal, especially in areas with high evaporation rates and low land costs. Disadvantages include the need for large tracts of land when the evaporation rate is low or the disposal rate is high, the need for impervious liners of clay or synthetic membranes such as PVC or Hypalon, and the potential for contaminating underlying potable water aquifers through seepage from poorly-constructed evaporation ponds (Ahmed et al. 2002).

A very recent development in brine management is the practice of brine concentration using forward osmosis (FO). In this process, a highly concentrated solution (draw solution) is used along with membrane freshwater from the brine. The current commercial process uses an ammonia-carbon dioxide mixture in the draw solution. Ammonia and carbon dioxide readily dissociate into gases when heated and the draw solutes can be effectively recovered and reused in a closed loop system. However, very little information is available on the cost of disposal from the plants. Analysis with limited cost data makes it clear that the unit cost of construction is reduced as the pond size is increased although there are other factors involved, such as remoteness of the plant location, distance from nearby towns, availability of local construction materials and labor, etc.

1.5 Water Produced by Desalination Process and Environment

At present, the combined production capacity of all seawater desalination plants worldwide provide about 66.4 Mm^3/day, where the share of agriculture is about 1.9 % (Desalination Yearbook 2011–2012). Desalination is becoming the sole source of drinking water in some countries in the GCC, with a total capacity of about 5,000 Mm^3/year by 2015 (Dawoud and Al Mulla 2012). It is expected that this capacity will double in the next decade. 61 % of the water is produced by thermal processes, mainly in the Gulf region, while 34 % is produced by RO, which is the first choice in many countries that have started to use desalination. Worldwide, RO desalination capacity for both sea and brackish water represents 60 % of total desalination capacity. Because all desalination plants use chemicals which cause pollution and also due to the large volume of brine discharge through many types of outfall systems into the sea, from simple surface discharge through an open-channel to modern submerged multiport outfall systems, desalination plants have been included in the list of major sources of land-based marine pollution in the Gulf by the United Nations Environment Programme (UNEP/MAP/MEDPOL 2003). Other main environmental concerns are the intakes, which may impinge upon and entrain marine organisms. Furthermore energy use may cause air pollution and greenhouse gas emissions.

The need for resource-saving, low-impact green desalination technologies is clearly evident, as the use of desalination rapidly increases in many parts of the world. The concept of best available techniques would be required at various stages of these technologies, starting with the identification of state-of-the-art technologies. Processes, or methods of operation, also need careful scrutiny. Furthermore, policies and programmes need to be implemented which specify the practical applicability of methods that may prevent or reduce pollution of the atmosphere, sea and land, as well as reduce the quantity of waste. These best available techniques would therefore be used through all the stages to help reduce any deleterious impact on the environment as a whole. The design and siting of submarine intakes and outfalls are a complex task that relies on many disciplines including oceanography, civil and environmental engineering, marine biology, construction, economics, and public relations.

1.6 Part I: Desalination Systems

There are three basic categories of water purification technologies that are used for desalination: membrane technologies, distillation processes (thermal technologies), and chemical approaches (Younos and Tulou 2005), with some water purification plants using a combination of all three technologies. A membrane is a thin film of porous material that allows water molecules to pass through it, but simultaneously prevents the passage of larger and undesirable molecules such as viruses, bacteria, metals, and salts (AWWA 1999). Thermal technologies are based on the concept of using evaporation and distillation processes (Van der Bruggen and Vandecasteele 2002) and modern thermal-based technologies are mostly developed as dual-purpose power and water desalination systems. Chemical approaches include processes such as ion exchange, which however, is considered impractical for treating water with high levels of dissolved solids (El-Dessouky and Ettouney 2002). Several new technologies are being researched with potential for future applications in desalination.

In Part I, there are 10 chapters which present different techniques and designs that are being used, not only to minimize the impact of desalination, but also to save energy and use natural resources to maximize the output of integrated desalination systems. Al-Haddabi et al. have presented the use of ceramic membrane technology for sustainable oil water production in the Chap. 2. In Chap. 3, Saeed et al. have analysed the thermal performance of single-slope solar water stills with enhanced solar heating systems, with a case study from Oman. In Chap. 4, Tahir et al. discuss the effect of fouling on the performance and design aspects of desalination systems. In order to achieve near-zero liquid discharge, an integrated membrane system has been discussed in Chap. 5 by Al Obaidani et al. Schmaeh has presented innovative trenchless methods for the installation of shore approaches and sealines with case studies from Australia and Germany in Chap. 6. In Chap. 7, Rashed has presented information regarding the photocatalytic degradation of divalent metals under sunlight irradiation, using nanoparticle TiO_2 modified concrete materials which can be used for desalination processes. Bruijs provides insights on cost-effective fouling control in (cooling) water intake systems with environmental and operational benefits in Chap. 8. Al-Kalbani et al., in Chap. 9, provide an investigation of the presence of chlorination by-products and other quality parameters in desalinated drinking water systems, using a case study from Oman. A review on the viability of renewable energy and energy storage as the power source for municipal-scale reverse osmosis desalination has been presented by Dansoh in Chap. 10, and Hasan discusses, in Chap. 11, the integration of desalination technology with renewable energy for climate change abatement in the Middle East and North Africa region.

1.7 Part II: Environmental Systems

Although desalination of seawater offers a range of human health, socio-economic, and environmental benefits by providing a seemingly unlimited, constant supply of high quality drinking water without impairing natural freshwater ecosystems, concerns have been raised due to potential negative impacts. These are mainly attributed to the concentrate and chemical discharges, which may impair coastal water quality and affect marine life, and to air pollutant emissions, which may be linked to the energy demand of the processes (Lattemann and Hopner 2008). The list of potential impacts can be further extended, but the information available on marine discharge alone (Lattemann and Hopner 2003) indicates the need for a comprehensive environmental evaluation of all major desalination projects. In order to avoid disorganised and unsustainable development of coastal areas, desalination activity needs to be integrated into management plans that regulate the use of water resources and desalination technology on a regional scale (UNEP 2003). Therefore, the potential environmental impacts of desalination activities need to be evaluated, adverse effects mitigated as far as possible, and the remaining concerns balanced towards the impact of alternative water supply and water management options in order to safeguard a sustainable use of the technology.

In the first chapter in Section II (Chap. 12), Palmer provides an overview of reducing the carbon footprint of desalination, with examples from Australia. The feasibility of using desalination for irrigation in the Souss Massa region in the south of Morocco is discussed in Chap. 13 by Abdelaziz et al., with a cost-benefit analysis of the scheme. Furthermore, desalination has been discussed in Chap. 14 from the viewpoint of integrated water resources management, highlighting a case study from Iran by Yazdandoost. Chap. 15 by Barau provides an overview of strategies for addressing sustainability risks of seawater desalination plants in the Arabian Gulf using the DPSIR framework. A study by Elshorbagy and Basioni, in Chap. 16, presented the impact of the coastal intake environment on the operating conditions of thermal desalination plants in the United Arab Emirates. Phytoplankton species and associated bacterial populations in the coastal waters of the Emirates are discussed by Khan et al. in Chap. 17. Purnama has provided environmental quality standards for brine discharge from desalination plants using the outfall discharge models in Chap. 18. In Chap. 19, a hydrodynamic and thermal dispersion modelling of the effluent in a coastal channel, with a case study from Oman, is presented by Sana. In Chap. 20, the formation of emerging disinfection by-products by chlorination/chloramination of seawater that is impacted by algal organic matter is discussed by Nihemaiti et al. and in Chap. 21, Dhage et al. discuss a functional appraisal of marine outfall for domestic waste disposal through a tracer technique used in a case study from India. Hydrodynamics modeling is presented in Chap. 22 by Baptistelli with the application of the Delft3D-FLOW in Santos Bay, São Paulo State, Brazil, and a mathematical model study of the effluent disposal from a desalination plant in the marine environment at Tuticorin in India is presented by Danish et al. in Chap. 23.

In summary, desalination is an important technology whose application addresses global water issues this century. Its use is expected to increase in various parts of the world and the desalination industry will continue to develop new technologies, taking environmental concerns into consideration, as well as aiming for low costs that will result from new financing models.

In conclusion, the objectives of this volume are to provide stakeholders with knowledge and technological practices focused on:

- international and local-level case studies that contribute significantly at the global level;
- concerns related to various environmental aspects that need to be considered beyond desalination; and
- initiatives that can simultaneously increase integrated planning and minimize environmental impacts in order to underscore the need for sustainable desalination practices and use of technologies.

References

Ahmed M, Shayya W, Hoey D, Al-Handhaly J (2001) Brine disposal from RO plants in Oman and the United Arab Emirates. Desalination 133:135–147

Ahmed M, Shayya WH, Hoey D, Al-Handaly J (2002) Brine disposal from inland desalination plants: research needs assessment. Water Int 27(2):194–201

Ahmed M, Hoey D, Shayya W, Goosen MFA (2004) Brine disposal from inland desalination plants: current status, problems, and opportunities. Published in Volume II of Environmental Sciences and Environmental Computing (Electronic Book Series). EnviroComp Consulting, Inc., Fremont, California, USA

American Water Works Association (AWWA) (1999) Manual of water supply practices: reverse Osmosis and nanofiltration. M46, AWWA, Denver

Birkett JD (1984) A brief illustrated history of desalination: from the Bible to 1940. Desalination 50:17–52

Brandt DC (1985) Seawater reverse osmosis: an economic alternative to distillation. Desalination 52:177–186

Burke PA (2013) Understanding desalination: the well that won't run dry. Water Technology Magazine 36 (Edition 4), 01 April 2013

Dawoud MA, Al Mulla MM (2012) Environmental impacts of seawater desalination: Arabian Gulf case study. Int J Environ Sustain 1(3):22–37

Desalination Yearbook (2011–2012) Water desalination report. International Desalination Association (IDA). Desalination Year Book, IDA, Australia

El-Dessouky HT, Ettouney HM (2002) Fundamentals of salt water desalination. Elsevier, Amsterdam

Global Water Intelligence (2012) Industrial desalination and water reuse report. Global Water Intelligence Office, Waterhub, Singapore

Henthorne L (2009) Desalination: a critical element of water solutions for the 21st century. In: Forare J (ed) Drinking water – sources, sanitation and safeguarding. The Swedish Research Council Formas, Sweden

International Desalination Associates (IDA) (1987) IDA worldwide desalting plants inventory. Wangnick Consulting Engineers, Gnarrenburg

International Desalination Associates (IDA) (2011) Desalination and the Gulf: the relationship between the environment and meeting the region's water needs. Environmental symposium on Desalination and the Gulf, Manama, 6–7 December 2010

Lattemann S, Hopner T (2003) Seawater desalination impacts of brine and chemical discharges on the marine environment. Desalination Publications, L'Aquila

Lattemann S, Hopner T (2008) Environmental impact and impact assessment of seawater desalination. Desalination 220(1–3):1–15

Maalouf S (2014) Planning and design of desalination plants effluent systems. PhD thesis, UCLA Electronic Theses and Disserations, University of California, USA

Sanz MA (2012) A look to desalination world. A presentation at the 3rd International Seminar on Desalination, Hotel Enjoy, Antofagasta, 1–October 2012, International Desalination Association, Australia

Sood A, Smakhtin V (2014) Can desalination and clean energy combined help to alleviate global water scarcity? J Am Water Resour Assoc 50(5):1111–1123

UNEP/MAP/MEDPOL (2003) Sea water desalination in the Mediterranean: assessment and guidelines, MAP technical reports series no. 139, United Nations Environment Programme (UNEP), Mediterranean Action Plan (MAP), Programme for the Assessment and Control of Pollution in the Mediterranean region (MEDPOL), Athens

Van der Bruggen B, Vandecasteele C (2002) Distillation vs. membrane filtration: overview of process evolutions in seawater desalination. Desalination 143(3):207–218

Younos T, Tulou KE (2005) Overview of desalination techniques. J Contemp Water Res Educ 132(1):3–10

Chapter 2
Use of Ceramic Membrane Technology for Sustainable Management of Oil Production Water: A Review

Mansour Al-Haddabi, Hari Vuthaluru, Mushtaque Ahmed, and Hussein Znad

Abstract The huge quantities of water produced along with oil make production of water one of the main challenges in the oil and gas industry. In the past, water produced in oil production was considered a tiresome by-product which represented a significant liability and cost to oil and gas production. Recently this attitude has changed and this water is now seen more as a resource than a by-product. By 2025, 2.8 billion people (from 48 countries) will be living in water-scarce and water-stressed countries. The Sultanate of Oman is considered a semi-arid country where the average annual rainfall is about 100 mm. Petroleum Development Oman (PDO) produces around 700,000 m^3/day of water associated with hydrocarbon production. Currently under half of this amount is injected back into reservoirs as water for reservoir management. The production of excessive quantities of water is the reason behind abandoning oil and gas wells, leaving huge quantities of hydrocarbons behind. Upgrading of low quality water (i.e. oil production water) for greening the desert or growing biofuels is becoming a strategic enabler for the sustainable development of remote oil fields. Although opportunities exist for the beneficial use of oil production water, there may be situations where treatment may not be economically feasible. In addition to volume, water quality is the other key determinant of suitable management options. Due to the poor quality of oil production water which contains a complex

---

M. Al-Haddabi (✉)
School of Chemical and Petroleum Engineering, Curtin University, GPO Box U1987, Perth, WA 6845, Australia

College of Agricultural and Marine Sciences, Sultan Qaboos University, Muscat, Oman
e-mail: mans99@squ.edu.om

H. Vuthaluru • H. Znad
School of Chemical and Petroleum Engineering, Curtin University, GPO Box U1987, Perth, WA 6845, Australia
e-mail: H.Vuthaluru@exchange.curtin.edu.au; H.Znad@curtin.edu.au

M. Ahmed
College of Agricultural and Marine Sciences, Sultan Qaboos University, Muscat, Oman
e-mail: ahmedm@squ.edu.om

M. Baawain et al. (eds.), *Recent Progress in Desalination, Environmental and Marine Outfall Systems*, DOI 10.1007/978-3-319-19123-2_2

mixture of organic and inorganic materials similar to those found in crude oil and natural gas, treatment might introduce potential economic, technological and environmental challenges. Another constraint for the management of oil production water is the regulatory framework which might cause restrictions to management options imposed by legislation. Economic constraints imposed by the costs associated with management options can influence the feasibility of particular management techniques. The application of ceramic membranes has shown great potential for oil in water separation and purification due to their superior mechanical, thermal and chemical stability and ease of generation after fouling. Very promising results were achieved by many researchers when using ceramic membranes for oily water treatment. In some studies the removal of oil, COD and TOC reach up to 99 %, 96 % and 94 %, respectively.

Keywords Oil production water • Hydrocarbon • Management • Ceramic membranes

2.1 Introduction

Large volumes of water produced during oil and gas production form the largest waste stream in the oil and gas industry. Formation water or produced water is a natural water layer that, being denser, lies under the hydrocarbons in oil and gas reservoirs (Ekins et al. 2007). For a long time, the produced water was considered as a waste and an obstacle to continued economic development and therefore it had to be handled and disposed of in a feasible and acceptable environmentally-friendly way which met local regulations (Khatib 2007; Zara 2002; Reynolds Rodney 2003; Chan et al. 2002). Many oil and gas companies, motivated by high oil and gas prices, started to extend the life of existing fields through secondary and tertiary recovery processes [Enhanced Water Recovery (EOR)]. Normally 30 % of oil can be extracted from a reservoir but this percentage can be increased through injecting water into the reservoir which will extend the life of the production well (Reynolds Rodney 2003). These activities require huge amounts of good quality water to be injected into the reservoir or for steam generation which will ultimately increase the quantities of extracted produced water. The water-oil-ratio (WOR) or the share of water in oil produced (water cut) continues to rise throughout the lifetime of a production well. On average, 3–5 barrels of water are produced with every barrel of oil but in mature and old oilfields the WOR can reach up to 10–14 barrels. The WOR in Saudi Arabia and Qatar is still below one. In Oman, it is approaching 0.9 barrels. The operational philosophy for most energy companies is to meet OPEC oil quotas rather than producing oil according to their technical or commercial limits. The availability of and access to fresh water is one of the most critical challenges facing the world. By 2025, 2.8 billion people, from 48 countries, will be living in water- scarce and water-stressed countries. Therefore, many countries have started to recognize the water produced because of oil production as a resource rather than as a by-product.

2.2 Composition of Produced Water

Oily produced water contains a complex mixture of organic and inorganic materials similar to those found in crude oil and natural gas (Ashaghi et al. 2007). The composition can significantly alter from well to well and from field to field depending on the nature of the geological location of the field, the geology of the reservoir, the chemicals involved in the production process and the operating conditions (Veil et al. 2004; Hansen and Davies 1994). Table 2.1 represents the complex composition of produced water.

Table 2.1 Composition of oil field-produced water (Fakhru'l-Razi et al. 2009)

Parameter	Minimum value	Maximum value	Heavy metal	Minimum value (mg/L)	Maximum value (mg/L)
Density (kg/m^3)	1,014	1,140	Calcium	13	25,800
Conductivity (μS/cm)	4,200	58,600	Sodium	132	97,000
Surface tension (dyn/cm)	43	78	Potassium	24	4,300
pH	4.3	10	Magnesium	8	6,000
TOC (mg/L)	0	1,500	Iron	<0.1	100
TSS (mg/L)	1.2	1,000	Aluminium	310	410
Total oil (IR; mg/L)	2	565	Boron	5	95
Volatile (BTEX; mg/L)	0.39	35	Barium	1.3	650
Base/neutrals (mg/L)	-	<140	Cadmium	<0.005	0.2
Chloride (mg/L)	80	200,000	Copper	<0.02	1.5
Bicarbonate (mg/L)	77	3,990	Chromium	0.02	1.1
Sulphate (mg/L)	<2	1,650	Lithium	3	50
Ammonium nitrogen (mg/L)	10	300	Manganese	<0.004	175
Sulphite (mg/L)	-	10	Lead	0.002	8.8
Total polar (mg/L)	9.7	600	Strontium	0.02	1,000
Higher acids (mg/L)	<1	63	Titanium	<0.01	0.7
Phenol (mg/L)	0.009	23	Zinc	0.01	35
Volatile fatty acids (mg/L)	2	4,900	Arsenic	<0.005	0.3
			Mercury	<0.005	0.3
			Silver	<0.001	0.15
			Beryllium	<0.001	0.004

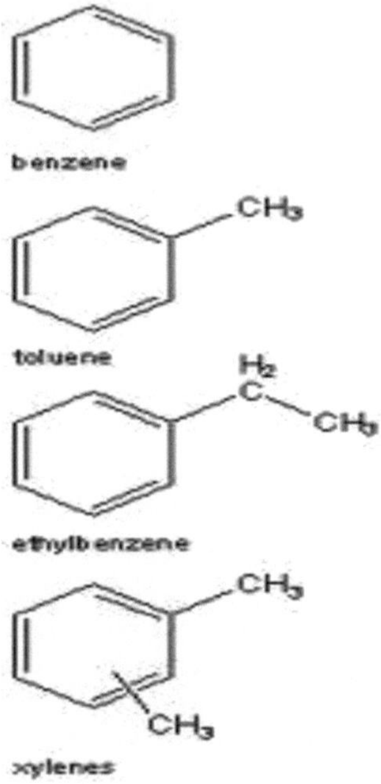

Fig. 2.1 Chemical composition of BTEX

2.2.1 *Dissolved and Dispersed Oil Components*

Dispersed and dissolved oil components are a mixture of hydrocarbons including benzene, toluene, ethylbenzene and xylene (BTEX), polyaromatic hydrocarbons (PAHs), naphthalene, phenanthrene, dibenzothiophene (NPD) and phenols. The polar organic compounds in produced water are known as dissolved oil or soluble oil, while the dispersed oil is the small droplets of oil suspended in the solution (Khosravi and Alamdari 2009; Stephenson 1992; Hayes and Arthur 2004). Produced water from gas/condensate production shows higher concentrations of dissolved oil which contains the groups of BTEX (Fig. 2.1), aliphatic hydrocarbons, carboxylic acid and low molecular aromatic compounds, while PAHs and heavy alkyl phenols are classified as dispersed oil (Veil et al. 2004; Ekins et al. 2007). The presences of dissolved and dispersed oil in the discharge water can have adverse effects on the environment. Many complex and interrelated factors govern the amount of oil present in produced water (Hudgins 1994; Hayes and Arthur 2004; Veil et al. 2004). The following factors affect the level of dissolved and dispersed oil present in the produced water (prior to treatment) (Fakhru'l-Razi et al. 2009):

- Oil composition
- pH, salinity, total dissolved solids (TDS), temperature
- oil/water ratio
- type and concentration of chemicals used in oil production

Temperature changes the relative ratio of carbon ranges and also affects the solubility of organics present in the produced water. Pressure enhances the solubility of organic compounds slightly. Salinity has no significant effect on the dissolved organics. Soluble compounds do not increase total dissolved organics in produced water.

The density of oil, the shear history of the droplet, the amount of oil precipitation and interfacial tension between the water and the oil, all determine the level of the dispersed oil in produced water (Stephenson 1992).

2.2.2 *Dissolved Minerals*

The dissolved inorganic minerals in produced water are usually present in very high levels. They consist of cations, anions, naturally occurring radioactive materials (NORM) and heavy metals. Cations and anions play a significant role in the chemistry of produced water. Salinity, which is mainly caused by the presence of Na^+ and Cl^-, can vary from a few milligrams per liter to ~300,000 mg/L (Roach et al. 1993). Heavy metals such as cadmium, chromium, copper, lead, mercury, nickel, silver and zinc are present in trace levels, depending on the formation geology and the age of oil wells (Fakhru'l-Razi et al. 2009; Hansen and Davies 1994; Hudgins 1994; Leifer 2010; Reynolds Rodney 2003; Sirivedhin et al. 2004).

NORM (mainly ^{226}Radium and ^{228}Radium) might be present in the produced water. The concentration of barium ions in produced water could give a strong indication of radium isotopes present in it (Fakhru'l-Razi et al. 2009; Jacobs et al. 1992; Jerez Vegueria et al. 2002; Lee 2011). Radium that is co-precipitated with barium sulphate (scale) or other types of scales is the primarily cause of radioactivity of produced water.

2.2.3 *Production Chemicals*

Through the oil and gas production process, different types of chemicals are used to mitigate or prevent operational problems. Treatment chemicals (production treating, gas processing, and simulation) and production treating chemicals (scale and corrosion inhibitors, biocides, emulsion breakers, antifoam and water treatment chemicals) are used in the oil and gas production processes (Stephenson 1992). The level of production chemicals in produced water is found to be as low as 0.1 mg/L.

The production chemicals can be used as pure compounds, or as compounds containing active ingredients dissolved as a solvent or a co-solvent and used for inhibition of corrosion, hydrate formation, scale deposition, foam production, wax deposition, bacterial growth, gas dehydration and emulsion breaking in order to enhance the process of oil and water separation (Hansen and Davies 1994). It is difficult to determine the fate of these chemicals as some active ingredients are consumed within the process (Hudgins 1994).

2.2.4 *Production Solids*

Produced solids include clays, precipitated solids, waxes, bacteria, carbonates, sand and silt, corrosion and scale products, formation solids and asphaltenes (Fakhru'l-Razi et al. 2009; Igunnu and Chen 2012; Neff 2002). Few microorganisms can survive in the produced water due to the presence of different toxic chemicals. Bacteria might clog and cause corrosion of equipment and pipelines (Veil et al. 2004).

2.2.5 *Dissolved Gases*

The most abundant gases present in the produced water are carbon dioxide (CO_2), oxygen (O_2) and hydrogen sulphide (H_2S).

2.3 Production Water Management

For many years, produced water was considered as an inconvenient by-product. This attitude has changed in the past few years and nowadays most oil companies regard it as a valuable resource. The management of produced water is considered to be a key component of the overall production costs. The reasons for that are:

- Environmental regulations are becoming more stringent. BAT (Best Available Technology) effluent limitations specify a maximum of 29 mg/L of oil in water, averaged over 30 days. Offshore BCT (Best Conventional Technology) regulations specify 48 mg/L, averaged over 30 days (Islam 2006). The effluent from onshore petroleum activities in the European standard is set to be less than 5 mg/L Total Hydrocarbons (HC) and less than 10 mg/L of suspended solids (Ashaghi et al. 2007).
- The quantities of produced water rise with time.
- There is a great need to save water and to look for new resources to match the great demand by human population growth and expansion in industry.

2.3.1 *Management Possibilities*

Most oil companies have started adopting management options when dealing with the huge quantities of produced water. These options are:

- Minimizing the volumes of water produced by adopting proper reservoir and well management.

- Injecting the water into oil-producing formations or reservoirs for water floods which maintains hydraulic pressure and enhances oil recovery.
- Discharging it into the environment after it complies with the discharge regulations.
- Reusing it for beneficial purposes where once it is treated properly, it can be applied in many agricultural and industrial applications (Clark and Veil 2009).

The ultimate objective of produced water treatment is to produce water which complies with stringent environmental regulations for disposal purposes or to meet the specifications of further applications. In order to meet these regulations the following components need to be removed:

- Dispersed and dissolved oil
- Suspended solids
- Dissolved salts and metals
- Dissolved gases like CO_2 and H_2S
- Excess water hardness
- Natural Occurring Radioactive Materials (NORM)

2.4 Membrane Separation Technology

A wide range of technologies exist for the treatment of produced water. Different physical, chemical, biological and combinations of the three processes were applied and tested (Deng et al. 2005; Zouboulis and Avranas 2000; Mostefa and Tir 2004; Chen 2004; van den Broek et al. 1998; Knudsen et al. 2004; Thoma et al. 1999). Over the last 30 years, membrane separation has gained momentum and popularity (Sonune and Ghate 2004). The membrane acts as a selective barrier to retain certain components while allowing other components to pass through (Burnett and Siddiqui 2006). There are four recognized membrane separation processes (Fig. 2.2), including microfiltration (MF), ultrafiltration (UF), nanofiltration (NF) and reverse osmosis (RO), (Madaeni et al. 2013; Xu and Drewes 2006; Bilstad and Espedal 1996).

2.4.1 Ceramic Membrane

Ceramic membranes (Fig. 2.3) are a new class of materials which can be made from different materials and processed in different ways to produce products with wide ranges of physical and chemical advantages and applications (Ashaghi et al. 2007). They can be classified according to their pore size distribution as microporous membranes (pore size $<$ 2.0 nm), mesoporous membranes (2.0 nm $\leq$ pore

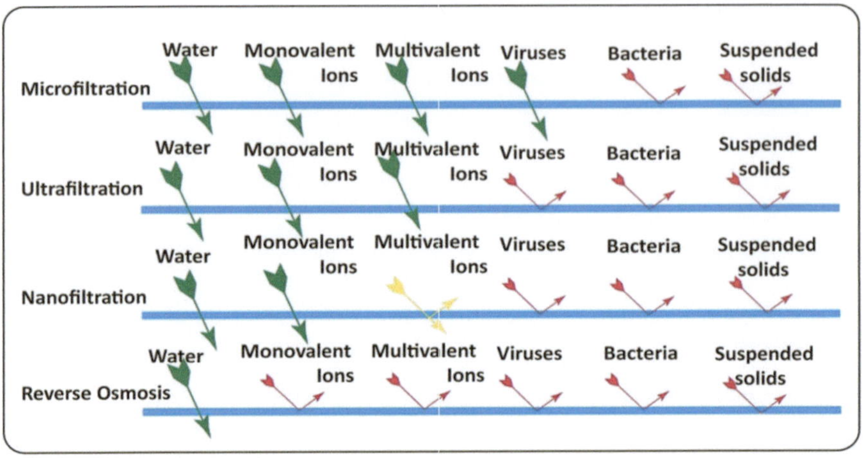

Fig. 2.2 Application of membrane filtration technologies

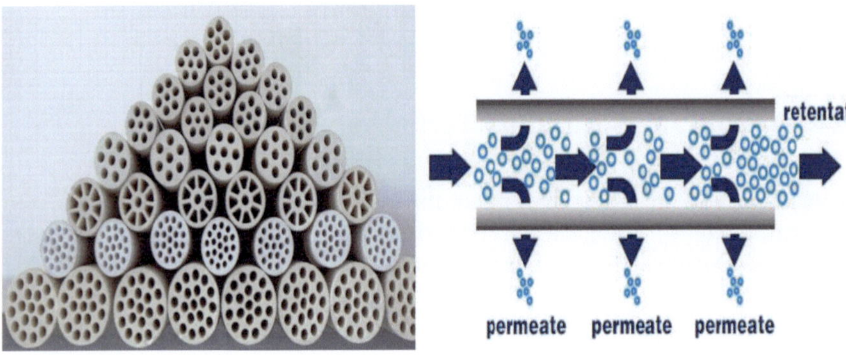

Fig. 2.3 Cross-flow ceramic membrane

size ≥ 50 nm) and macroporous membranes (pore size > 50 nm) (Li and Lee 2009). Due to their extraordinary properties, such as thermal stability, high mechanical strength and ease of regeneration after fouling, they become a good choice to treat oily produced water (Tsuru 2001). Other advantages are they require low energy and they are operational in a pH range of 1–14. Tubular ceramic membranes are produced by a porous support (generally α-alumina), one or more layers of decreasing pore diameter and an active or separating layer (α-alumina, zirconia, etc.) covering the internal surface of the tube (Ashaghi et al. 2007).

Many researchers tested the application of ceramic membranes for the purification of oily produced water (Ciora Jr and Liu 2003; Higgins et al. 1994; Yang et al. 1998; Zhong et al. 2003; Madaeni et al. 2012; Abadi et al. 2011). They investigated the effect of different operating conditions on the efficiency of ceramic membranes in terms of oil removal efficiency, flux and resistance to fouling. Very promising results were obtained. In some studies the oil removal reached more than 90 % (Mueller et al. 1997; Hua et al. 2007; Mehrdad Ebrahimi et al. 2010; Abbasi et al. 2010). Zirconia ceramic membrane is found to perform better compared to alumina ceramic membrane. Fouling and concentration polarization is found to be the main challenge in the application of ceramic membranes.

Mueller et al. (1997) in their study investigated the treatment of oily water containing constituents ranging from 250 to 1,000 ppm by using two α-alumina ceramic membranes (0.2–0.8 μm pore sizes) and a surface-modified polyacrylonitrile membrane (0.1 μm pore size). The results showed that 99 % of oil removal can be achieved producing permeate having <6 ppm total hydrocarbon. Hua et al. (2007) investigated the application of ceramic membranes(α-Al_2O_3) with 50 nm to treat edible oil mixed with water and surfactant. Different operational conditions were tested. From Table 2.2 we can observe that the increase in Transmembrane Pressure (TMP) has a positive effect on permeate flux but it has a negative effect on the quality of permeate because the efficiency of oil removal has decreased. This is due to the fact that higher TMP will allow droplets (both solvent and solute) to pass quickly through the membrane pores.

Abadi et al. (2011) tested the application of tubular ceramic MF (α-Al2O3) membrane to treat the outlet of the American Petroleum Institute (API) unit of Tehran refinery. The process achieved 85 % reduction in oil and grease and 100 % in TSS (Table 2.3).

Lehman and Liu (2009) introduced the application of ozone as a pretreatment with ceramic membrane to treat secondary wastewater effluent (Table 2.4). The

Table 2.2 Effects of Transmembrane Pressure (TMP) and Cross Flow Velocity (CFV) on permeate flux (Hua et al. 2007; Ashaghi et al. 2007)

Effect of TMP on permeate flux (CFV: 1.68 mS^{-1})			Effect of CFV on permeate flux (TMP: 0.2 MPa)		
TMP (MPa)	Flux ($Lm^{-2}h^{-1}$)	TOC removal efficiency (%)	CFV (mS^{-1})	Flux ($Lm^{-2}h^{-1}$)	TOC removal efficiency (%)
0.05	30	97.3	0.17	122	97.5
0.1	70	97	0.42	135	97.4
0.15	110	95.2	0.8	140	97.5
0.2	170	93	1.7	165	97.4

Table 2.3 Properties of the feed and treated wastewater (TMP: 1.25 bar, CFV: 2.25 mS^{-1} and T: 32.5 °C) (Abadi et al. 2011)

Parameter	Unit	Feed	Treated wastewater by using tubular ceramic MF (α-Al2O3)
Total Suspended Solids (TSS)	mg/L	92	Trace
Oil and grease content	mg/L	26	4
Total Organic Carbon (TOC)	mg/L	141	7
Turbidity	mg/L	21	0.3

Table 2.4 Membrane performance and ozone residual in train study. $J_{sp}/J_{sp,0}$ is the normalized specific flux (Lehman and Liu 2009)

Ozone dose (mg/L)	Ozone residual (mg/L)	$J_{sp}/J_{sp,0}$ at day 3	$J_{sp}/J_{sp,0}$ at day 5
6	0.89	0.98	0.98
5	0.61	0.88	0.82
3.5	0.15	0.72	0.70
2.7	<0.05	0.42	0.36

Table 2.5 Summary of the results obtained from different ceramic membrane measured TMP, 1 bar, Temperature, 60 °C (Ebrahimi et al. 2009)

Membrane cut-off	Coil, feed (ppm)	Oil removal (%)	CTOC, feed (%)	TOC removal (%)
MF-0.2 μm	113	67	94	13
UF-0.05 μm	565	99	582	14
	56	71	28	36
UF-20,000 Da	113	84	94	27
NF-1,000 Da	113	80	94	13

feed water was subjected to a pressurized ozone contact system and a poly aluminum chloride (PACL) was added as a coagulant. The study demonstrated steady performance at a filtration flux of 100 gallons per square foot per day (170 L/m^2h) at 20 °C by using ozone (4 mg/L) and PACL (1 mg/L) as a pretreatment. The study found that the ceramic membrane could operate with minimum fouling at fluxes as high as 125 gfd (212.5 L/m^2h).

Ebrahimi et al. (2009) characterized the permeate flux using different ceramic MF, UF and NF. Tubular ceramic membranes with pore sizes ranging from 0.2 μm to 750 Da for MF, UF and NF in different stainless steel housing were investigated. The CFV through membrane was in the range of 0.6–1.3 m/s. PF varied from 3.4 to 3,300 L/m^2h. Permeate flux declined from 1,150 to 200 L/m^2h after 5 h operation. The oil removal reached up to 99 % and TOC removal was around 39 % when UF (0.05 μm) was used. Table 2.5 summarizes the results obtained from different ceramic membranes.

Abbasi et al. (2010) studied the capabilities of the synthetic MF mullite ceramic membranes for the treatment of oily wastewater (Table 2.6). A condensate gas with distilled water and emulsifier was used as feed water. The effects of pressure, flow rate, temperature, oil content and salt concentration on the separation process were tested. The rejection of TOC for the synthetic feed was found to be more than 94 %. The study shows that the PF increases with the increase in volumetric flow rate and temperature. By increasing the oil content and salt concentration, the PF decreases. The MF treatment is very effective in reducing COD and TDS. The salt reduction is found to be around 5–10 %.

Table 2.6 Performance of the mullite-alumina ceramic membrane. PF, FR and R stands for Permeation Flux, Fouling Resistance and Rejection respectively (Abbasi et al. 2010)

Membrane type	PF (L/m^2 h)	FR (%)	R (%)
Mullite	41.3	46.83	84
Mullite-alumina (25 % alumina)	46.8	62.31	82.2
Mullite-alumina (50 % alumina)	58.1	72.63	78.7
Mullite-alumina (75 % alumina)	91.5	96.27	70.8

2.5 Conclusion

The production of huge quantities of water during oil and gas production is found to be the greatest challenge facing the oil and gas industry. These quantities increase with time as the oilfield mature. New technologies need to be implemented in order to meet the stringent environmental regulations. Many physical, chemical and biological processes have failed to produce a permeate that complies with regulations. Ceramic membrane is found to be a promising technology with a high potential to purify oil production water. There is a great need to come up with low cost materials for manufacturing ceramic membranes and to focus on reducing membrane fouling and concentration polarization.

List of Figures

References

Abadi SRH, Sebzari MR, Hemati M, Rekabdar F, Mohammadi T (2011) Ceramic membrane performance in microfiltration of oily wastewater. Desalination 265(1–3):222–228. doi:10.1016/j.desal.2010.07.055

Abbasi M, Mirfendereski M, Nikbakht M, Golshenas M, Mohammadi T (2010) Performance study of mullite and mullite-alumina ceramic MF membranes for oily wastewaters treatment. Desalination 259(1–3):169–178. doi:10.1016/j.desal.2010.04.013

Ashaghi KS, Ebrahimi M, Czermak P (2007) Ceramic ultra- and nanofiltration membranes for oilfield produced water treatment: a mini review. Open Environ J 1:1–8

Bilstad T, Espedal E (1996) Membrane separation of produced water. Water Sci Technol 34 (9):239–246

Burnett DB, Siddiqui M (2006) Recovery of fresh water resources from desalination of brine produced during oil and gas production operations. Texas Engineering Experimental Station, Washington, DC

Chan L-H, Starinsky A, Katz A (2002) The behaviour of lithium and its isotopes in oilfield brines: evidence from the Heletz-Kokhav field, Israel. Geochim caetcosmochim acta 66(4):615–623

Chen G (2004) Electrochemical technologies in wastewater treatment. Sep Purif Technol 38(1):11–41
Ciora RJ Jr, Liu PK (2003) Ceramic membranes for environmental related applications. Fluid/Part Sep J 15:51
Clark C, Veil J (2009) Produced water volumes and management practices in the United States. Argonne National Laboratory (ANL), Washington, DC
Deng S, Yu G, Jiang Z, Zhang R, Ting YP (2005) Destabilization of oil droplets in produced water from ASP flooding. Colloids Surf A Physicochem Eng Asp 252(2):113–119
Ebrahimi M, Ashaghi KS, Engel L, Willershausen D, Mund P, Bolduan P et al (2009) Characterization and application of different ceramic membranes for the oil-field produced water treatment. Desalination 245(1):533–540
Ebrahimi M, Willershausen D, Ashaghi KS, Engel L, Placido L, Mund P et al (2010) Investigations on the use of different ceramic membranes for efficient oil-field produced water treatment. Desalination 250(3):991–996
Ekins P, Vanner R, Firebrace J (2007) Zero emissions of oil in water from offshore oil and gas installations: economic and environmental implications. J Clean Prod 15(13):1302–1315
Fakhru'l-Razi A, Pendashteh A, Abdullah LC, Biak DRA, Madaeni SS, Abidin ZZ (2009) Review of technologies for oil and gas produced water treatment. J Hazard Mater 170(2):530–551
Hansen B, Davies S (1994) Review of potential technologies for the removal of dissolved components from produced water: oil and natural gas production. Chem Eng Res Des 72 (2):176–188
Hayes T, Arthur D (2004) Overview of emerging produced water treatment technologies. In: 11th Annual international petroleum conference, Albuquerque
Higgins R, Bishop B, Goldsmith R (1994) Reclamation of waste lubricating oil using ceramic membranes. In: Proceeding of the third international conference on inorganic membranes, Worcester Polytechnic Institute, Worcester, pp 447–463
Hua F, Tsang Y, Wang Y, Chan S, Chua H, Sin S (2007) Performance study of ceramic microfiltration membrane for oily wastewater treatment. Chem Eng J 128(2):169–175
Hudgins C (1994) Chemical use in North Sea oil and gas E&P. J Pet Technol 46(1):67–71
Igunnu ET, Chen GZ (2012) Produced water treatment technologies. Int J Low-Carbon Technol 1–21
Islam S (2006) Investigation of oil adsorption capacity of granular organoclay media and the kinetics of oil removal from oil-in-water emulsions. Texas A&M University, College Station
Jacobs R, Grant R, Kwant J, Marquenie J, Mentzer E (1992) The composition of produced water from shell operated oil and gas production in the North Sea. In: Produced water, Springer, U.S., pp 13–21
Jerez Vegueria S, Godoy J, Miekeley N (2002) Environmental impact studies of barium and radium discharges by produced waters from the "Bacia de Campos" oil-field offshore platforms, Brazil. J Environ Radioact 62(1):29–38
Khatib Z (2007) Produced water management: is it a future legacy or a business opportunity for field development. In: International Petroleum Technology Conference (IPTC), Dubai, U.A.E.
Khosravi J, Alamdari A (2009) Copper removal from oil-field brine by coprecipitation. J Hazard Mater 166(2):695–700
Knudsen B, Hjelsvold M, Frost T, Svarstad M, Grini P, Willumsen C et al (2004) Meeting the zero discharge challenge for produced water. In: SPE international conference on health, safety, and environment in oil and gas exploration and production, Dubai, U.A.E.
Lee K (2011) Produced water. Springer, New York
Lehman SG, Liu L (2009) Application of ceramic membranes with pre-ozonation for treatment of secondary wastewater effluent. Water Res 43(7):2020–2028
Leifer I (2010) Characteristics and scaling of bubble plumes from marine hydrocarbon seepage in the coal oil point seep field. J Geophys Res-Ocean 115, Artn C11014. doi: 10.1029/2009jc005844

Li L, Lee R (2009) Purification of produced water by ceramic membranes: material screening, process design and economics. Sep Sci Technol 44(15):3455–3484

Madaeni SS, Monfared HA, Vatanpour V, Shamsabadi AA, Salehi E, Daraei P et al (2012) Coke removal from petrochemical oily wastewater using gamma-A1203 based ceramic microfiltration membrane. Desalination 293:87–93. doi:10.1016/J.DESAL.2012.02.028

Madaeni SS, Gheshlaghi A, Rekabdar F (2013) Membrane treatment of oily wastewater from refinery processes. Asia Pac J Chem Eng 8(1):45–53. doi:10.1002/Apj.1619

Mostefa NM, Tir M (2004) Coupling flocculation with electroflotation for water oil/water emulsion treatment. Optimization of the operating conditions. Desalination 161(2):115–121

Mueller J, Cen YW, Davis RH (1997) Crossflow microfiltration of oily water. J Membr Sci 129(2):221–235. doi:10.1016/S0376-7388(96)00344-4

Neff JM (2002) Bioaccumulation in marine organisms: effect of contaminants from oil well produced water. Elsevier, Amsterdam

Reynolds Rodney R (2003) Produced water and associated issues. A manual for independent operator. Petroleum Technology Transfer Council, Tulsa

Roach RW, Carr RS, Howard CL (1993) An assessment of produced water impacts at two sites in the Galveston Bay System, U.S. Fish and Wildlife Service

Sirivedhin T, McCue J, Dallbauman L (2004) Reclaiming produced water for beneficial use: salt removal by electrodialysis. J Membr Sci 243(1):335–343

Sonune A, Ghate R (2004) Developments in wastewater treatment methods. Desalination 167:55–63

Stephenson M (1992) A survey of produced water studies. In: Ray JP, Engelhardt FR (eds) Produced water, Springer, pp 1–11

Thoma G, Bowen M, Hollensworth D (1999) Dissolved air precipitation/solvent sublation for oil-field produced water treatment. Sep Purif Technol 16(2):101–107

Tsuru T (2001) Inorganic porous membranes for liquid phase separation. Sep Purif Rev 30(2):191–220

van den Broek W, Plat R, van der Zande M (1998) Comparison of plate separator, centrifuge and hydrocyclone. In: SPE international oil and gas conference and exhibition in China, Dubai, U.A.E.

Veil JA, Puder MG, Elcock D, Redweik Jr RJ (2004) A white paper describing produced water from production of crude oil, natural gas, and coal bed methane. Prepared by Argonne National Laboratory for the US Department of Energy, National Energy Technology Laboratory. January Available at http://www.ead.anl.gov/pub/dsp_detail.cfm

Xu P, Drewes JE (2006) Viability of nanofiltration and ultra-low pressure reverse osmosis membranes for multi-beneficial use of methane produced water. Sep Purif Technol 52 (1):67–76

Yang C, Zhang G, Xu N, Shi J (1998) Preparation and application in oil-water separation of $ZrO_2/\alpha\text{-}Al_2O_3$/MF membrane. J Membr Sci 142(2):235–243

Zara K (2002) Water to value-produced water management for sustainable field development of mature and green fields. In: SPE international conference on health, safety and environment in oil and gas exploration and production, Dubai, U.A.E.

Zhong J, Sun XJ, Wang CL (2003) Treatment of oily wastewater produced from refinery processes using flocculation and ceramic membrane filtration. Sep Purif Technol 32 (1–3):93–98. doi:10.1016/S1383-5866(03)00067-4

Zouboulis A, Avranas A (2000) Treatment of oil-in-water emulsions by coagulation and dissolved-air flotation. Colloids Surf A Physicochem Eng Asp 172(1):153–161

Chapter 3
Single Slope Solar Water Still with Enhanced Solar Heating System

Abdullah M. Al-Shabibi and M. Tahat

Abstract This paper experimentally investigates the thermal performance of a conventional solar water still with an enhanced solar heating system in Oman. A number of variables have been considered, including the water depth inside the still and the saline water temperature inlet to the solar still from the preheater solar collector system. A single slope, single effect conventional solar still with a solar preheating unit was constructed and experimentally tested under different Omani weather conditions. The still had been modified to include a preheating solar energy system so that saline water would be preheated before entering the solar still and this preheating would enhance hourly or daily yield of pure water. Different quantities of water in the solar still basin were tested to find the effect of water quantity on the hourly yield and also to assess the thermal efficiency of the still.

It was found that 1 cm depth gives the best performance in terms of fresh water yield and thermal efficiency. The addition of the solar water preheater to the system significantly increases the inlet basin saline water temperature to almost saturated temperature. The saline water in the basin needed only a small amount of heat to be vaporized, hence increasing the production of fresh water and enhancing the solar still's thermal efficiency.

Keywords Single slope solar water still • Solar collector • Solar desalination • Solar energy • Thermal performance

3.1 Introduction

The main source of fresh water in Oman is the underground water supply which is limited and cannot meet the increasing demands on fresh water. Oman's population over the last four decades has almost trebled and the availability of freshwater is

A.M. Al-Shabibi (✉) • M. Tahat
Department of Mechanical and Industrial Engineering, Sultan Qaboos University, Muscat, Oman
e-mail: ashabibi@squ.edu.om; tahat@squ.edu.om

M. Baawain et al. (eds.), *Recent Progress in Desalination, Environmental and Marine Outfall Systems*, DOI 10.1007/978-3-319-19123-2_3

becoming critical. Furthermore, there is increased demand for water from the expansion of the industrial sector in Oman. Water desalination represents a very attractive alternative to produce enough fresh water to maintain sufficient supply and Oman has a very long coast line of about 1700 km.

Oman is an oil-producing country but the oil is expected to run out in the near future. Furthermore, water desalination using fossil fuel is known to be costly and harmful to the environment. The demand for, as well as the unit cost of, conventional energy is increasing every year, so alternative sources of energy need to be utilized. Solar energy is a renewable source of energy which can be utilized effectively in many applications in Oman such as water heating, cooking and solar desalination.

Oman is located in the southeastern quarter of the Arabian Peninsula and according to official estimates, covers a total land area of approximately 300,000 km^2. The land area is composed of varying topographic features where valleys and desert account for 82 % of the land mass, mountain ranges, 15 % and the coastal plain, 3 %.

The climate of the Sultanate of Oman is dry and tropical and is characterized by extreme heat in the summer around June and by coolness in the winter around January (Fig. 3.1a). The Sultanate receives a high degree of solar radiation thought the year. It is therefore advisable for the country to use solar energy which is renewable and readily available. Oman has on average 9.49 h of sunshine per day with a standard deviation of 1.78 h. The average number of sunshine hours ranges from 7.88 h in Salalah to 10.24 h in Buraimi. The solar radiation averaged 18.71 MJ/m^2/day with a standard deviation of 4 MJ/m^2/day for Oman over the period. Marmoul has the highest average solar radiation of 21.86 MJ/m^2/day while Sur and Salalah have the lowest average of 15.92 MJ/m^2/day and 16.22 MJ/m2/day respectively. The greatest amount of radiation occurs in April, May and June with the highest level of radiation occurring in May (Fig. 3.1b). The May average is 23.11 MJ/m^2/day. The least amount of monthly radiation is in December and January (Seeb Meteorological 2006).

A single slope water still consists of a thermally-insulated box with a glass cover. Solar energy penetrates the glass and causes a greenhouse effect inside the box. The water inside the box heats up and evaporates leaving behind any impurities such as the salt. The water vapor then condenses when it is in contact with the inside of the glass cover. The pure water condensate can then be collected. The single slope still was studied both theoretically and experimentally by a number of researchers. Nafey et al. (2000) investigated the effect of solar radiation, wind speed, brine depth and glass tilting angle on still productivity. It was found that solar radiation is the greatest affecting parameter on condensate productivity. Productivity decreases with the increase of the wind speed due to the decrease in the fractional energy of evaporation. Furthermore, when the brine depth was increased, the productivity decreased. Moreover, it was found that in summer the tilt angle of the glass cover should be maintained at as acute an angle as possible because the horizontal plane receives more radiation than the inclined plane does, while in winter the productivity increases when the inclination is increased. The

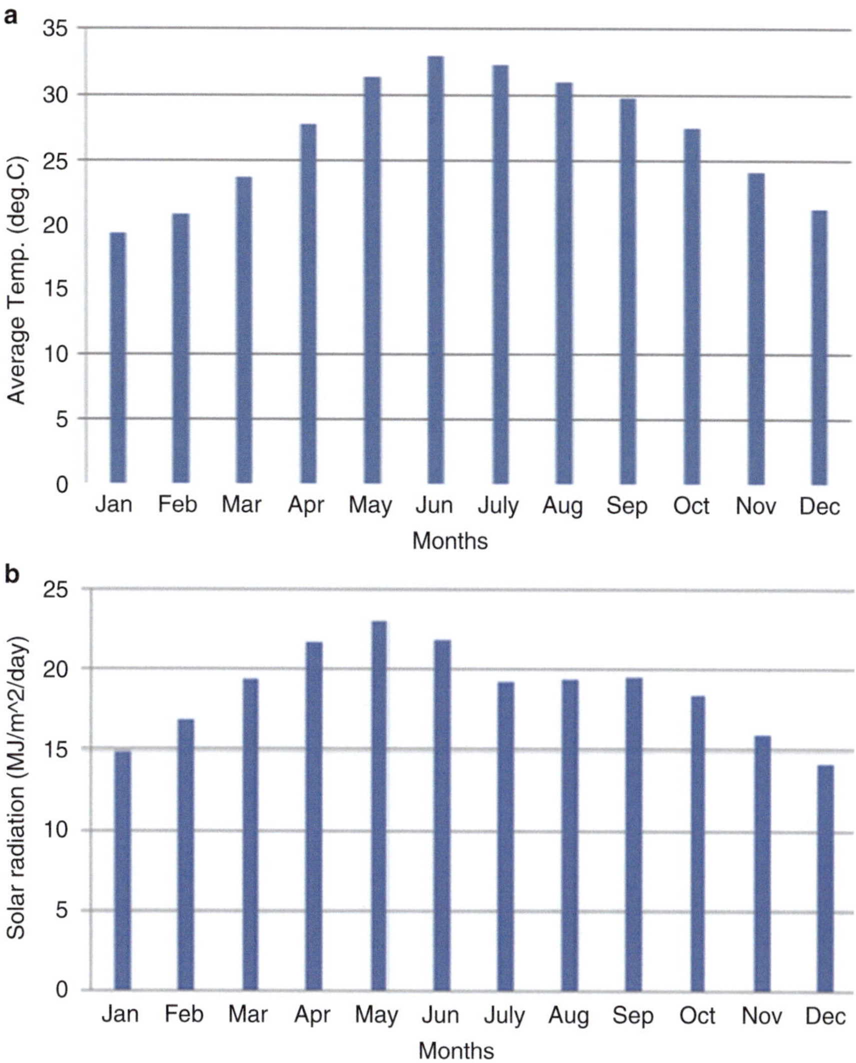

Fig. 3.1 (**a**) Average temperature in Oman during 2006 (Seeb Meteorological Center). (**b**) Average solar radiation in Oman during 2006 (Seeb Meteorological Center)

solar still was tested in different climatic conditions in many places around the world. Singh and Tiwari (2004) studied the monthly performance of passive and active solar stills for different Indian climatic conditions. Medugu and Ndatuwong (2009) designed and tested a solar still under the actual environmental condition of Mubi, Nigeria. They developed a theoretical analysis of heat and mass transfer mechanisms inside the still. They did experimental and theoretical investigations on the distillation performance of the solar still and found that the instantaneous efficiency increases with the increase of solar radiation and the feed water temperature.

Radwan et al. (2009) conducted a study on the single slope solar still for sea water distillation. The investigation was carried out under the open environmental conditions of Egypt on a single slope still inclined at 20 %. The investigation addressed the following parameters: still productivity, distilled water salinity and still performance in term of the still efficiency and the coefficient of performance. They found that still productivity and efficiency increases when solar radiation increases. The highest productivity was 0.226 L/m^2 obtained during the month of July. Also, they found that still performance increases gradually from sunrise until it reaches the maximum value at noon and an hour later and then decreases until it reaches the minimum value at the time of sunrise.

Arjunan et al. (2009) conducted an experimental study on a solar still which had a sponge liner. Two types of measurements were performed for the same climatic conditions: one with the sponge sheet placed on the inner wall surfaces (the back and the side wall) and one without the sponge sheet. It was found that the productivity of the solar still was 15 % higher than the conventional still. They also found that decreasing the water depth increases the productivity of the still.

Afrand et al. (2010) carried out a theoretical study of solar distillation in a single basin under the open environmental condition of Chabahar, Iran. In this research, they investigated still productivity, distilled water salinity and still performance in terms of still efficiency. They found that the maximum efficiency of the solar still was at noon due to the high radiation at that time.

Another study about a single solar still was conducted in India by Prasad et al. (2011). The still was modified with graphite powder to maximize absorptivity. In this study, the effects of four parameters were investigated; the amounts of silicate, acid and graphite powder as well as water depth. The maximum productivity of the still was 1.6 L/m^2. It was found that the productivity of the solar still decreases as the amount of water is increased. When the amount of sodium silicate is increased, the productivity increases, but when more than 150 g was added, the productivity decreased. The same result was obtained when increasing the amount of acid and graphite. Peak performance was obtained with 150 g of sodium silicate, 100 ml of 2NHCl, and 50 g of graphite.

The main aim of the present work is to design and test a single slope water still with a presolar water heating system. The effect of the amount of water in the basin, and of the inlet water temperature to the basin from the presolar water heater on the amount of fresh water yield and on its thermal performance will be investigated.

3.2 Test Rig

A single slope solar water still with a solar preheater system was designed and constructed. The still consisted mainly of a base unit, made from galvanized steel, and a glass cover. The surface area of the still is 1 m^2. The bottom inner surface of the base unit is painted black and the inner side walls are painted white. The side walls of the base unit are insulated from outside with Styrofoam. At the bottom end of the

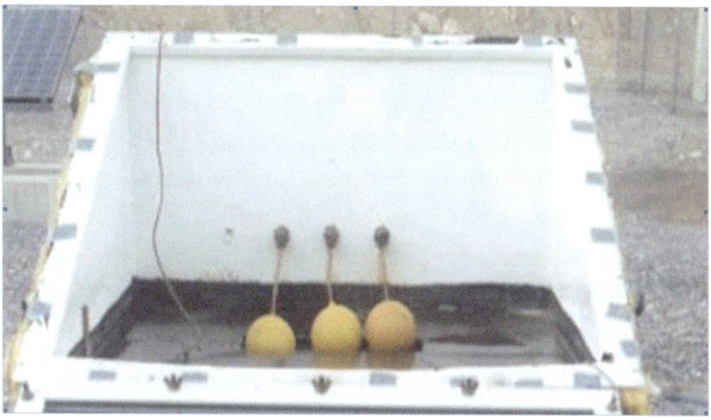

Fig. 3.2 Float valves attached to the still

Fig. 3.3 Experimental set up of single slope solar water still with solar energy heating system

inclined surface of the base unit, a passage is made to collect the fresh water coming down from the cover glass. Three floating valves are attached to the back of the still to allow the adjustment of the water depth inside the basin of the still (see Fig. 3.2). A simple solar collector was constructed and made of a single glass cover with a water loop to preheat the saline water before it entered the still. The whole system, the still, the collector and the supply tank, was assembled as shown in Fig. 3.3.

3.3 Experimental Procedure

The experiments were conducted to study the effect of water level in the basin and the effect of using a solar preheater on the yielded fresh water and on the solar still thermal performance. Thermocouples were positioned at different locations in the still to measure the temperatures throughout the day and were connected to a data logger to record and save the temperature readings every 15 minutes.

First, the solar collector was connected to the inlet of the solar still basin and the correct amount of saline water was introduced into the basin. The basin inlet and the solar collector outlet temperatures were measured by the thermocouples and recorded every 15 minutes in order to find the mean average hourly temperatures. The same procedure was followed for measuring the still water, the cover glass, the vapor and the ambient temperatures. The fresh water yield was collected and recorded using sealed graduated glass containers on an hourly basis.

3.4 Thermal Efficiency of the Water Still

To assess the performance of the still, it is important to identify its thermal efficiency. The still efficiency can be calculated using the following formula

$$\eta = \frac{evaporation\ heat}{total\ input\ solar\ energy} = \frac{\dot{Q}_{ev\,p}}{\dot{Q}_{in}}$$

The heat of vaporization can be calculated using the following formula

$$\dot{Q}_{ev\,p} = \dot{m}h_{fg}$$

Where $\dot{m}$ is the mass flow rate and h_{fg} is the latent heat of vaporization. The energy input due to solar radiation can be calculated as

$$\dot{Q}_{in} = \alpha\tau A_p I$$

Where α, τ, A_p, and I are the absorptivity coefficient, the transmission coefficient of the glass cover, the area of the absorbing plate and the solar intensity on the horizontal surface respectively.

3.5 Results

The results of the experimental tests show that the production rate of fresh water is directly related to the water level height in the basin and to the solar radiation. As the water level in the basin decreases and solar radiation increases, the production rate increases. For a flat bottom surface, the hourly rate yielded and the hourly-daily accumulated fresh water for three different water depths are shown in Figs. 3.4 and 3.5. It was observed that a depth of 1 cm outperforms 2 cm and 3 cm of water depth in the solar still basin. This is due to the small heat capacitance of water at 1 cm

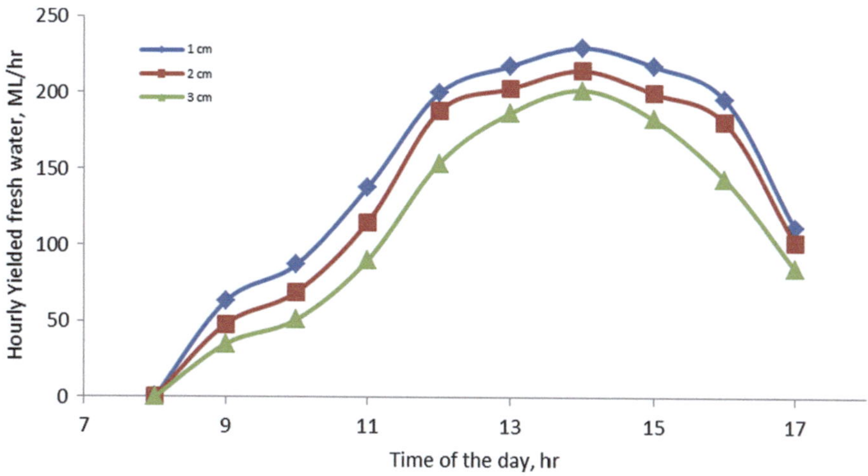

Fig. 3.4 Hourly yielded fresh water for flat surface

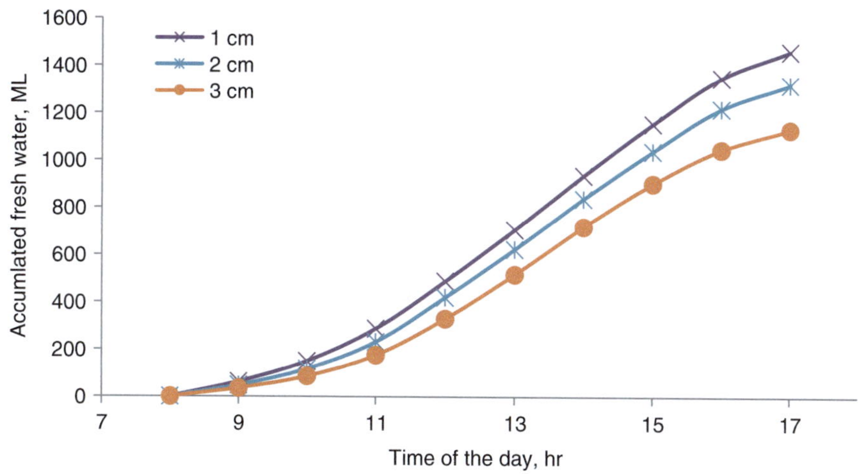

Fig. 3.5 Fresh water cumulative yield

depth compared to the other two depths and alos because it needs less energy to be vaporized. The maximum hourly amount of fresh water yield is at noon and the total hourly accumulation is 1500 ml per meter squared for a water depth of 1 cm, where the solar radiation is at a maximum (See Figs. 3.4 and 3.5).

The average efficiency is calculated over the testing time which lasted between 7 am and 5 pm and was found to be 40.8 % for a flat basin surface. The addition of the solar of water preheater to the system significantly increases the inlet basin saline water temperature to almost saturated temperature and the saline water in the basin needed only a small amount of heat to be vaporized as shown in Figs. 3.6, 3.7 and 3.8 (Table 3.1).

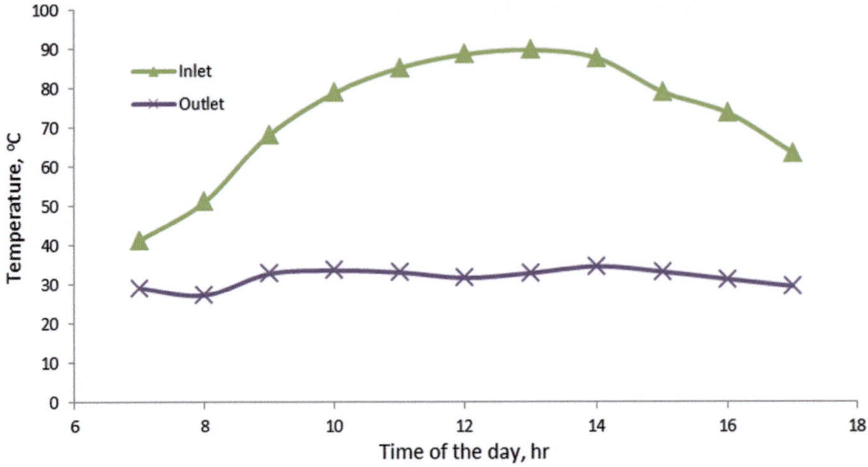

Fig. 3.6 Temperature readings at the solar preheater's inlet and outlet

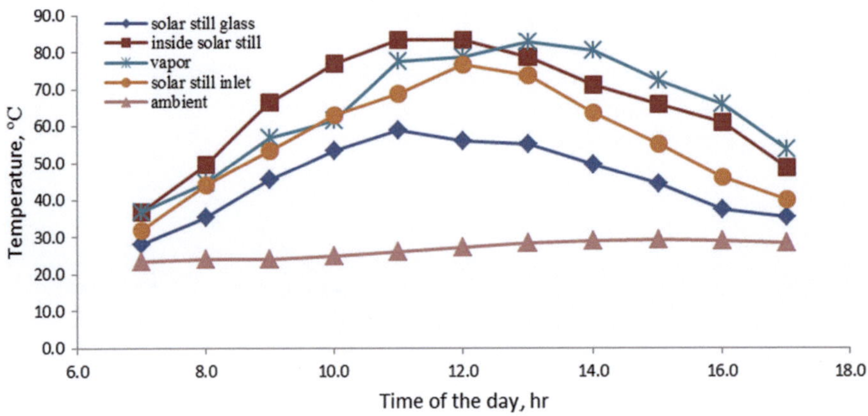

Fig. 3.7 Temperature readings at different locations in the still

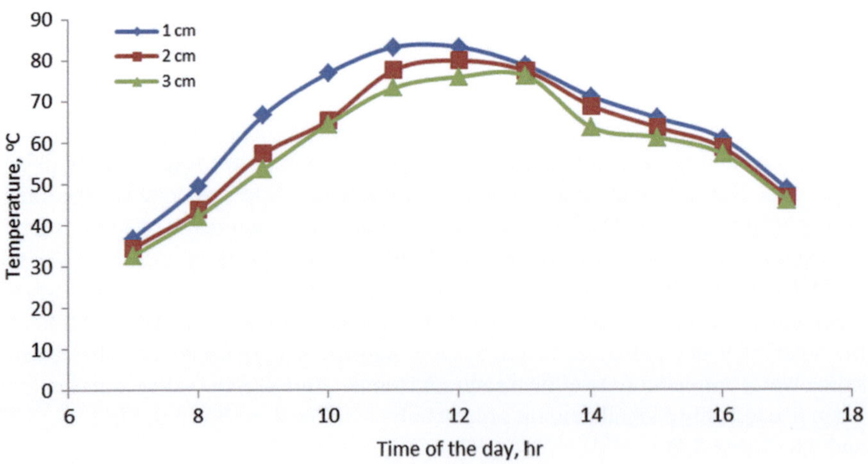

Fig. 3.8 Temperature readings inside the still for three different water depths

Table 3.1 Still thermal efficiency for different tested cases

Tested case	Average thermal efficiency (%)
1 cm depth	40.8
2 cm depth	35.1
3 cm depth	27.9

3.6 Conclusion

This paper has investigates the thermal performance of a solar water still with an enhanced solar heating system. A number of variables have been considered, including the water depth inside the still and the inlet saline water temperature to the solar still from the preheater solar collector system. It was found that a depth of 1 cm gives the best performance in terms of fresh water yield and thermal efficiency. The addition of the solar water preheater to the system significantly increased the inlet basin saline water temperature to almost saturated temperature and the saline water in the basin needed only a small amount of heat to be vaporized, thus increasing the production of fresh water and enhancing the thermal efficiency of the solar still.

List of Figures

References

Afrand M, Behzadmehr A, Karimipour A (2010) A numerical simulation of solar distillation for installation in Chabahar-Iran. World Acad Sci Eng Technol 47:469–474

Arjunan TV, Aybar HS, Nedunchezhian N (2009) A study on effects of water capacity on the performance of a simple solar still. Int J Appl Eng Res 4(11):2223–2234

Medugu DW, Ndatuwong LG (2009) Theoretical analysis of water distillation using solar still. Int J Phys Sci 4(11):705–712

Nafey AS, Abdelkader M, Abdelmotalib A, Mabrouk AA (2000) Parameters affecting solar still productivity. Energy Convers Manag 41:1797–1809

Prasad PR, Padma P, Rajeev GV, Vikky K (2011) Energy efficient solar water still. Int J ChemTech Res 3(4):1781–1787

Radwan SM, Hassanin AA, Abu-zeid MA (2009) Single slope solar still for sea water distillation. World Appl Sci J 7(4):485–497
Seeb Meteorological Center (2006) Seeb Airport, Sultanate of Oman
Singh HN, Tiwari GN (2004) Monthly performance of passive and active solar stills for different Indian climatic conditions. Desalination 168:145–150

Chapter 4
The Effect of Fouling on Performance and Design Aspects of Multiple-Effect Desalination Systems

F. Tahir, M. Atif, and M.A. Antar

Abstract Multiple-effect desalination systems are attractive water desalination units for large capacity production since the brine temperature does not exceed 65 °C and there is less power consumption compared to other thermal desalination systems. The system is modeled for performance prediction as well as for the assessment of design aspects such as the total heat transfer area in the effects, the preheaters and the system down condenser. Then, the system degradation with time due to fouling is considered for both rating and design aspects. It is therefore believed that the results presented can provide guidelines for designers and engineers using multiple-effect desalination systems for specifying system sizes and/or scheduling maintenance of the existing systems.

Keywords MED • Model • Fouling • Design • Performance analysis

4.1 Introduction

Scarcity of drinkable water has become a critical issue because the demand for fresh water is increasing exponentially especially in those areas where sweet and clean water resources are not present in the form of rivers, lakes etc.

Among various water desalination systems, multiple-effect desalination (MED) systems are receiving increasing attention so that they can be considered as the future of large-scale thermal desalination systems. Although the majority of thermal desalination systems in the Gulf region are multi-stage flash (MSF) systems, the potential of increased attention directed to MED is due to higher productivity and lower power consumption (around 2.75 kWh/m^3) compared with MSF (more than 4 kWh/m^3) systems (Darwish and Al-Sairafi 2004; Alasfour

F. Tahir • M. Atif • M.A. Antar (✉)
Department of Mechanical Engineering, King Fahd University of Petroleum and Minerals, Dhahran 31261, Saudi Arabia
e-mail: g201202820@kfupm.edu.sa; maimoonatif@kfupm.edu.sa; antar@kfupm.edu.sa

M. Baawain et al. (eds.), *Recent Progress in Desalination, Environmental and Marine Outfall Systems*, DOI 10.1007/978-3-319-19123-2_4

et al. 2005). MED systems have received considerable attention in the literature as several desalination units were built in Saudi Arabia based on MED technology within the past two to three decades, including three 1500 m^3/day units in Azizia, a remote area, ten units with a capacity of 4500 m^3/day in the southwest coast of the Kingdom and two units of 9000 m^3/day in Rabigh. The unit that was built in Yanbu has a capacity of 68,000 m^3/day and about three years ago, a 800,000 m^3/day unit with thermal vapor compression was built on the east coast. MED units are characterized by a high overall heat transfer coefficient and low pumping power. The top brine temperature (TBT) is in the order of 65 °C, thus allowing low grade energy sources for supplying heat to the desalination units. In addition, the low TBT allows for the use of low grade material such as aluminum alloys and carbon steel shells coated with an epoxy layer (Ophir and Lokiec 2005).

Several attempts were made to model MED systems for better understanding of the system and for possible improvements to the system performance. Models apply the first law of thermodynamics to simulate energy balances for the effects and the condenser and to estimate the required surface area needed as a system design parameter. The second law analysis is used to pinpoint the possible room for improvement through reduction of exergy loss.

El-Dessouky and Ettouney (1999, 2002) and El-Dessouky et al. (2000) have performed performance and design modeling of various layouts of MED systems, including forward feed (FF), parallel feed and parallel cross feed. They also considered the effect of thermal vapor compression (TVC) and mechanical vapor compression (MVC). The major findings of their work are related to the decrease of the required surface area at higher steam temperature and to the significant increase in the performance ratio with the addition of TVC while the performance ratio decreases slightly with increasing steam temperature. They also reported that although a parallel flow arrangement is simpler to build and operate compared to a parallel cross feed, it has a lower performance ratio.

Various assumptions were also considered in the developed models that affected the results. Examples include constant temperature drop between stages, heat transfer coefficient and specific heat in a non-iterative model (Al-Sahali and Ettouney 2007; Darwish and Alsairafi 2004).

Performance as well as design parameters that were reported include the dependence of performance on the number of effects, water salinity and feed water temperature (Ameri et al. 2009). Increasing the condenser area by 32 % will result in an increase in the performance ratio of about 15 % as indicated by Kamali and Mohebinia (2008).

Similar results were reported by Aly and El-Fiqi (2003), through their analysis of the forward feed MED (FF-MED) system. A more detailed model of the FF-MED system was reported by Mistry et al. (2013), where more updated seawater properties were considered in the model. They directed the vapor leaving an effect to preheat the feed seawater before being admitted to the next effect. They indicated the relative importance of the boiling point elevation on the calculated surface area compared to other minor losses like non-equilibrium allowance.

The limitation imposed on the TBT due to scale formation was examined by Minnich et al. (1995), who reported that optimum value of gain output ratio of 14 may be obtained at TBT of 110 °C with a constant heat flux of 12.6 kW/m^2 to take advantage of the high heat transfer coefficient at such a high temperature in addition to the assumption of a constant temperature difference. In another study, El-Allawy (2003) modeled the MED-TVC performance with TBT that varies from 58 to 70 °C and for a number of effects that varied from 3 to 6. Increasing the number of effects was a key factor in improving the performance ratio. Increasing TBT while the bottom temperature is kept unchanged was reported to increase the heat transfer area (Darwish and Al-Najem 1987) whereas fixing both the temperature and increasing the number of effects increases both the heat transfer area and the performance ratio. This result has also been confirmed by (Hamed 1992). Computational fluid dynamics simulation of a thermocompressor were presented by Park et al. (2005), where they reported that these are sensitive to design parameters including mixing process flow rates and dimensions. A formula for calculating the entrainment ratio was proposed by Power (1994) based on operating parameters. However, it has a limited validity range and cannot be used for high motive steam pressures (Ameri et al. 2009).

Second law analysis has been a useful tool to indicate the exergy losses that would, when reduced, improve system performance. The major portion of exergy destruction was found in the thermocompressor and the effects. Choi et al. (2005), Darwish and Al-Najem (1987), and Alasfour et al.(2005) indicated that lower energy consumption can be realized through decreasing the effect-to-effect temperature difference. MED-TVC desalination unit has the lowest exergy destruction compared to MED and MED-MVC units as reported by Hamed et al. (1996) who added that a better performance ratio can be obtained with less TBT and more effects. Lower exergy losses were responsible for the superior performance of parallel flow MED with preheaters as compared with forward feed MED with preheaters (Greogorzewski and Genthner 1995).

Keeping in view the thermodynamic superiority of MED plants and the growing need of thermal desalination, a MED forward feed plant in the present study has been modeled where the performance is studied, design considerations are proposed and in the end the fouling effect on the plant has been discussed.

4.2 Model

A detailed mathematical model of FF-MED systems has been developed by El-Dessouky and Ettouney (2002). It is important to state that updated seawater properties are used for a more accurate prediction of the results. The system layout is shown in Fig. 4.1. Steam from a power plant is used as a primary source of heat for MED. In the forward feed type, the flow of vapor and the flow of brine in the evaporators is in the same direction.

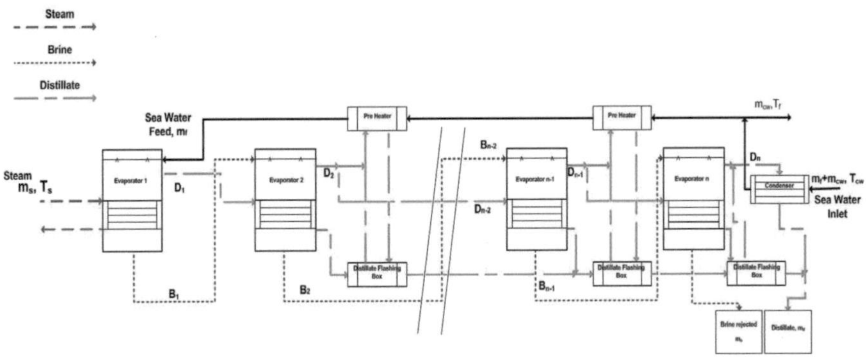

Fig. 4.1 Layout of forward feed MED plant

The pre-heaters and flash boxes are used for heat recovery. Each effect contains heat exchanger tubes, vapor space, brine spray nozzles, a mist eliminator and a brine collecting box.

The intake feed ($m_{cw} + m_f$) flows into the condenser of the last effect where it is pre-heated. Then it passes through several preheaters where the condensed distillate raises the feed temperature. After pre-heating, the feed (m_f) is sprayed and further heated in the first effect by condensing steam flowing into the tubes. Part of the feed is evaporated and the rest (brine) flows into the next effect where the pressure is reduced to decrease the saturation temperature.

A part of this brine is evaporated using the heat liberated from condensing the vapor formed in the previous effect and this process is repeated in the following effects. The condensed vapors are collected as desalinated water.

Assumptions used in this model are as follows:

(a) The operation is steady state.
(b) The plant is designed for a distillate flow of 1 kg/s.
(c) Steam temperature is taken as $T_s = 70$ °C, cooling water temperature $T_{cw} = 25$ °C, the last effect temperature $T_n = 40$ °C and the feed temperature $T_f = 35$ °C.
(d) Temperature losses in piping and during condensation are taken as 1 °C.
(e) The intake and rejected salinity of seawater are taken $X_f = 42{,}000$ ppm and $X_n = 70{,}000$ ppm, respectively.
(f) The salinity of distillate water is zero.
(g) The distillate formed in the effects is pure vapor condensate.
(h) Non-equilibrium allowances for flashing and boiling point elevation (BPE) are considered in the model.
(i) Total areas for effects ($A_e = \sum_{j=1}^{n} A_{ej}$) are constant; and total pre-heater areas ($A_p = \sum_{j=1}^{n} A_{pj}$) are constant.

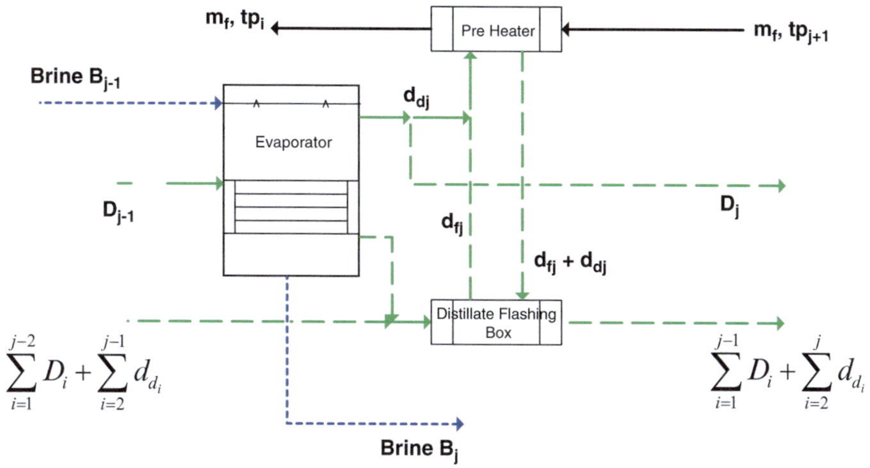

Fig. 4.2 Sectional view of *j*th effect (from 2nd effect to (n-1) effect)

(j) Time-dependent fouling is considered in estimating the overall heat transfer coefficient.
(k) The outer (r_o) and inner (r_i) radii of tubes are taken as 0.015 m and 0.0125 m respectively.

For material balances and energy balances, *j*th effect is considered as shown in Fig. 4.2. Although the model equations are listed in other references, they are written here for the sake of completeness.

4.2.1 Mass Balances

The overall liquid mass and salt balances are given by

$$m_f = m_d + m_b \tag{4.1}$$

$$m_b X_b = m_f X_f \tag{4.2}$$

where X is the salt concentration, m is the mass flow rate, and the subscripts d, f and b represent the distillate, feed seawater and the rejected brine.

The total distillate flow rate as the final product can be expressed as

$$m_d = \sum_{j=1}^{n} D_j + \sum_{j=2}^{n} d_{dj} \tag{4.3}$$

where D_j and d_{dj} are the amounts of vapor formed by boiling in *j*th effect and flashing in the flash boxes respectively and the subscripts j and n define the effect number and the total number of effects in the plant. On the other hand, the subscript i denotes the counter for the summation in Fig. 4.2.

The total seawater feed is divided into the distillate produced and the brine formed in the first effect

$$m_f = B_1 + D_1 \tag{4.4}$$

Similarly for effects 2 to n, the mass balance of each effect is given by

$$m_f = B_j + \sum_{i=1}^{j} D_i + \sum_{i=2}^{j} d_{d_i} \tag{4.5}$$

The salt balance for each effect can be expressed as

$$X_j B_j = X_f m_f \tag{4.6}$$

4.2.2 Energy Balances

In the 1st effect, the latent heat of steam condensation used to raise the feed seawater temperature from t_{p2} to the boiling temperature T_1 and to evaporate some of the feed seawater into vapor (D_1) at T_1 is given as

$$q = m_s h_{fgs} \tag{4.7}$$

$$q = m_f C_p (T_1 - t_{p2}) + D_1 h_{fgv1} \tag{4.8}$$

where q represents the heat required for each effect. This value is taken as a constant for all of the effects as supposed by El-Dessouky et al. (2000).

The heat required to produce the distillate from effect 2 to n is given by

$$q = D_2 h_{fgv2} \tag{4.9}$$

$$q = D_j h_{fgvj} \tag{4.10}$$

The latent heat of the produced vapors by boiling in an effect j-1 is used to evaporate a small amount of brine feed in the next effect j. The decrease in the distillate amount is caused by the increase in the vapor latent heat upon the decrease in effect temperature.

$$D_j h_{fgvj} = D_{j-1} h_{fgcj-1} \tag{4.11}$$

The boiling process occurs on the outer surface of the evaporator tubes. The condensation temperature T_{cj} is lower than the effect temperature T_j by (boiling point elevation) BPE_j and the saturation temperature depressions associated with pressure losses ΔT_{cj}. h_{fgcj} is the latent heat after condensation at temperature T_{cj}. The resulting condensation temperature after the thermodynamic losses is given by

$$T_{cj} = T_j - BPE_j - \Delta T_{cj} \tag{4.12}$$

The energy balance in the flash boxes in the 2nd effect and from the effects 3 to n which give the flow rate of distillate is

$$h_{fgv2}{}'' d_{f2} = D_1 C_p \left(T_{c1} - T_2{}'' \right) \tag{4.13}$$

$$h_{fgvj}{}'' d_{fj} = \left(\sum_{i=1}^{j-1} D_i + \sum_{i=2}^{j-1} d_{d_i} \right) C_p \left(T_{cj-1} - T_j{}'' \right) \tag{4.14}$$

where $h_{fgvj}{}''$ is the latent heat of vaporization at temperature $T_{vj}{}''$. The amount of condensed vapors entering the flash boxes in the 2nd effect is equal to D_1. This amount increases in the subsequent flash boxes by the amount of vapor formed by boiling and flashing within these effects.

$$T_j{}'' = T_{vj}{}'' + NEA_j{}'' \tag{4.15}$$

$$T_{cj}{}'' = T_{vj}{}'' - \Delta T_{cj}{}'' \tag{4.16}$$

The energy balances for the condenser are given as

$$q_c = \eta_c \left(D_n h_{fgcn} + d_{dn} h_{fgcn}{}' + d_{fn} h_{fgcn}{}'' \right) \tag{4.17}$$

$$q_c = \left(m_f + m_{cw} \right) C_p \left(T_f - T_{cw} \right) \tag{4.18}$$

where q_c is the heating energy of the condenser, η_c and is the heat exchanger efficiency of the condenser. The feed stream is heated in the condenser and energy is provided by the condensation of vapors which are formed by flashing and boiling in the last effect and by flashing in the associated last flash box.

Energy balance of the preheaters is given by

$$\eta_p \left(d_{dj} h_{fgcj}{}' + d_{fj} h_{fgcj}{}'' \right) = m_f C_p \left(t_{pj} - t_{pj+1} \right) \tag{4.19}$$

$$d_{dj} h_{fgvj}{}' = B_{j-1} C_p \left(T_{j-1} - T_j{}' \right) \tag{4.20}$$

In all the effects, the boiling temperature is higher than the vapor saturation temperature by the boiling point elevation, where the boiling point elevation is due to the difference in the concentration of salt.

$$T_{vj}^{'} = T_j^{'} - BPE_j^{'} \tag{4.21}$$

$$T_j^{'} = T_j + NEA_j^{'} \tag{4.22}$$

$$T_{cj}^{'} = T_{vj}^{'} - \Delta T_{cj}^{'} \tag{4.23}$$

In the above equations, $h_{fgcj}{'}$ and $h_{fgcj}{''}$ are the latent heat of condensation of flashed vapors in the pre-heaters at $T_{cj}{'}$ and $T_{cj}{''}$.

4.2.3 Heat Transfer Areas

The heat transfer surface area of any effect is given by

$$A_{e1} = \frac{q}{U_1(T_s - T_1)} \tag{4.24}$$

$$A_{ej} = \frac{q}{U_j \Delta T_j} \tag{4.25}$$

where $j = 2$ to n. The pre-heater heat transfer area is given by

$$A_{pj} = \frac{m_f C_p (t_{pj} - t_{pj+1})}{U_p LMTD_{pj}} \tag{4.26}$$

where the log mean temperature difference across the preheaters is given by

$$LMTD_{pj} = \frac{t_{pj} - t_{pj+1}}{\ln\left[\frac{T_{cj}^{'} - t_{pj+1}}{T_{cj}^{'} - t_{pj}}\right]} \tag{4.27}$$

The area of the condenser is given by

$$A_c = \frac{q_c}{U_c LMTD_c} \tag{4.28}$$

The log mean temperature difference across the condenser is given by

$$LMTD_c = \frac{T_f - T_{cw}}{\ln\left[\frac{T_{cn}^{'} - T_{cw}}{T_{cn}^{'} - T_f}\right]} \tag{4.29}$$

4.2.4 Performance

The performance ratio (PR) is defined as the ratio of the distillate formed to the amount of steam required and is given by

$$PR = \frac{m_d}{m_s} \tag{4.30}$$

The specific mass of cooling flow rate is defined as follows

$$sM_{cw} = \frac{m_{cw}}{m_d} \tag{4.31}$$

The specific area (sA) is the total sum of areas and includes the area of the evaporators, the pre-heaters and the condenser per unit distillate formed and is given as

$$sA = \frac{\sum_{j=1}^{n} A_{ej} + \sum_{j=2}^{n-1} A_{pj} + A_c}{m_d} \tag{4.32}$$

4.2.5 Heat Transfer Coefficients

The generalized overall heat transfer coefficient for the evaporators can be expressed as

$$\frac{1}{U} = \frac{1}{h_{in}} \frac{r_o}{r_i} + R_{fo} + r_o \frac{\ln\left({}^{r_o}/_{r_i}\right)}{k_w} + \frac{1}{h_o} \tag{4.33}$$

The generalized overall heat transfer coefficient for the pre-heaters and the condenser is given by

$$\frac{1}{U} = \frac{1}{h_{in}} \frac{r_o}{r_i} + R_f \frac{r_o}{r_i} + r_o \frac{\ln\left({}^{r_o}/_{r_i}\right)}{k_w} + \frac{1}{h_o} \tag{4.34}$$

The fouling resistance in the inner tube surface of the pre-heaters and the condenser and the fouling resistance in the outer tube surface of the evaporators are taken into account. The surface with which the desalinated vapors/water is in contact, and the fouling are assumed to be negligible. The same asymptotic expression for the fouling factor has been used for all the equipment as reported by Antar and Zubair (2007).

$$R_f = 0.199\left(1 - \exp\left(\frac{-time}{61.55}\right)\right) \tag{4.35}$$

Han and Fletcher (1985) proposed a heat transfer coefficient on the tube outside for the falling film evaporator

$$h_{out} = 0.0004\left(\rho_{sw}^2 g \frac{k_{sw}^3}{\mu_{sw}^2}\right)^{1/3} \mathrm{Re}^{0.2}\mathrm{Pr}^{0.65} q''^{0.4} \tag{4.36}$$

Shah (1978) proposed heat transfer coefficients for vapor condensation inside tubes

$$h_l = 0.023\mathrm{Re}_i^{0.8}\mathrm{Pr}_v^{0.4}\frac{k_{LC}}{2r_i} \tag{4.37}$$

$$h_u = h_l(1-x)^{0.8} \tag{4.38}$$

$$h_{in} = \left(1 + \frac{3.8}{Z^{0.95}}\right)h_u \tag{4.39}$$

$$Z = \left(\frac{1}{x} - 1\right)^{0.8} \mathrm{Pr}_v^{0.4} \tag{4.40}$$

Wangnick (1995) proposed an expression for the heat transfer coefficient of seawater flowing inside tubes. This expression will be used for condenser and pre-heaters.

$$h_{in} = \frac{3293.5 + T_f(84.24 - 0.1714T_f) - X_f(8.471 + 0.01161X_f + 0.2716T_f)}{1000\left(\frac{d_i}{0.017272}\right)^{0.2}(0.656V_c)^{0.8}\frac{d_i}{d_o}} \tag{4.41}$$

Henning and Wangnick (1995) proposed expressions for the heat transfer coefficient of vapor condensation outside the tube surface. These expressions will be used for condensers and pre-heaters.

$$h_{out} = 0.725\left(\frac{k_{LC}^3\rho_{LC}(\rho_{LC} - \rho_{VC})gh_{fgv}}{2r_o\mu_L\Delta T_c}\right)^{0.25} C_{1c}C_{2c} \tag{4.42}$$

$$C_{1c} = 1.23795 + 0.353808N_{1c} - 0.0017035N_{1c}^2 \tag{4.43}$$

$$C_{2c} = 1 - 34.313X_{nc} + 1226.8X_{nc}^2 - 14923X_{nc}^3 \tag{4.44}$$

$$N_{tc} = \frac{m_f}{\pi r_i^2 \rho_{sw} V_c} \tag{4.45}$$

$$N_{1c} = 0.564\sqrt{N_{tc}} \tag{4.46}$$

4.3 Discussion of Results

4.3.1 Comparisons and Validation

The model developed is compared with the models of El-Dessouky and Ettouney (1999), Darwish and Alsairafi (2004) and Mistry et al. (2013) as shown in Figs. 4.3

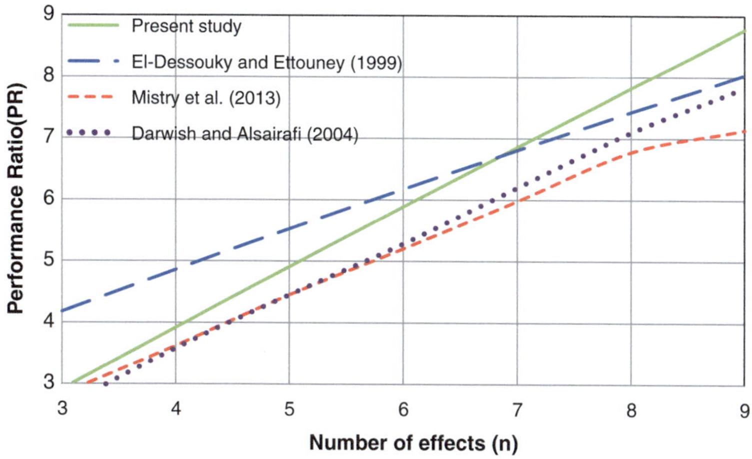

Fig. 4.3 Comparison of performance ratio for different models

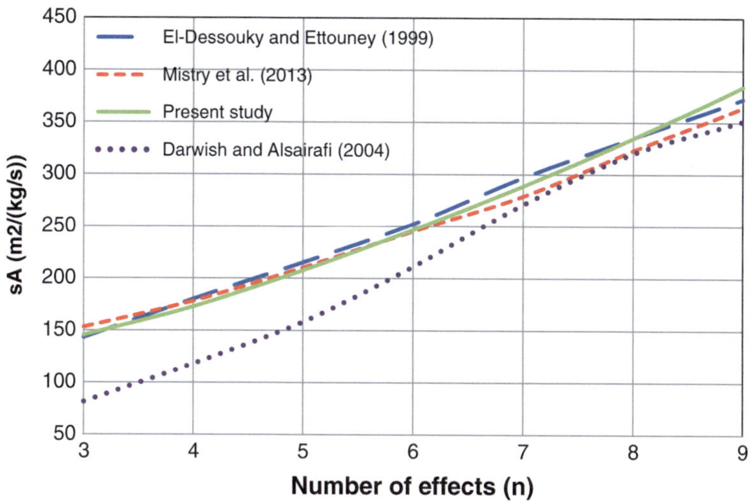

Fig. 4.4 Comparison of specific areas for different models

and 4.4. The performance ratio of the developed model is higher than that of Mistry et al. (2013) and Darwish and Alsairafi (2004), but it is comparable to the El-Dessouky and Ettouney (1999) detailed model. The difference in PR to the El-Dessouky and Ettouney (1999) model is due to a difference in input parameters and updated seawater properties used in this model. The difference to Mistry et al. (2013) is due to the flow direction of the vapor leaving each effect. The specific areas are in good agreement with the Mistry et al. (2013) and El-Dessouky and Ettouney (1999) models. However, the Darwish and Alsairafi (2004) model predicts a less specific area due to simplified assumptions.

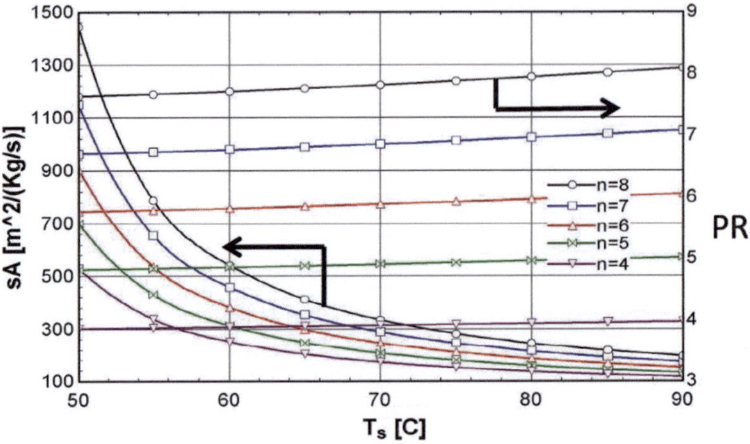

Fig. 4.5 Variation of specific area and performance ratio with steam temperature and number of effects

4.3.2 Design Results

Figure 4.5 shows the effect of steam temperature on sA and PR. Due to an increase in the steam temperature and flashing in flash boxes and evaporator increases, the overall performance ratio improves. Increasing the steam temperature reduces the required specific heat transfer areas, hence reducing the overall cost. As a design option, the selection of high steam temperature enhances plant performance but it is limited by the scaling problem.

Cooling water flow rates and condenser areas may vary to keep performance unchanged in case the temperature of seawater/cooling water changes as shown in Figs. 4.6 and 4.7. As the temperature increases, a larger condenser area and a higher cooling water flow rate would be required to condense the vapor leaving the last effect. Increasing the number of effects decreases the specific cooling water flow rate since the amount of vapor coming from the last effect decreases which needs less energy to condense. An increase in the cooling water temperature, while keeping T_f fixed, results in a low temperature difference and the area of the condenser must be increased in order to keep the same performance.

Practically, seawater temperature fluctuates due to seasonal change. This factor should be taken into account in the design phase. Either a condenser area has to be selected based on the maximum cooling water temperature or the plant may be equipped with an additional condenser.

The fouling factor has an asymptotic profile. Initially, the resistance increases sharply, and then the rate of increase is reduced with time and the overall heat transfer coefficient decreases as shown Fig. 4.8 for the 2nd effect (the trend is similar for the other effects). Since the fouling resistance is an exponential function, the specific area required to keep the same performance increases with time. Hence, the specific area at the time of designing will be selected on the basis of scheduled overhauling, say after every 5 years (60 months).

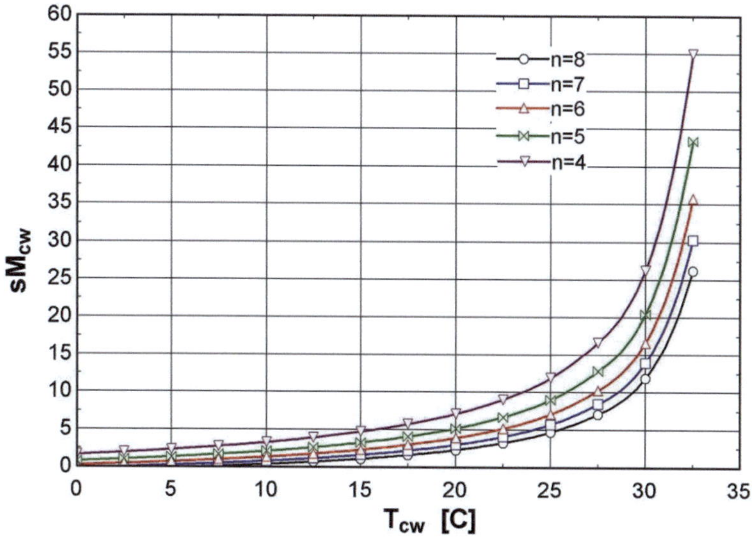

Fig. 4.6 Variation of specific cooling water flow rate with cooling water temperature and number of effects

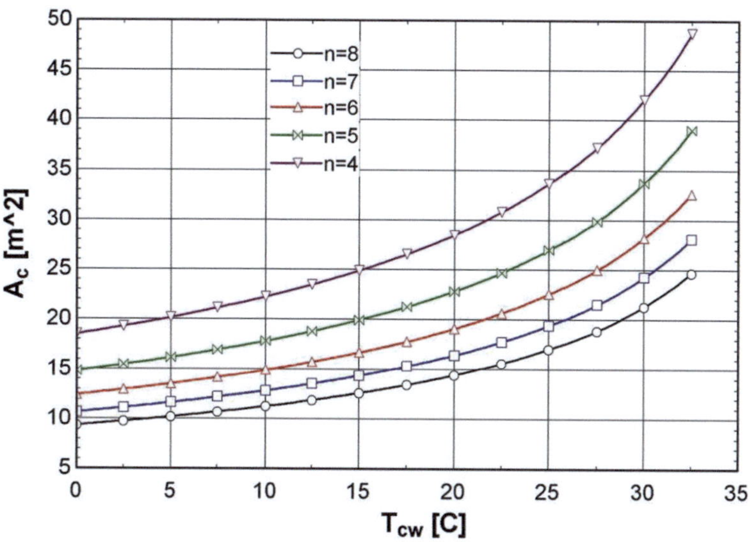

Fig. 4.7 Required condenser area at cooling water temperature and number of effects for same plant performance

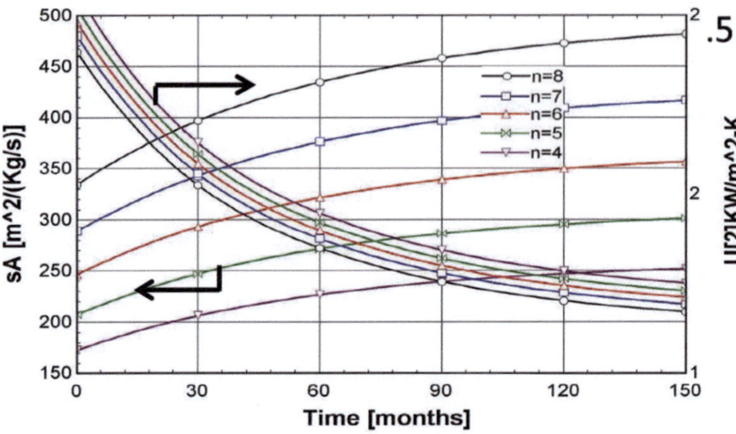

Fig. 4.8 Variation of overall heat transfer coefficient and required specific area with time and number of effects for same performance

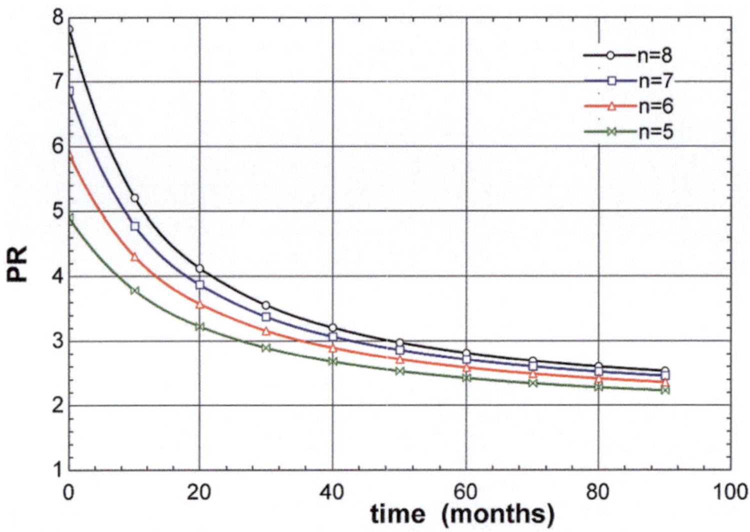

Fig. 4.9 Degradation of plant performance over time for plant having different number of effects

4.3.3 Rating Results

Once the plant is designed and commissioned, its performance will not be the same as time passes. The equipment degrades due to fouling that occurs in the effects, in the pre-heaters and in the condenser tubes which will reduce the heat transfer rate. Therefore, it is important to predict the performance of the plant over time to decide on a major overhauling time to remove scales or replace the faulty tubes so that the plant can retain its original performance. The performance ratio of the plant over time for a different number of effects is shown in Fig. 4.9. PR starts around

8 (for $n = 8$) and decreases with time and after 90 months the PR is approximately 2.5 which is around one third of the initial performance. Due to the fouling asymptotic profile, the decline of PR at the start is sharp but is negligible after 80 months.

Although the MED system considered in this work is a forward feed system, it is believed that qualitatively similar results are applicable to other layouts such as the parallel and parallel cross arrangements.

4.4 Conclusion

Detailed forward feed MED plant modeling has been carried out and compared with other models present in the literature. The performance of the plant and the design considerations have been discussed for different parameters such as steam temperature, cooling water temperature and fouling resistance. Rating of the plant's performance ratio is analyzed over time so that suggestions for overhauling in order to optimize the performance and economics can be made.

Design parameters are chosen so that the surface areas are selected based on predicting the overall heat transfer coefficient as a function of time. Oversizing can be avoided once fouling resistance is estimated and incorporated into the model so that time intervals between shutdowns can be determined.

Acknowledgement The support provided by KFUPM to complete this study is gratefully acknowledged.

Nomenclature

A_c	Condenser area (m^2)
A_{ej}	Area of *j*th effect, $j = 1$ to n(m^2)
A_{pj}	Area of *j*th pre-heater, $j = 2$ to n-1 (m^2)
B_j	Brine leaving *j*th effect, $j = 1$ to n (kg/s)
BPE_j	Boiling point elevation in *j*th effect, $j = 1$ to n (°C)
BPE_j'	Boiling point elevation of flashed vapour in *j*th effect, $j = 2$ to n (°C)
C_p	Specific heat of water (kJ/kg-K)
d_i	Internal diameter of tube (m)
d_o	Outer diameter of tube (m)
D_j	Distillate formed in *j*th effect, $j = 1$ to n (kg/s)
ΔT_j	Temperature difference across *j*th effect, $j = 1$ to n(°C)
d_{dj}	Distillate formed due to flashing in *j*th effect, $j = 2$ to n(kg/s)
d_{fj}	Flashed distillate in the *j*th flashing box, $j = 2$ to n(kg/s)
g	acceleration due to gravity, m/s^2
η_c	Efficiency of condenser
η_p	Efficiency of pre-heater
h_{in}	Heat transfer co-efficient inside the tube (kW/m^2-K)
h_o	Heat transfer co-efficient outside the tube (kW/m^2-K)
h_{fgcj}	Latent heat of condensation inside the tubes at T_{cj} of the *jth* effect, $j = 2$ to n(kJ/kg)
h_{fgcj}'	Latent heat of condensation of the flashed vapor at T_{cj}' in the *j*th effect, $j = 2$ to n (kJ/kg)

$h_{fgcj}^{''}$	Latent heat of condensation of the flashed vapor at $T_{cj}^{''}$ in the *jth* effect, $j=2$ to n (kJ/kg)
h_{fgs}	Latent heat of steam (kJ/kg)
h_{fgvj}	Latent heat of vaporization in *j*th effect, $j=1$ to n(kJ/kg)
$h_{fgvj}^{'}$	Latent heat of vaporization of the flashed vapour in the *j*th effect, $j=2$ to n(kJ/kg)
$h_{fgvj}^{''}$	Latent heat of vaporization of the flashed vapor from *j*th flashing box, $j=2$ to n (kJ/kg)
k_w	Thermal conductivity of the tubes (W/m-K)
k_{sw}	Thermal conductivity of the seawater (W/m-K)
k_{LC}	Thermal conductivity in the liquid phase (W/m-K)
$LMTD_c$	Log mean temperature difference across the condenser (°C)
$LMTD_{pj}$	Log mean temperature difference across the *jth* pre-heater, $j=2$ *to n-1* (°C)
m_b	Total mass flow rate of brine (kg/s)
m_{cw}	Mass flow rate of cooling water (kg/s)
m_d	Total mass flow rate of distillate (kg/s)
m_s	Mass flow rate of steam (kg/s)
m_f	Feed flow rate (kg/s)
μ_{sw}	Seawater viscosity, kg/m-s
μ_L	Viscosity in the liquid phase, kg/m-s
n	Total number of effects
$NEA_j^{'}$	Non equilibrium allowance within the *jth* effect, $j=2$ to n (°C)
$NEA_j^{''}$	Non equilibrium allowance for the flashing box for the *jth* effect, $j=3$ to n (°C)
PR	Performance ratio of the plant
Pr	Prandtl number outside the tubes
Pr_v	Prandtl number inside the tubes
q	Heat transfer within each effect (kW)
q_c	Heat transfer within the condenser (kW)
q''	Heat flux (kW/m^2)
ρ_{sw}	Density of seawater (kg/m^3)
ρ_{LC}	Density in the liquid phase (kg/m^3)
ρ_{VC}	Density in the vapor phase (kg/m^3)
r_o	Outside radius of the tubes (m)
r_i	Inside radius of the tubes (m)
Re	Reynolds number outside the tubes
Re_i	Reynolds number inside the tubes
R_{fo}	Fouling factor outside the tubes (kW/m^2-K)
R_f	Fouling factor inside the tubes (kW/m^2-K)
sA	Total specific area of the system (m^2/(kg/s))
sM_{cw}	Specific mass of cooling water flow rate
t_{pj}	Temperature at the outlet of the *j*th pre-heater, $j=2$ to n-1 (°C)
T_{cw}	Temperature of cooling water (°C)
T_{cj}	Condensation temperature of the distillate in *j*th effect, $j=1$ to n (°C)
$T_{cj}^{'}$	Condensation temperature of the flashed vapor in *j*th effect, $j=2$ to n (°C)
$T_{cj}^{''}$	Condensation temperature of flash distillate from the flashing box, $j=2$ to n (°C)
ΔT_{cj}	Vapor condensation losses inside the tubes of *jth* effect, $j=2$ to n (°C)
$\Delta T_{cj}^{'}$	Vapor condensation losses associated with the flashed vapor within the *jth* effect, $j=2$ to n (°C)
$\Delta T_{cj}^{''}$	Vapor condensation losses associated with the flashing box of *jth* effect, $j=2$ to n (°C)
T_f	Temperature of feed (°C)
T_j	Temperature of brine at *j*th effect, $j=1$ to n (°C)
$T_j^{'}$	Temperature of flashing brine within the *jth* effect, $j=2$ to n (°C)
$T_j^{''}$	Temperature of flashing brine in flash boxes at the *jth* effect, $j=2$ to n (°C)

T_s	Temperature of steam (°C)
T_{vj}	Vapor saturation temperature in the *jth* effect, $j = 1$ to n (°C)
T_{vj}'	Vaporization temperature of flashed vapor in the *j*th effect, $j = 2$ to n (°C)
T_{vj}''	Vaporization temperature of flashed vapor within the flashing box at the *jth* effect, $j = 2$ to n (°C)
U	Generalized overall heat transfer coefficient (kW/m^2-K)
U_c	Overall heat transfer coefficient of the condenser (kW/m^2-K)
U_j	Overall heat transfer coefficient of *j*th effect, $j = 1$ to n (kW/m^2-K)
U_p	Overall heat transfer coefficient of the pre-heaters, $j = 2$ to n-1 (kW/m^2-K)
V_c	Seawater velocity inside the tubes (m/s)
x	vapor phase mass fraction
X_f	Feed salinity (ppm)
X_j	Salinity of brine leaving each effect, $j = 1$ to n (ppm)
$X_b(X_n)$	Salinity of brine leaving the last effect (ppm)

List of Figures

References

Alasfour FN, Darwish MA, Bin Amer AO (2005) Thermal analysis of MEE-Tvc + MEE desalination systems. Desalination 174(1):39–61

Aly NH, El-Fiqi AK (2003) Thermal performance of seawater desalination systems. Desalination 158:127–142

Al-Sahali M, Ettouney H (2007) Developments in thermal desalination processes: design, energy, and costing aspects. Desalination 214(1–3):227–240

Ameri M, Mohammadi SS, Hosseini M, Seifi M (2009) Effect of design parameters on multi-effect desalination system specifications. Desalination 245(1–3):266–283

Antar MA, Zubair SM (2007) The impact of fouling on performance evaluation of multi-zone feedwater heaters. Appl Therm Eng 27:2505–2513

Choi H-S, LeeT-J KV-G, Song S-L (2005) Performance improvement of multiple-effect distiller with thermal vapor compression system by exergy analysis. Desalination 182(1–3):239–249

Darwish MA, Alsairafi A (2004) Technical comparison between TVC/MEB and MSF. Desalination 170(3):223–239
Darwish MA, Al-Najem NM (1987) Energy consumptions and costs of different desalting systems. Desalination 64:83–96
El-Allawy M (2003) Predictive simulation of the performance of MED/TVC desalination distiller. In: IDA conference, Bahamas
El-Dessouky HT, Ettouney HM (1999) Multiple-effect evaporation desalination systems. Thermal analysis. Desalination 125(1–3):259–276
El-Dessouky HT, Ettouney HM, Mandani F (2000) Performance of parallel feed multiple effect evaporation system for seawater desalination. Appl Therm Eng 20(17):1679–1706
El-Dessouky HT, Ettouney HM (2002) Principles of seawater desalination. Elsevier Science BV, Amsterdam
Greogorzewski A, Genthner K (1995) Multi-effect distillation: a study and comparison of different process configuration. In: Proceedings of the IDA World congress on desalination and water science, Abu Dhabi, UAE, November 1995
Hamed OA (1992) Thermal assessment of a multiple effect boiling (Meb) desalination system. Desalination 86(3):325–339
Hamed OA et al (1996) Thermal performance and exergy analysis of a thermal vapor compression desalination system. Energy Convers Manag 37(4):379–387
Han J, Fletcher L (1985) Falling film evaporation and boiling in circumferential and axial grooves on horizontal tubes. Ind Eng Chem Process Des Dev 24:570–597
Henning S, Wangnick K (1995) Comparison of different equations for the calculation of heat transfer coefficients in MSF multi-stage flash evaporators. In: Proceedings of the IDA World congress on desalination and water sciences, Abu Dhabi, UAE, November 1995
Kamali RK, Mohebinia S (2008) Experience of design and optimization of multi-effects desalination systems in Iran. Desalination 222(1–3):639–645
Mistry KH, Antar MA, Lienhard VJH (2013) An improved model for multiple effect distillation. Desalin Water Treat 51(4–6):807–821
Minnich K, Tonner J, Neu D (1995) A comparison for heat transfer requirement and evaporator cost for MED/TVC and MSF. In: Proceedings of the IDA World congress on desalination and water science, Abu Dhabi, UAE, November 1995
Ophir A, Lokiec F (2005) Advanced med process for most economical sea water desalination. Desalination 182(1–3):187–198
Park IS, Park SM, HaJS (2005) Design and application of thermal vapor compressor for multi-effect desalination plant. Desalination 182(1–3):199–208
Power RB (1994) Steam jet ejectors for the process industries. McGraw-Hill, New York
Shah MM (1978) Heat transfer, pressure drop, visual observations, test data for ammonia evaporating inside tubes. ASHRAE Trans 84 Part 2, pp 239–256
Wangnick K (1995) How incorrectly determined physical and constructional properties in the seawater and brine regimes influence the design and size of an MSF desalination plant – Stimulus for further thoughts. In: Proceedings of the IDA World congress on desalination and water science, Abu Dhabi, UAE, November 1995

Chapter 5
Achieving the Zero-Liquid-Discharge Target Using the Integrated Membrane System for Seawater Desalination

Sulaiman Al Obaidani, Mohammed Al-Abri, and Nabeel Al-Rawahi

Abstract Membrane desalination technology has emerged in recent years as the most viable solution to water shortage. However, despite the enormous improvement in membrane desalination technology, some critical developments are still necessary in order to accomplish possible improvements in the process efficiency (to increase recovery), operational stability (to reduce fouling and scaling problems), environmental impact (to reduce brine disposal), water quality (to remove harmful substances) and costs. In particular, cost- effective and environmentally-sensitive concentrate management is today recognized as a significant obstacle to extensive implementation of desalination technologies. As a result of the significant impact of desalination plants on the environment, the requirements for concentrate management are brine disposal minimization and zero liquid discharge (ZLD), both being the demanding targets for several applications. Conventional pressure-driven membranes such as MF, NF and RO were integrated with the innovative units of membrane contactors such as Membrane Distillation/Crystallization (MD/MC). The integration of different membrane units represents an interesting way for achieving the ZLD goal due to the possibility of overcoming the limits of the single units, thus improving the performance of the overall operation.

The present research study focuses on the evaluation of the integrated membrane system which merges membrane contactor technology with conventional pressure-driven membrane operations for seawater desalination. Sensitivity studies were performed for several configurations of the integrated system to obtain the most sensitive parameter in the total water cost and optimal design of the system.

The results revealed that the pressure-driven membrane operations were very sensitive to the feed concentration and the cost of electricity consumption. On the other hand, MD processes were not sensitive to the variation of the feed concentration or the electricity costs. The most sensitive parameter in the total water cost of the MD plant was the cost of steam which contributed to values as high as high as 11.4 % in case of MD without a heat recovery system. The best tolerance to the

S. Al Obaidani (✉) • M. Al-Abri • N. Al-Rawahi
Department of Mechanical and Industrial Engineering, College of Engineering, Sultan Qaboos University, Al-Khod, P.O. Box 33, Muscat P.C. 123, Sultanate of Oman
e-mail: sobeidani@squ.edu.om; alabri@squ.edu.om; alrawahi@squ.edu.om

M. Baawain et al. (eds.), *Recent Progress in Desalination, Environmental and Marine Outfall Systems*, DOI 10.1007/978-3-319-19123-2_5

variation of these parameters was obtained when using the integrated membrane system of pressure-driven membranes and MC processes.

Keywords Membrane distillation • Integrated membrane system • Sensitivity analysis

5.1 Introduction

Water shortage problem is now becoming more and more evident worldwide due to limited water resources and increased population growth. Rational utilization and sustainable water resources management in combination with waste water treatment and the development of high efficiency desalination technologies are the only solutions to face the water shortage problem (Drioli and Macedonio 2007). Desalination is no longer a marginal water resource for municipal and industrial use as it is in some countries like Qatar, Saudi Arabia and Kuwait. Two-third of the world's desalination plants are located in Gulf Cooperation Council (GCC) countries: Saudi Arabia, Kuwait, Qatar, Bahrain, UAE, and Oman (Al-Hajri and Al-Misned 1994; Mehdizadeh 2006). Among different technologies available, multi-stage flash (MSF) distillation and reverse osmosis (RO) dominate the existing plants. Other technologies include multi-effect distillation (MED), vapor compression (VC), and electrodialysis (ED) (Wangnick 2000).

Membrane distillation (MD) is a process that has great potential to replace conventional desalination processes since it is a concentration process carried out at low temperature and is not affected by concentration polarization phenomena like in RO. MD requires lower operating temperatures and a smaller vapor space than the MSF and MED processes (Curcio and Drioli 2005). Recently, the interest in using MD process for desalination is increasing world-wide due to these attractive features, especially when coupled with using solar energy or a low-grade heat source (Ding et al. 2005).

Although desalination has offered a key solution for water shortage problems, one of the main obstacles is the management of concentrated brine (wastes). Considering a typical RO desalination plant with a water recovery factor of 45–60 % (seawater) or 75–85 % (brackish water) would result in a significant excess of high concentrated solutions to be disposed of. Currently, the most frequent disposal practice for brine is direct discharge into lakes, lagoons, rivers, ocean and sanitary drains. However, the requirements of more and more rigid environmental protection regulations will stop this low-cost brine disposal in the near future (Drioli et al. 2006).

In the past, several process engineering strategies have been implemented in order to accomplish the concept of the zero-liquid discharge (ZLD) in seawater desalination (Van der Bruggen 2003). In this context, the combination of conventional pressure-driven membrane operations such as MF, NF and RO with membrane contactors technology such as Membrane Distillation/Crystallization (MD/MC) is expected to offer alternative design-pathways for brine management. The integration of different membrane units represents an interesting way of achieving the ZLD goal due to the possibility of overcoming the limits of single units, thus improving the performance of the overall operation (Drioli et al. 2006).

The purpose of this work is to perform a sensitivity analysis for different combinations of pressure-driven membrane operations with membrane contactor technologies, thus realizing the role of MD/MC technologies in an integrated membrane system in order for seawater desalination to approach ZLD requirements.

5.2 Scheme and Techniques

Four configurations of desalination plants were considered in this study. More details of these plants are given below.

5.2.1 *MF-RO*

The first system considered in the study was the commonly used MF-RO desalination plant. The microfiltration (MF) membranes are used as pretreatment for the reverse osmosis (RO) desalination plant instead of conventional pretreatment methods like sand and multi-media filtration. A schematic diagram of such a plant is presented in Fig. 5.1.

The overall water recovery of the plant in this case was 47.5 %, assuming 95 % and 50 % water recovery rates of the MF and the RO, respectively.

5.2.2 *MF-NF-RO*

In this case a nanofiltration (NF) unit was installed between the MF and the RO as shown in Fig. 5.2. The water recovery was 95 %, 70 % and 60 % for the MF, NF and RO, respectively. The overall plant recovery was 39.9 %.

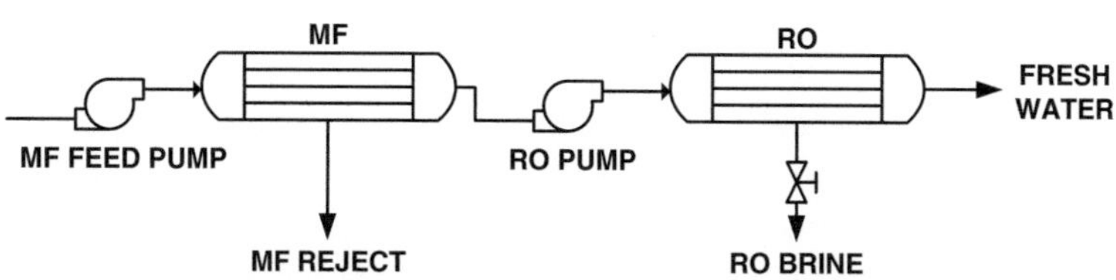

Fig. 5.1 Pressure-driven membrane operations (MF-RO)

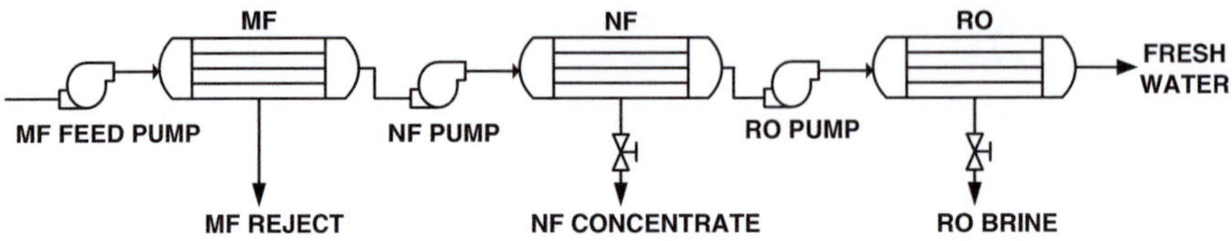

Fig. 5.2 Pressure-driven membrane operations (MF-RO)

Fig. 5.3 Schematic diagram of the MD plant

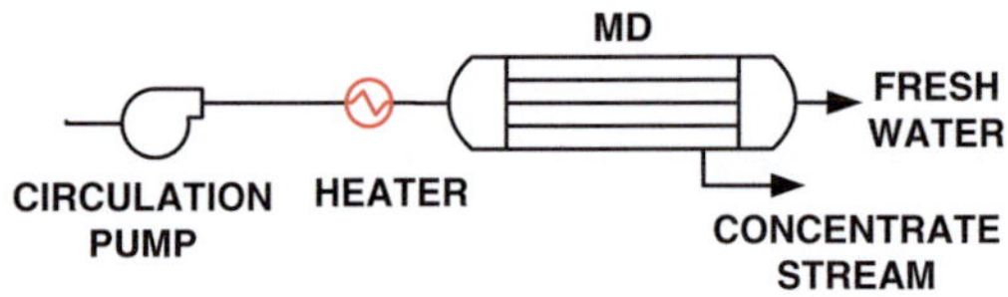

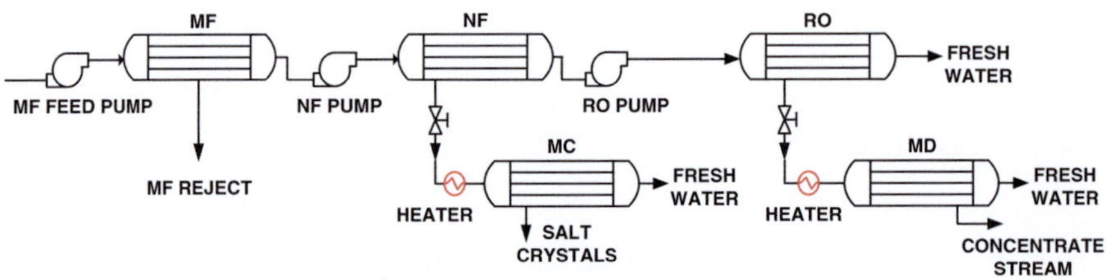

Fig. 5.4 Schematic representation of the integrated membrane system

5.2.3 MD Alone

Here the MD plant was operated as a stand-alone plant. The water recovery was 80 % and the MD units were operated under the conditions of 30° C temperature difference, producing a flux of 8.2 kg/m^{2}h in case of the MD, and a flux of 6.8 kg/m^{2}h in the case of the MC as obtained from the MD simulation program using MATLAB studied earlier (Al Obaidani et al. 2008). A schematic diagram of the MD plant is illustrated in Fig. 5.3.

5.2.4 MF-NF-RO-MC_NF-MD_RO Plants

In this case, the desalination plant consisted of pressure-driven membranes (MF-NF-RO) in combination with membrane crystallization (MC) units operated in the rejected stream of the NF units and membrane distillation units operated in the brine stream of the RO units as shown in Fig. 5.4. The overall fresh water recovery of the whole integrated system in this configuration was 89 %, assuming the same recoveries of MF, NF, RO and MD as above. The water recovery of the MC units was assumed as 97.6 %.

5.3 Results and Discussion

The sensitivity of changing variables of the desalination process on the product water cost was studied in order to identify the most sensitive parameters on water cost and to establish optimal conditions for minimizing total water cost. A water production capacity of 24,000 m^3/day was assumed in all cases.

At the beginning, the reference water cost of each plant configuration was calculated based on the reference cost values of membranes, electricity and steam as shown in Table 5.1. The estimated water cost of desalination plants obtained by this study was comparable to the cost of water produced by conventional process at around US $0.5/$m^3$ for RO (Fritzmann et al. 2007), US $1.00/$m^3$ for MED and US $1.40/$m^3$ for MSF (Van der Bruggen 2003).

5.3.1 *Water Recovery (Yield)*

The effects of changing the water recovery on the total water cost for the MF-RO, MF-NF-RO and MD plants were considered and the results are discussed below.

5.3.1.1 RO Recovery in MF-RO Plants

The first system considered in the study was the commonly used MF-RO desalination plant. The effect of changing the RO water recovery on the total water cost was investigated and the results are shown in Fig. 5.5. The results showed that when increasing the RO recovery from 40 % to 60 %, the total water cost decreased from US $0.58 to US $0.46/$m^3$ in case of using the energy recovery system and from US $0.64 to US $0.50/$m^3$ in case not using the energy recovery system. Since an RO recovery of 50 % typically used, it has been taken as a reference value. An increment of 20 % in this value will contribute to an average reduction of 10 % in

Table 5.1 Reference values and estimated water cost for each plant configuration

Cost of MF membranes	90 $/$m^2$ (Macedonio et al. 2007)
Cost of NF and RO membranes	30 $/$m^2$ (Malek et al. 1996)
Membrane life	6.7 years Al Obaidani et al.
Electricity cost	0.03 $/kWh (Helal et al. 2004)
Steam cost	7 $/t (Ettouney et al. 2002)
Plant configuration	Estimated reference water cost ($/$m^3$)
MF-RO	0.51
MF-NF-RO	0.62
MD alone	1.13
MF-NF-RO-MC_NF-MD_RO	1.28

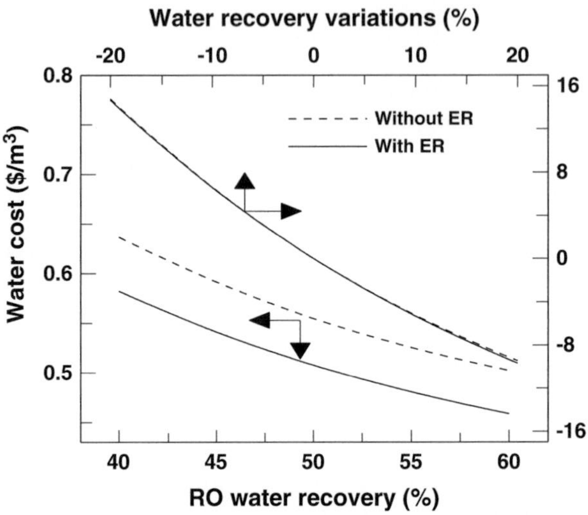

Fig. 5.5 Effects of the RO water recovery on the total water cost of MF-RO plants

the total water cost for both cases with and without energy recovery system as obtained from the slope of the water cost variations graph in Fig. 5.5.

5.3.1.2 RO Recovery in MF-NF-RO Plants

In this case, higher RO recovery values can be achieved at lower osmotic pressure since most of the bivalent ions were removed by the NF membranes. As shown in Fig. 5.6, the RO recovery was changed within the range 48–72 % and the results showed that the total water cost decreased from US \$0.71 to US \$0.56/m^3 and decreased from US \$0.75 to US \$0.59/m^3 for systems with and without energy recovery devices, respectively. Considering a recovery value of 60 % as a reference, the total water cost was reduced by an average value of 10 % as the RO recovery was increased by 20 % for both cases with and without energy recovery systems as deduced from the slope of the water cost variations graph in Fig. 5.6.

5.3.1.3 NF Recovery in MF-NF-RO Plants

The NF recovery varied from 56 % to 84 % and the results showed that total water cost reduced from US \$0.69 to US \$0.57/m^3 in case of the plant with the energy recovery system, and reduced from US \$0.73 to US \$0.61/m^3 in case of the plant without the energy recovery system as shown in Fig. 5.7. Taking a reference recovery value of 70 %, an increase of 20 % in this value will result in an average reduction of 8 % and 7.8 % respectively in the total water cost for the plants with and without the energy recovery system, as given from the slope of the water cost variations graph in Fig. 5.7.

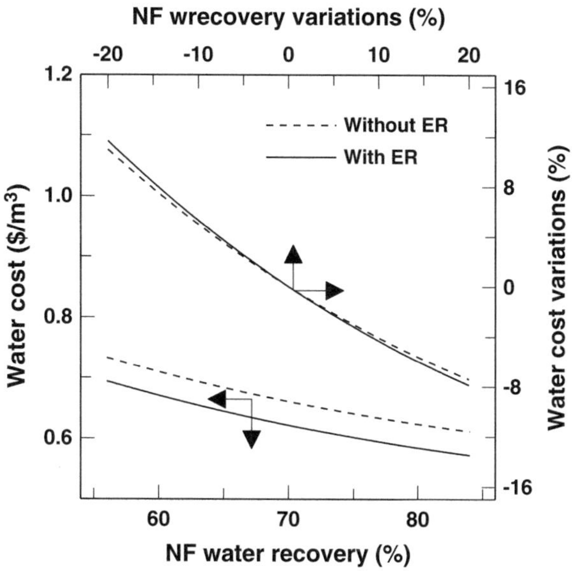

Fig. 5.6 Effects of the RO water recovery on the total water cost of MF-NF-RO plants

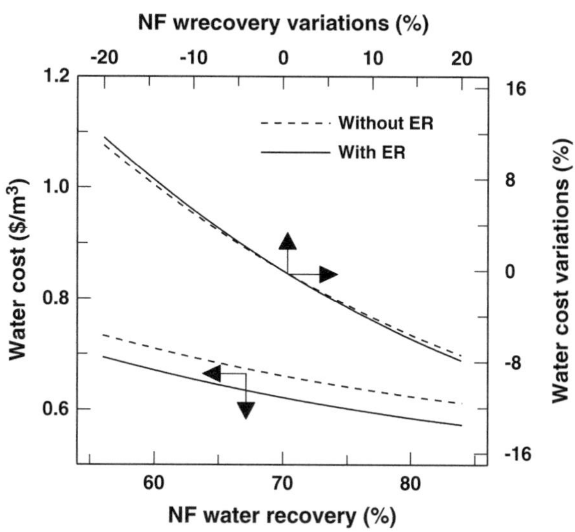

Fig. 5.7 Effects of the NF water recovery on the total water cost of MF-NF-RO plants

5.3.2 *MD Recovery in MD Plants*

High MD recovery values can be achieved since the MD plant is a thermal process and not limited by osmotic pressure. The MD recovery was increased from 64 to 96 % and the total water cost was reduced from US $0.24 to US $1.08/m^3 and from US $1.39 to US $1.09/m^3 for both cases with and without the heat recovery systems respectively as illustrated in Fig. 5.8. The difference in the water cost between MD with and without heat recovery was high (12 %) when operating MD at low

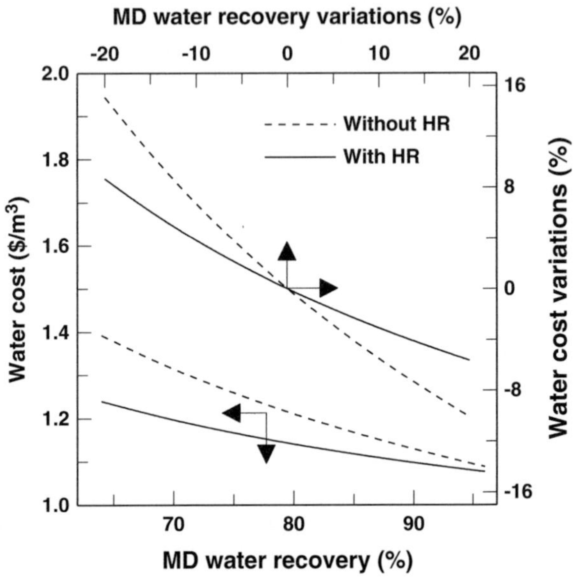

Fig. 5.8 Effects of the MD water recovery on the total water cost of MD plants

recovery values. However, at high water recovery values, the water cost of MD without the heat recovery system became very close to the one of MD with heat recovery with a difference of about 1 % only. This was due to the fact that as the recovery was increased, the amount of hot brine was reduced and hence the amount of heat that could be recovered using the heat recovery system was reduced too. Increasing the MD recovery value by 20 %, based on a reference recovery value of 80 %, will lead to a reduction of 6.2 % and 10.3 % in the total water cost for cases with and without heat recovery systems respectively as taken from the slope of the water cost variations graph in Fig. 5.8.

Generally, the total water cost decreased as the water recovery was increased, in all cases. This was due to the fact that smaller intake/discharge facilities and lower membrane surface areas were required since higher flux values were obtained at higher recoveries. The pressure-driven membrane desalination plants showed similar sensitivity to the changes in the water recovery values for both cases with and without energy recovery systems. However, in case of MD, the sensitivity of the MD plant without heat recovery system was higher than the one with a heat recovery system especially at low recovery values. This was due to the dependency of the amount of heat to be recovered by the heat recovery system on the amount of the rejected brine which varied with the recovery value.

5.3.2.1 Feed Water Concentration

The desalination plants were evaluated at situations where the feed concentration might vary. The total water cost was considered when the feed concentration changed between 10 and 25 g/L for brackish water and between 30 and 50 g/L for seawater.

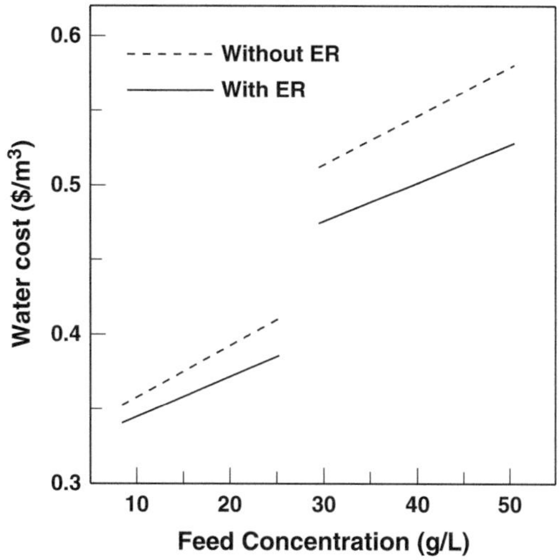

Fig. 5.9 Effects of the feed concentration on the total water cost of MF-RO plants

5.3.2.2 MF-RO Plants

The results showed that when the concentration was increased from 10 to 25 g/L in case of brackish water, the water cost increased from US \$0.34 to US \$0.39/m^3 for plants with the energy recovery system and increased from US \$0.35 to US \$0.41/m^3 for plants without the energy recovery system as shown in Fig. 5.9. In case of seawater plants, the concentration was increased from 30 to 50 g/L and accordingly the water cost increased from US \$0.47 to US \$0.53/m^3 and from US \$0.51 to US \$0.58/m^3 in the case of plants with and without energy recovery systems respectively as shown in Fig. 5.9.

5.3.2.3 MF-NF-RO Plants

In this case, the results showed that when increasing the concentration from 10 to 25 g/L for brackish water plants, the water cost increased from US \$0.34 to US \$0.41/m^3 for plants with the energy recovery system and increased from US \$0.35 to US \$0.42/m^{3}for plants without the energy recovery system as shown in Fig. 5.10. The water cost increased from US \$0.59 to US \$0.65/m^3 and increased from US \$0.62 to US \$0.70/m^3 for seawater plants with and without energy recovery systems respectively as shown in Fig. 5.10.

5.3.2.4 MD Plants

In case of thermal processes, the total water cost was not affected by the change in the feed concentration for brackish and seawater plants as shown in Fig. 5.11. The water cost was US \$1.13/m^3 and US \$1.20/m^3 for brackish water plants with and

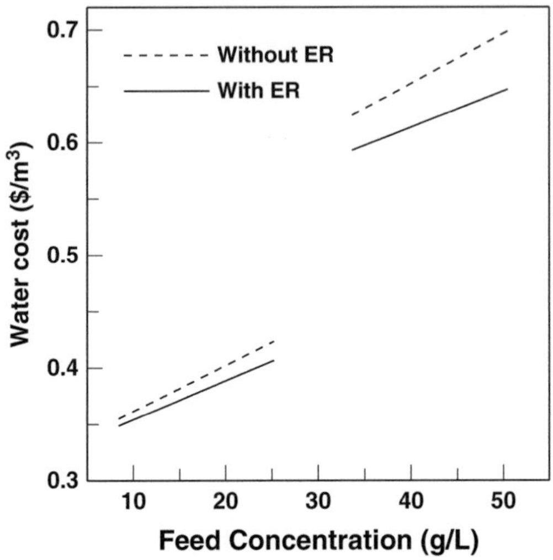

Fig. 5.10 Effects of the feed concentration on the total water cost of MF-NF-RO plants

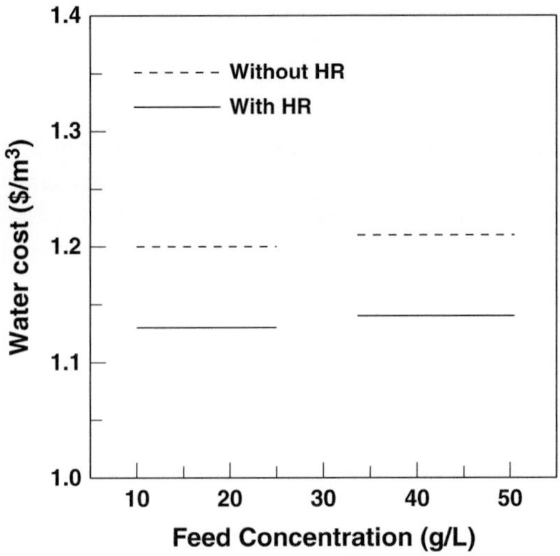

Fig. 5.11 Effects of the feed concentration on the total water cost of MD plants

without the heat recovery system respectively. For seawater plants, the water cost was US $\$1.14/m^3$ and US $\$1.21/m^3$ for plants with and without the heat recovery system, respectively.

5.3.2.5 MF-NF-RO-MC_NF-MD_RO Plants

These plants consisted of an integrated system which includes a membrane crystallization unit operated in the rejected stream of the NF unit and a membrane

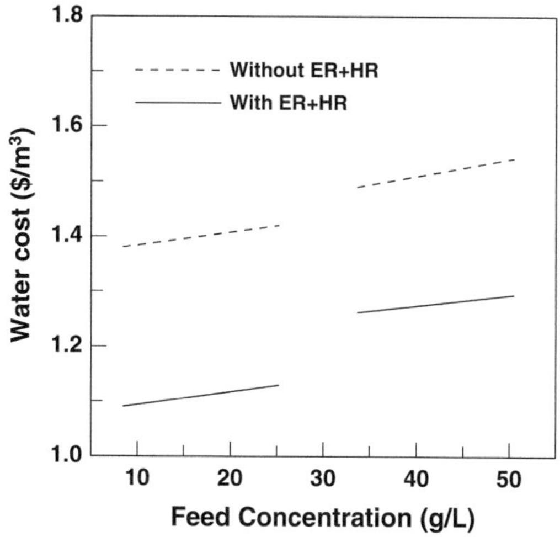

Fig. 5.12 Effects of the feed concentration on the total water cost of MF-NF-RO-MC_NF-MD_RO plants

distillation unit operated in the brine stream of the RO unit. In this case, the results showed that when increasing the concentration from 10 to 25 g/L for brackish water plants, the water cost increased gradually from US $1.09 to US $1.13/m^3 for plants with the energy recovery system and increased steadily from US $1.38 to US $1.42/m^3 for plants without the energy recovery system as shown in Fig. 5.12. The water cost increased slightly from US $1.26 to US $1.29/m^3 and increased from US $1.50 to US $1.53/m^3 for seawater plants with and without the energy recovery systems respectively as shown in Fig. 5.12.

As shown above, the total water cost increased when the feed concentration increased in the case of pressure-driven membranes. This rise in the total water cost was related to the additional requirement of a membrane surface area due to the flux reduction as well as the additional power requirement due to higher pressure operations to overcome the additional resistance of osmotic pressure at higher concentrations. On the other hand, MD plants were not greatly affected by the feed concentration. This was expected due to the fact that MD plants are thermal processes and are not limited by osmotic pressure or concentration polarization phenomena. In the case of integrated plants which contain pressure-driven membrane operations and membrane contactors (MD and MC), the total water cost was less sensitive to the changes in the feed concentration than the other cases.

5.3.3 Membrane Cost

The sensitivity of changing the membrane cost on the total water cost of desalination plants was studied. The reference values of the membrane cost was taken as US $90/m^2 for MF membranes (Macedonio et al. 2007) and US $3/m^2 for NF and RO

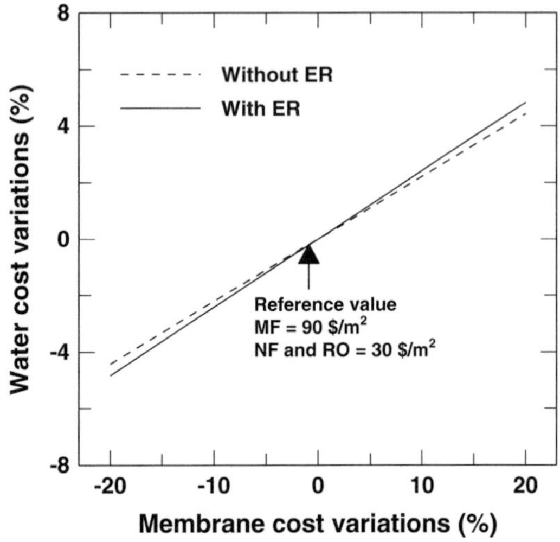

Fig. 5.13 Effects of the membrane cost on the total water cost of MF-RO plants

membranes (Malek et al. 1996). For study purposes, the membrane cost was changed by a factor of ±20 % of these reference values in all cases.

5.3.3.1 MF-RO Plants

An increase of 20 % in membrane cost contributed to an increase of 4.8 and 4.4 % in the total water cost for plants with and without energy recovery system as shown in Fig. 5.13.

5.3.3.2 MF-NF-RO Plants

In this case, the results showed that the total water cost was increased by 4.8 % for plants with the energy recovery system and increased by 5.6 % for plants without the energy recovery system when the membrane cost was increased by 20 % as shown in Fig. 5.14.

5.3.3.3 MD Plants

In case of MD processes, the total water cost was increased by 5.6 % and 5.4 % for plants with and without heat recovery system, respectively when the MD membrane cost was increased by 20 % as shown in Fig. 5.15.

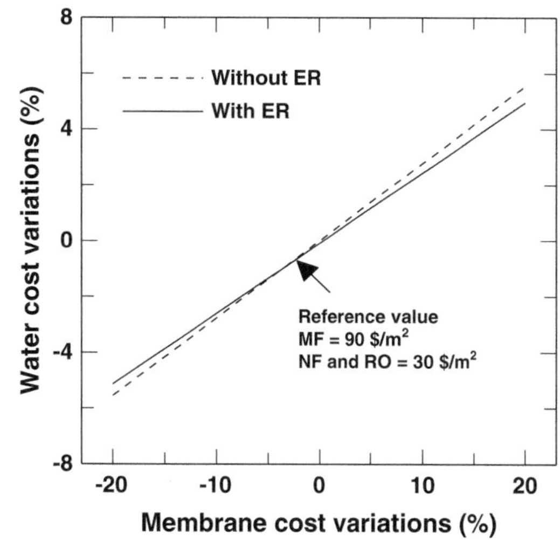

Fig. 5.14 Effects of the membrane cost on the total water cost of MF-NF-RO plants

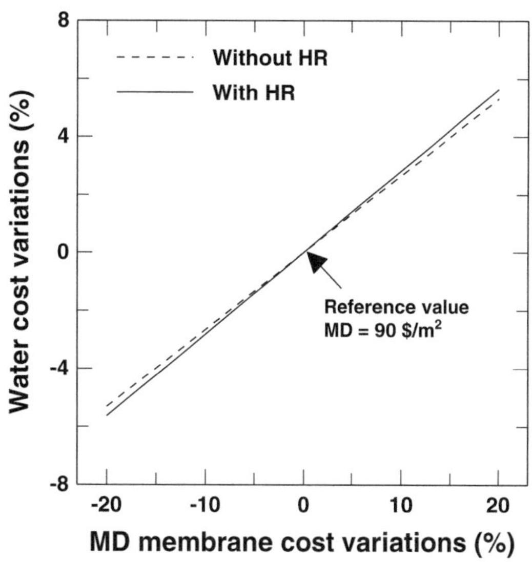

Fig. 5.15 Effects of the membrane cost on the total water cost of MD plants

5.3.3.4 MF-NF-RO-MC_NF-MD_RO Plants

In case of the integrated system, the changes in the total water cost were less evident than the other cases. The total water cost was increased only by 1.5 % and 1.6 % for plants with and without energy and heat recovery systems respectively when the membrane cost was increased by 20 % as shown in Fig. 5.16. The results showed that when increasing the membrane cost per unit area by 20 %, the total water cost was increased within the range 4.4–5.6 % in all cases except for the plant with the

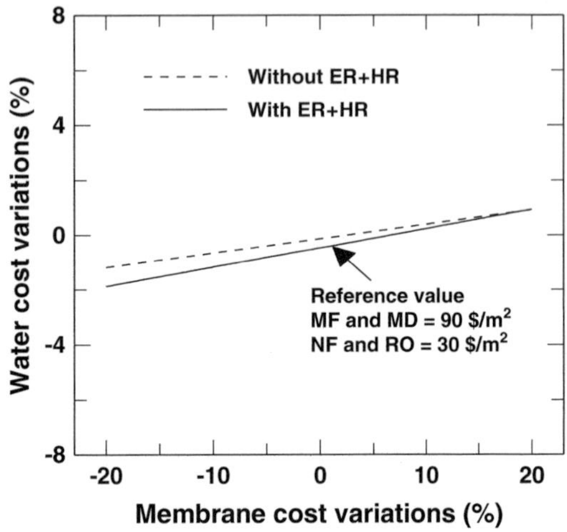

Fig. 5.16 Effects of the membrane cost on the total water cost of MF-NF-RO-MC_NF-MD_RO plants

integrated systems which showed better cost stability for the changes in the total water cost.

5.3.4 *Electricity Cost*

The effects of the variations in the electricity cost were studied. The reference value was taken as US $0.03/kWh (Helal et al. 2004). The electricity cost was varied by a factor of ±20 % of this reference value in all cases.

5.3.4.1 MF-RO Plants

The total water cost was increased by 3.4 % and 5.6 % for plants with and without energy recovery system respectively when the electricity cost was increased by 20 % as shown in Fig. 5.17.

5.3.4.2 MF-NF-RO Plants

An increase of 20 % in the electricity cost will result in increasing the total water costs by 3.6 % and 5.2 % for plants with and without energy recovery systems respectively as shown in Fig. 5.18.

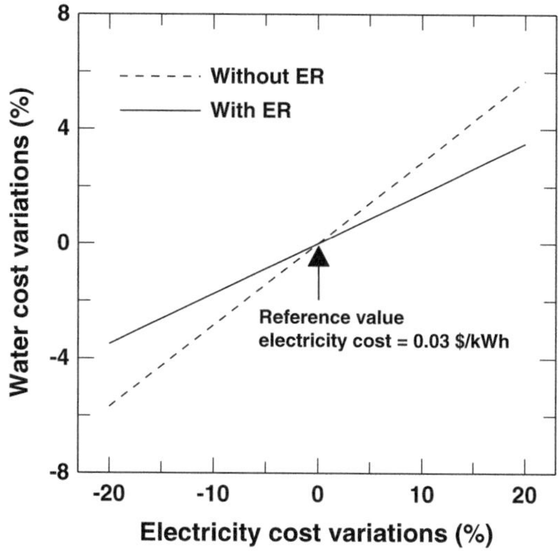

Fig. 5.17 Effects of the electricity cost on the total water cost of MF-RO plants

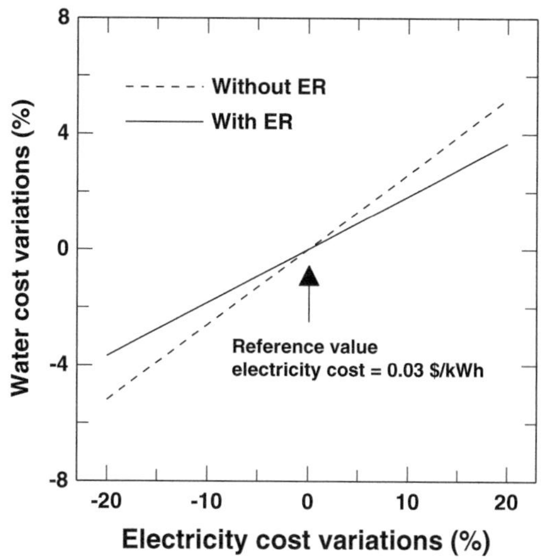

Fig. 5.18 Effects of the electricity cost on the total water cost of MF-NF-RO plants

5.3.4.3 MD Plants

In case of MD processes, the total water cost was not affected by the change in the electricity costs since the electricity power was used only as an auxiliary power supply for running the circulation pumps. The main energy input was the thermal energy provided as hot steam for heating the feed water. The total water cost was constant at US $1.14 and US $1.21/m^3 for plants with and without heat recovery systems, respectively.

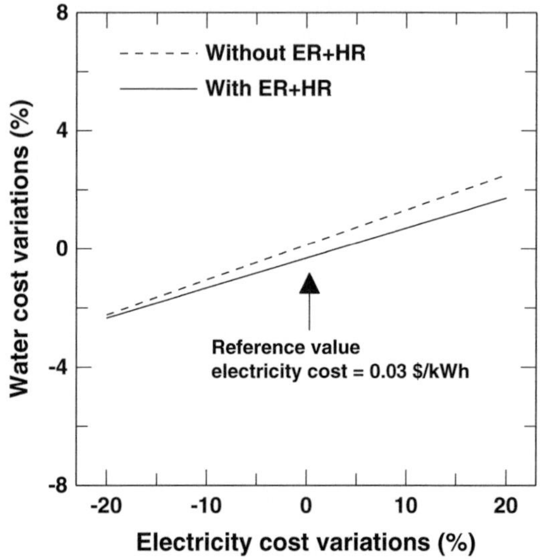

Fig. 5.19 Effects of the electricity cost on the total water cost of MF-NF-RO-MC_NF-MD_RO plants

5.3.4.4 MF-NF-RO-MC_NF-MD_RO Plants

In case of the integrated system, the total water cost was increased only by 2.0 % and 2.4 % for plants with and without energy and heat recovery systems respectively when the membrane cost was increased by 20 % as shown in Fig. 5.19.

The results showed that the plants with energy recovery systems showed less sensitivity to the changes in the electricity cost than the ones without energy recovery systems in case of pressure-driven membrane operations. This means that the energy recovery system was useful in making the plants less sensitive to the changes in the electricity cost. The thermal process (MD) showed constant water costs against the changes in the electricity costs since the electrical power was used only to operate auxiliary pumps and systems. Again, the integrated membrane system showed better stability to the changes in the electricity cost than the pressure-driven membrane operations alone.

5.3.5 *Steam Cost*

The cost of energy has a very wide variation in different countries and it might be even different in the same country depending on the location of the plant. In thermal plants, the main energy input is heat in the form of steam. This is applicable only for cases where the steam was used as the heat energy input like in MD plants and the integrated system when using MD and MC processes. The reference value of the steam was taken as US $7/t (Ettouney et al. 2002). The steam cost was varied by a factor of ±20 % of this reference value in all cases.

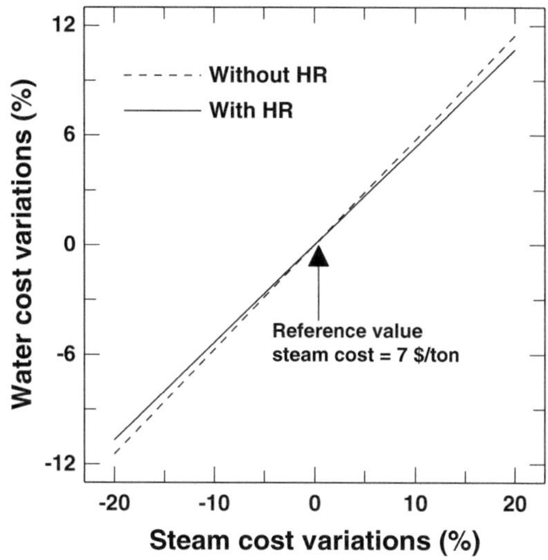

Fig. 5.20 Effects of the steam cost on the total water cost of MD plants

5.3.5.1 MD Plants

In case of operating MD process as a stand-alone desalination plant, the total water cost increased by 10.6 % for plants with the heat recovery system and 11.4 % for plants without the heat recovery system when the steam cost was increased by 20 % as shown in Fig. 5.20.

5.3.5.2 MF-NF-RO-MC_NF-MD_RO Plants

The integrated system showed less sensitivity to the changes in the steam cost. The total water cost increased by 4.7 % for plants with energy and heat recovery systems and by 4.1 % for plants without energy and heat recovery systems when the steam cost was increased by 20 % as shown in Fig. 5.21.

The results showed that the MD process was very sensitive to any changes in the steam costs when operated as a stand-alone desalination plant. However, when integrating MD and MC plants with the pressure-driven membrane operations, the system confirmed better stability against the changes in the steam costs.

5.3.6 Membrane Life

The operating lifetime of the membrane is a very important factor affecting the total water cost of desalination plants. In this study, the shortest lifetime was considered as 2 years and the longest lifetime was 8 years. The total water cost was evaluated when changing the lifetime of the membrane between these values for all cases. The

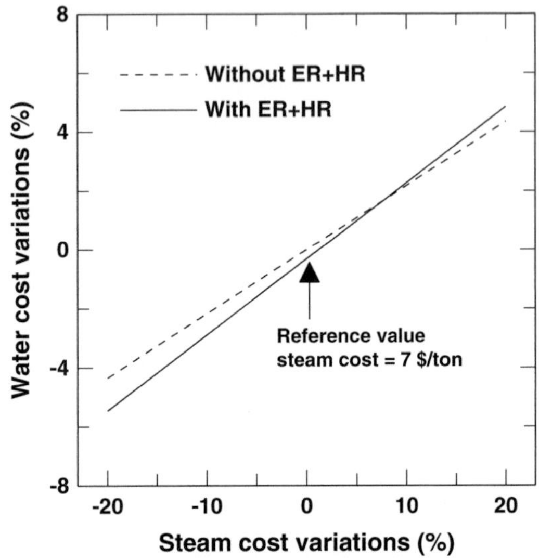

Fig. 5.21 Effects of the steam cost on the total water cost of MF-NF-RO-MC_NF-MD_RO plants

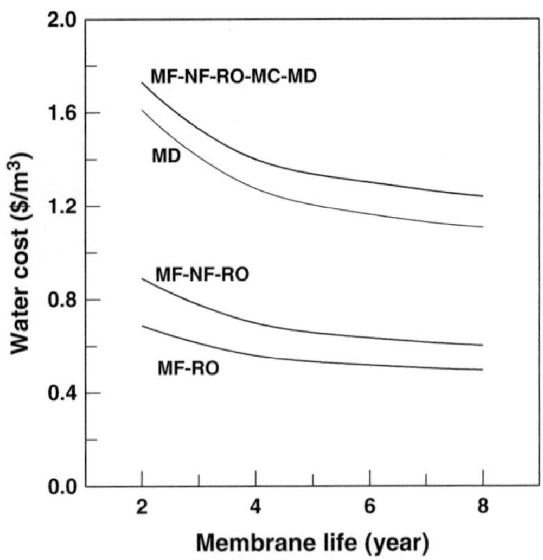

Fig. 5.22 Effects of membrane lifetime on the total water cost of desalination plants

results showed that the pressure-driven membranes had lower dependency on the membrane life than the thermal and integrated processes. The total water cost was reduced from average values of about US $0.8/m^3$ for a membrane life of 2 years to reach average values of about US $0.5/m^3$ for a membrane life of 8 years for pressure-driven membrane operations as shown in Fig. 5.22. The MD plants showed similar dependency on the membrane life as the integrated system. In this case, the total water cost decreased from US $1.75 to US $1.15/m^3$ for a membrane life time of 2 and 8 years respectively. The total water cost showed higher sensitivity to the change in the membrane lifetime for values less than 4 years and the sensitivity started to be lower at membrane lifetimes of more than 4 years as shown in Fig. 5.22.

5.4 Conclusions

The sensitivity analysis showed that the pressure-driven membrane operations were very sensitive to the changes in the feed concentration and the cost of electricity per kilowatt-hour. On the other hand, MD processes were not sensitive to the variation in the feed concentration or the electricity costs. The most sensitive parameter in the total water cost of the MD plant was the cost of steam which contributed to values as high as 11.4 % in the case of MD without the heat recovery system. The best tolerance to the variation of these parameters was obtained when using the integrated membrane system of pressure-driven membranes and MC processes.

Concerning the membrane lifetime, the results showed that the pressure-driven membranes had lower dependency on membrane life than the thermal and the integrated processes. In addition, the total water cost showed higher sensitivity to the change in membrane lifetimes for values less than 4 years and the sensitivity started to be lower at membrane lifetimes more than 4 years.

Accordingly, it was demonstrated that combining several membrane processes in an integrated membrane system will offer essential improvements in efficiency, water cost and environmental impact which will maintain process sustainability and growth, thus meeting the process intensification targets. In addition, the integrated membrane system has great potentials to achieve the zero-liquid discharge goal since the amount of the rejected brine was very limited.

List of Figures

References

Al-Hajri KR, Al-Misned LA (1994) Water resources in the GCC countries: a strategic option. Renew Energy 5(1–4):524–528

Al-Obaidani S et al (2008) Potential of membrane distillation in seawater desalination: thermal efficiency, sensitivity study and cost estimation. J Membr Sci 323:85–98

Curcio E, Drioli E (2005) Membrane distillation and related operations – A review. Sep Purif Rev 34:35–86

Ding Z, Liu L, El-Bourawi MS, Ma R (2005) Analysis of a solar-powered membrane distillation system. Desalination 172:27–40

Drioli E, Macedonio F (2007) D 1.1.1 Report on raising public participation and awareness. pp 1–128

Drioli E et al (2006) Integrating membrane contactors technology and pressure-driven membrane operations for seawater desalination energy, exergy and costs analysis. Chem Eng Res Des 84:209–220

Ettouney HM, El-Dessouky HT, Gowing PJ (2002) Evaluating the economics of desalination. CEP Magazine, pp 32–39

Fritzmann C, Lowenberg J, Wintgens T, Melin T (2007) State-of-the-art of reverse osmosis desalination. Desalination 216:1–76

Helal AM, El-Nashar AM, Al-Katheeri E, Al-Maler S (2004) Optimal design of hybrid RO/MSF desalination plants. Part III. Sensitivity analysis. Desalination 169:43–60

Macedonio F, Curcio E, Drioli E (2007) Integrated membrane systems for seawater desalination: energetic and exergetic analysis, economic evaluation, experimental study. Desalination 203:260–276

Malek A, Hawlader MNA, Ho JC (1996) Design and economics of RO desalination. Desalination 105:245–261

Mehdizadeh H (2006) Membrane desalination plants from an energy-exergy point of view. Desalination 191:200–209

Van der Bruggen B (2003) Desalination by distillation and by reverse osmosis – trends towards the future. Membr Technol 2:6–9

Wangnick K (2000) IDA Worldwide desalting plants inventory 16, 2000

Chapter 6
Installation of Shore Approaches and Sea-Lines Using Trenchless Methods: Technologies and Case Studies

Peter Schmaeh

Abstract From the mainland into the sea or vice versa, for a sea outfall, an intake or landfall, a pipeline is laid from the mainland out to sea or into a lake or river. Such lines can be installed by Trenchless Technology using pipe jacking, segment lining or pipeline installation methods (HDD and Direct Pipe), according to the requested diameter, length and geological conditions. This paper shows different trenchless installation methods, including those for large diameter pipelines. Selected case histories show the multipurpose use of the trenchless outfall technology in various industrial sectors. Herrenknecht machines have installed in numerous marine outfall systems safely beneath the seabed, such as sewer and brine outlets or landfall constructions for oil and gas pipelines. Another application field which has been gaining importance in recent years are water intake and brine tunnels for desalination plants. With machine concepts for Offshore Foundations, e.g. wind-parks for renewable energy, Herrenknecht has also entered the energy sector. Tunnels and casings for offshore-onshore connection lines and cooling water intakes for coastal power plants complete its trenchless range of energy solutions.

This paper highlights the environmental and economic aspects of tunnelled outfall construction. Trenchless installation methods have less impact on the environment and on the existing infrastructure than methods applying open-cut trenching. They reduce environmental impact and can also be carried out in heavily built-up areas. Shipping and tourism remain unaffected. The pipeline is better protected against damage and therefore has a longer lifetime. Due to the fact that the tunneled pipeline is laid underground, it is safe and maintenance-free for decades.

Keywords Tunnelled outfalls • Intakes landfalls • Horizontal Directional Drilling (HDD) • Microtunnelling • Pipe-jacking • Segment lining • Direct pipe

P. Schmaeh (✉)
Herrenknecht AG, Schlehenweg 2, 77963 Schwanau, Germany
e-mail: schmaeh.peter@herrenknecht.de

M. Baawain et al. (eds.), *Recent Progress in Desalination, Environmental and Marine Outfall Systems*, DOI 10.1007/978-3-319-19123-2_6

6.1 Introduction

The situation in coastal areas is becoming critical as more and more people move to the world's urban centers. Big cities are turning into megacities with more than 5 million inhabitants. Most of these megacities, like Jakarta, Lagos, Manila, Mumbai, Bangkok, New York, Osaka-Kobe, Rio de Janeiro and Shanghai, all have one thing in common: they are all located on narrow stretches of coast line. More than 40 % of the world's population already live within 100 km of the sea and another 40 % live in cities along rivers. Thus, the majority of the world population lives close to water. The challenge to overcome in these places is to provide people with sufficient potable water and to handle sewage disposal to assure a healthy quality of life.

In the construction of utility infrastructures in coastal areas or in river regions, tunneled outfalls, intakes and landfalls are an effective and sustainable method. With the help of sea and river outfalls, wastewater can be transported away from the coastline or river bank and discharged at locations where diffusion, dispersion and decomposition are enhanced. The municipal wastewater may be fully treated, pre-treated or untreated.

Water intakes can be constructed, for example, to supply water to desalination or power plants. If no beach or sandy floor exists near the plant location, or if the site conditions are inadequate for infiltration, a tunnelled offshore intake system is the ideal choice. The worldwide growing demand for oil and gas makes the construction of pipelines on- and off-shore necessary. Please find hereafter an overview of application fields for tunnelled pipeline systems from the mainland into the sea and vice versa. Figure 6.1 below gives a summary of the application fields of outfalls, intakes and landfalls.

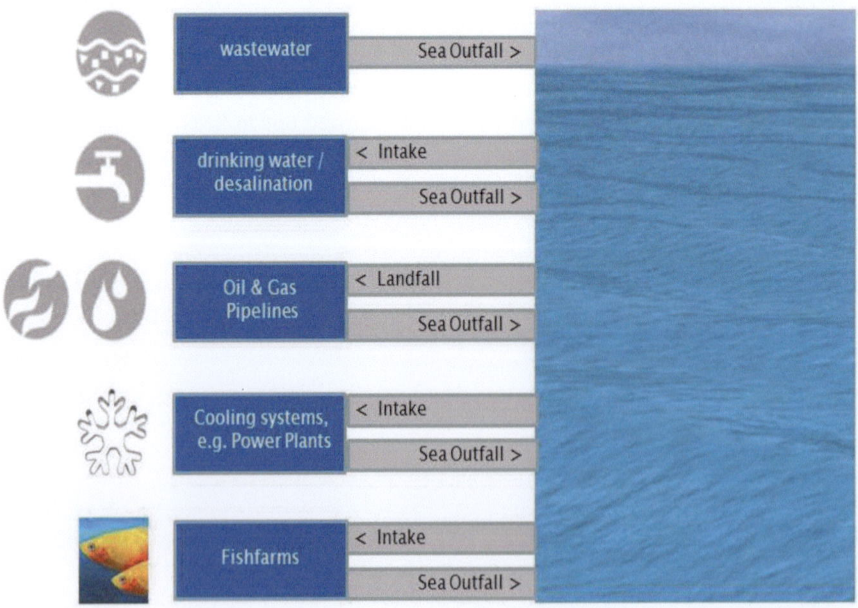

Fig. 6.1 Application fields of sea outfalls, intakes and landfalls

6.2 Advantages of Trenchless Installation Techniques

6.2.1 Minimized Impact on the Environment

Sea outfalls or seawater intakes may generally be installed by trenchless or open-cut trenching methods. In comparison to open-cut methods of pipe installation, trenchless installation techniques reduce impact on the environment to a minimum. The installation takes completely place underground without harming marine wildlife. Sea water quality remains untouched and emissions and vibrations caused by the pipeline installation are considerably reduced.

6.2.2 Minimized Impact on Existing Infrastructure

Trenchless marine pipeline installation can also be applied in densely populated areas where limited space is available for jobsite installation. Due to the displacement of the jobsite underground, life in coastal areas can continue so that tourism and shipping traffic is not affected. Furthermore, existing pipe networks can be conserved.

6.2.3 Benefits for Construction Time and Pipeline Life Cycle

The installation work of a pipeline underground is minimally affected by external conditions like the weather, storms, tidal ebb and flow or sediment transport. This makes trenchless technology a safe and reliable construction method. In the long run, the underground installation even extends the pipeline life cycle as the pipeline remains protected underground against damage by ships or sabotage, with a lower risk of settling as well as having higher seismic safety.

6.2.4 Influence of Currents on Pipelines Laid on the Sea Bed

Pipelines laid by open-cut techniques on the sea bottom are directly influenced by hydrodynamic forces resulting from currents. Typical currents caused by orbital movements of wave particles and near-coast currents can damage pipes especially during heavy sea or storms. Hydrodynamic forces cause erosion, transport and accretion of seabed material. Offshore buildings in general influence currents and may lead to heavy erosion or accretion near the buildings. Pipelines laid on the sea bottom may increase the current velocity and thus its turbulence. A two-phase flow of sand and water under the pipeline forms and scours may develop, which depend on the following parameters:

- Vertical current velocity profile
- Turbulence
- Wave reflexion
- Bed material
- Bed roughness

If the pipeline is laid directly on the sea bottom, scouring can lead to a large free span of the pipeline between two supports. If the pipeline is laid on concrete supports, the scouring may cause the supports and the pipeline to sink. The bending radius of the pipeline between two supports may be exceeded and pipeline may break.

6.3 Methods for Trenchless Installation of Offshore Lines

6.3.1 Pipe Jacking and Microtunnelling

6.3.1.1 Installation of Pipeline

A launch shaft is excavated at the land side of the pipeline. The dimensions and the design of this shaft may vary according to the specific requirements of the site with economics being a key factor. A thrust wall is constructed to provide a reaction against which to jack. The initial alignment of the pipe jack is obtained by accurately positioning guide rails within the thrust pit on which the pipes are laid.

Powerful hydraulic jacks are used to push the jacking pipes through the ground. At the same time, excavation at the tunnel face can take place within a steerable shield. Remote-controlled microtunnelling machines are operated from a control panel in a container which is located on the surface next to the launch shaft. This is an advantage regarding safety regulations, because no workers need to be in the tunnel during construction. The position of the remote-controlled machine is supervised by a guidance system.

Today, this developed tunnelling technique enables the realization of long-distance advances, even in difficult ground conditions. Figure 6.2 below shows a stylized rendition of pipe jacking marine outfall technology.

The following technological features reduce friction and jacking forces.

6.3.1.2 Automatic Lubrication System

During the pipe jacking process, the whole pipeline is pushed through the ground. Rising friction forces between the surrounding ground and the pipe string lead to increasing jacking forces. However the maximum jacking force is strictly limited

Fig. 6.2 Pipe jacking marine outfall technology

by the maximum permissible pipe load. To reduce friction, the pipeline should be lubricated continuously so Bentonite suspensions are used to provide lubrication during the pipe jacking process. They are mixed in bentonite plants at the job site and are pumped into the tunnel via hoses or pipes. Through injection nozzles within the jacking pipes, the lubricant is squeezed into the annular gap between the pipeline and the surrounding ground. Thus, the jacking forces can be reduced considerably which optimize the performance of the pipe jacking process in terms of lower pipe loads and thus longer jacking distances.

A new generation of our bentonite lubrication system enables the automatic distribution of bentonite suspensions along the alignment, while controlling and recording the injected bentonite volumes along each pipe at the same time. This permits the selection and adjustment of the desired bentonite volume along the tunnel route, while also considering changes in geology. The monitoring, control and recording of relevant data, such as bentonite volume, pressure and friction forces, is performed in the control container where the machine operator supervises the pipe jacking process.

6.3.1.3 Use of Interjacking Stations

Another possibility to keep jacking forces down in difficult ground or long drive lengths is the use of intermediate jacking stations. They are installed to reduce and distribute the jacking forces, thereby reducing the forces applied by the main hydraulic jacks in the start pit and secondly on the thrust wall. These interjack stations are integrated into several points in the pipeline, serving to separately advance the laying of the pipeline in sections.

6.3.1.4 Additional Push Force with Telescopic Station

To realize longer advance distances in accordance with the given maximum load, telescopic stations are a further possible option. The telescopic station consists of two steel pipes, the inner and the outer part, which can be pushed into each other. These steel pipes are connected by integrated jacking cylinders. The telescopic station is used as a jacking station following the machine. Situated behind the tunnelling machine, it controls the forces applied to the cutting head of the machine. Now, only the part in front of the telescopic station moves forward. When the telescopic station has reached its maximum stroke, the remaining pipeline is jacked forward with the use of the main jacking station in the launch shaft, whereby the telescopic station is pushed together again. The result is a divided jacking force. Consequently the allowed maximum loads of the jacking pipes will not be exceeded and longer distances can be realized. Furthermore, after a longer standstill, the tunnelling machine is able to be started up again without problems, because it is not necessary to move the whole pipeline at once.

6.3.1.5 Access to Cutting Wheel/Airlock System

A deciding factor for attainable drive length is the lifetime of cutting tools. Due to worn-out cutting tools, tunnelling performance will be decreased, so requiring their replacement. There are tunnelling machines (Ø ID > 1200 mm) which can enter the excavation chamber to inspect the cutting head, and to change worn cutting tools if required. Also, if necessary, an airlock can be used to insulate the tunnelling machine from possible ground water in front of it. Consequently, longer drives can be realized.

6.3.1.6 Machine Recovery

Tunnelling machines to be used for sea outfalls are equipped with an additional recovery module, consisting of a steel can with its bulkhead to close the machine and hydraulic cylinders to separate the tunnel from the machine. Hydraulic oil is put into these cylinders by divers and connected from the outside skin of the recovery module. After complete installation of the tunnel, the seaside end of the pipeline is mostly closed with a bulkhead equipped with a valve (Figs. 6.3 and 6.4).

In most cases, the tunnelling equipment has to be recovered and lifted up to the surface. Therefore, the jacking machine is equipped with lifting eyes on its upper side. There are two ways to lift the tunnelling machine: (1) A barge with a crane is moored at the position from which the jacking machine can be recovered. A jack-up platform with crane can be installed which is able to lift higher weights than a floating barge. The crane is connected to the lifting eyes of the jacking machine by means of a spreader beam. The connection has to be carried out with the help of divers. The jacking machine is lifted to surface by the crane. (2) Another possibility

Fig. 6.3 Machine is uncovered by special suction system excavator

Fig. 6.4 Divers fix the crane to lifting eyes

to lift the machine from the seabed to the water surface is the application of airbags. These are fixed by divers to the lifting eyes of the machine. A compressor installed on a ship or barge on the surface inflates the airbags required to lift the weight of the machine. Water level fluctuations caused by tidal ebb and flow may need to be

Fig. 6.5 Recovery of tunnelling machine via airbags and lifting by crane in harbor

taken into consideration to reduce the lifting height. The barge or a ship transports the jacking machine to the next harbour, where it can be taken out of the water by a high-capacity crane (Fig. 6.5).

Four case studies are given below highlighting the technology used for laying pipes.

6.3.1.7 Case Study 1 Pipe Jacking: Cooling Water Tunnel for Power Plant

Coal-fired power station GDF Suez, Germany	
Use of sealine	Cooling water tunnel for coal-fired power station
No. of sealines	4 parallel tunnels
Machine type	AVND 2000 (upsized OD 4100) ǀ AVND 3000 (OD 4100)
Method of installation	Pipe Jacking
Geology	Sand, clay, gravel
Ground water level max.	10 m

Wilhelmshaven is located in Jade Bay in the North Sea. Here, one of the world's cleanest and most efficient coal-fired power plants started operation in September 2013. The plant, with a capacity of 800 MW, will produce 5.5 billion kilowatts of electricity annually. Its location directly on the North Sea and/or close to the port of Bremerhaven has two advantages, the required coal can be delivered across short distances, and the seawater can be used for cooling so that no cooling tower is needed.

The cooling water from the North Sea is led to the power plant through two tunnels and returned by another two tunnels. Two AVN machines, named Katja

Fig. 6.6 Jacking cylinder pushing machine ahead from launch shaft

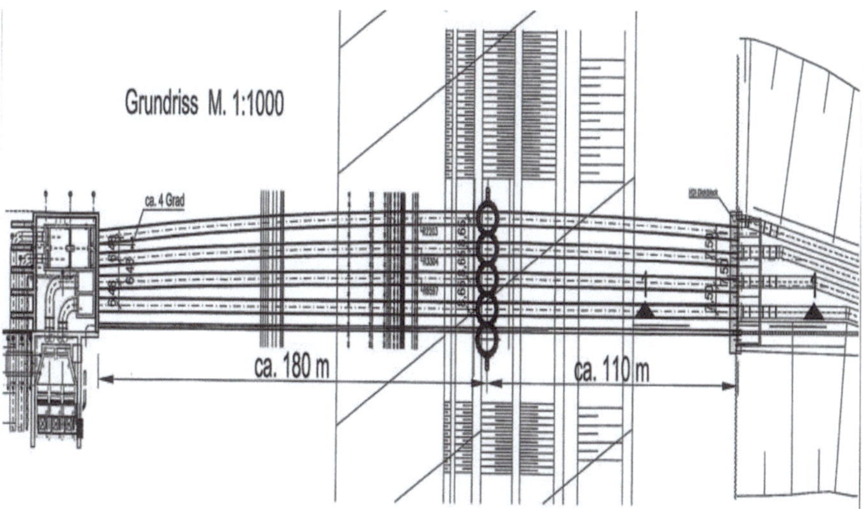

Fig. 6.7 Alignment of the tunnels crossing diaphragm wall in embankment area

(German acronym for "cooling water from the deep Jade Bay") and Wilma ("Wilhelmshaven *pipeline* drilling machine"), have built four tunnels with a length of 370 m each. Through these two tunnels, up to 30 m^3 of cooling water per second will be fed into the operating power plant (Figs. 6.6 and 6.7).

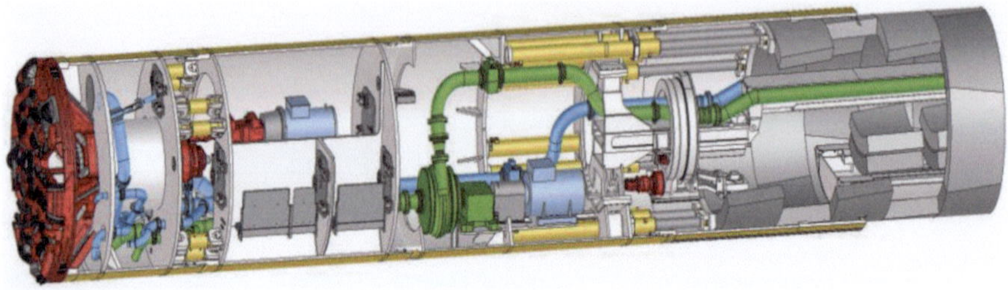

Fig. 6.8 Section through AVND segment lining tunnelling machine

6.3.2 Segment Lining

Segment lining is a tunnel construction method mainly used for marine tunnels with larger diameters (> ID 2000 mm) and/or longer distances. The tunnel lining is assembled in the tailskin of the tunnelling machine and consists of segments, which are assembled into a complete segment ring. The reinforced concrete segments are transported to the machine on a rail system. The excavated material is transported to the surface via a rail-bound or slurry system.

6.3.2.1 Function Principle of Segment Lining

Hydraulic cylinders are situated in the tunnelling machine within this support method. And represent the advance unit. After having completed an advance cycle, e.g. achievement of the length of segment, the required number of jacking cylinders is retracted in order to create sufficient space for the first segment. The other cylinders maintain their contact to the ring already built, in order to prevent the shield from moving backwards due to the pressure of the earth. A so-called erector picks the segments up and places them in the required position. After completion of one segment ring, all jacking cylinders are extended for a further advance cycle. By this method of tunnel lining, the prefabricated tunnel frame remains on the spot in the ground and serves as an abutment for the advance. Grout is injected into the overcut which is created by the cutting wheel during excavation (Fig. 6.8).

6.3.2.2 Segment Lining for Tunnelled Outfalls

For the construction of tunnelled outfalls, the segmental lining tunnelling machine is equipped with a recovery module integrated into the machine structure, similar to pipe jacking. Segment lining uses the jacking cylinders to separate the machine from the tunnel after completion of the last segment ring. Segment lining can be applied for inner diameters of a minimum of 2000 mm. The advantage of this lining technology is the longer drive distance where no interjacking stations are required, because the tunnel is erected behind the machine. Recovery of the machine takes place in the same way as the pipe jacking method.

6.3.2.3 Case Study 2 Segment Lining: Sydney Desalination Plant

Sydney desalination plant, Australia	
Use of sealine	Desalination Plant
No. of sealines	Brine outlet – 2579 m
	Feed line – 2579 m
Machine type	2 × Double Shield TBM 3400 (OD 4340)
Method of installation	Segment Lining
Geology	Sandstone (up to 70 MPa)
Ground water level max.	30 m

To ensure a secure water supply for the future, Australia's largest cities like Sydney, Melbourne, Brisbane, Perth and Adelaide, all of which are located on the coast, are building or planning seawater desalination plants. The desalination plant in Sydney went online at the end of 2009 and now feeds 250 million litres of fresh water into the municipal system every day. The plant is located on a spit of land separating Botany Bay from the open sea. Close to Australia's biggest airport in Sydney, two identical Double Shield TBMs excavated the underground intake and outfall structures in sandstone (max. 70 MPa). Lining was completed using segments, which were transported to the machine on a rail-bound system and following each drilling thrust, assembled to form a ring. The excavation material was transported out of the tunnel by a specially installed tunnel conveyor belt.

The control unit, with air conditioning, interface screens and a control panel is located in a cabin inside the TBM. As tunnelling progressed by around 30 m per day, the TBMs worked their way 2.5 km out to sea. Each installed a total of 1,904 rings with an overburden of 20 m.

Both machines followed parallel routes for about half their distance. The machine creating the outfall then changed direction, following two curves ($r = 500$ m), to separate the desalination brine outlet from the seawater intake point. After they reached their targets, the two machines were abandoned at the bottom of the sea, as the tight schedule left no room for their usual recovery. The machines were dismantled from the inside and any machine fluids, such as oil, were removed and encased in concrete.

A special in-feed sieve was mounted on the end of the intake pipe. A 180 m long and 1 m diameter piece of pipeline was attached with a flange onto the outfall pipe. This is where the brine is released through eight openings and is diluted in the sea (Figs. 6.9 and 6.10).

Connecting the potable water supply from the new seawater desalination plant to the municipal infrastructure was a particular challenge for the client. The pipeline had to run for several kilometres straight through Botany Bay, crossing ecologically important seagrass meadows. On the first, most sensitive section of 800 m, an AVND 2000 for pipe jacking was used.

Fig. 6.9 View into machine and segment ring construction with erector

Fig. 6.10 Launch shaft of AVND 2000

6.3.3 Horizontal Directional Drilling

6.3.3.1 General Function Principle

In the HDD method, pipelines are laid in three stages: First, a pilot drill is carried out from the launch point, using rotating drilling rods. The excavated material is transported to the surface by the drilling fluid which also gives the chisel extra drive. In the second phase, with the retraction of the drilling pipeline, the excavating diameter is gradually enlarged with a reamer. In most cases, the borehole is supported by a bentonite suspension which at the same time serves as the transport medium for the excavated material.

Finally, the pipeline is installed by pullback of the pipeline. This method is suitable for diameters of up to 56 in. (approx. 1.5 m) and for lengths of up to around 3000 m, depending on the diameter. Drilling in less stable geological surfaces, such as gravel, is not always possible because, unlike in the pipe jacking or segmental lining methods, the drill hole is not immediately stabilized.

6.3.3.2 HDD Application for Landfall – Shore Approach

The Horizontal Directional Drilling (HDD) method is applied for seawater intakes, outfall lines and landfalls for oil, gas or telecommunication pipelines. There are several options regarding construction method and drilling direction depending on the diameter and length of pipeline and geological conditions.

Horizontal Directional Drilling can be considered as a very flexible construction method for connection lines between the land and the sea bottom. The HDD-Rig can be positioned on the onshore jobsite or offshore on a barge, a jack-up platform or cofferdam. To increase flexibility, a pipe thruster can be used to generate additional thrust force to push longer or larger pipes. According to the circumstances onsite, the pipeline can be prepared onshore or floated on the water. The final design of a landfall installed by HDD depends on the project requirements and conditions (Fig. 6.11).

Fig. 6.11 HDD drilling steps for landfalls

6.3.3.3 Case Study 3 HDD: Landfalls Barrow Island

Landfalls Barrow Island, Australia	
Use of sealine	LNG flowlines and supply lines
No. of sealines	1 – 2 × Ø 8″ – supply lines – 470 + 520 m
	2 – 2 × Ø 10″ – supply lines – 460 + 512 m
	3 – 3 × Ø 18″ – supply lines – 453 – 497 m
	4 – 2 × Ø 34″ – LNG Flowline – 507 + 517 m
Machine type	HK 400 M + HK 250 T + HK500PT
Method of installation	HDD
Geology	Fractured limestone, calcarenite (up to 70 MPa)

By 2014, Australia will cover 8 % of the global demand for liquefied gas. The Gorgon Joint Venture is developing the Greater Gorgon Area gas fields, located approximately 130 km off the northwest coast of Western Australia in Commonwealth waters. These gas fields contain some 40 trillion cubic feet (1.12 trillion cubic meters) of gas, the nation's largest undeveloped gas resource.

The exploration of the huge deposits of Great Gorgon close to Barrow Island is currently the largest recovery project of liquefied gas worldwide. Offshore pipeline systems and industrial exploration plants have to be built. The Gorgon gas fields are linked by submarine pipelines to the northwest coast of Barrow Island. The Island is a Class 'A' Nature Reserve for the purpose of 'Conservation of Flora and Fauna', which represents the highest level of protection afforded under State legislation. Barrow Island is home to at least 22 unique terrestrial species and is also a significant nesting site for marine turtles. Therefore, the Gorgon Project underwent stringent environmental assessment prior to obtaining approval to proceed with development in 2009. The State and Federal Governments have put in place a range of measures to protect the environment and ensure the highest environmental standards are met throughout the life of the project. This entails strict environmental management practices for both the operation and mobilisation of personnel and equipment at the site.

To match these requirements in this sensitive area, the construction company in charge provided two HDD-Rigs (HK400M & HK250T) to execute the directional drilling of nine 450–520 m long HDD holes, including the stringing, welding and NDT testing of the pipelines. Besides three 34″ landfalls for high pressure LNG flowlines, the rigs installed 6 landfalls (6″–10″) for utility lines supplying water, electricity etc. to the plant. Due to drilling through fractured sandstone and calcarenite with up to 70 MPa compression strength, a 500 t Pipe-Thruster (HK500PT) has been used to install the pipes from the land to the sea. Offshore diving vessels were used to guide the pipelines beyond the exit of each HDD hole to the target laydown area (Figs. 6.12 and 6.13).

Fig. 6.12 View on jobsite Barrow Island

Fig. 6.13 Pipe Thrusters in operation to install supply line

6.3.4 *The Direct Pipe® Method*

The Direct Pipe®-method for trenchless installation of prefabricated steel pipelines combines microtunnelling technology of excavation with a thrust unit, the Pipe Thruster. Direct Pipe® incorporates the advantages of microtunnelling which enable application in difficult ground conditions, while reducing risks to a minimum. This opens up new application potentials. Originally developed as an auxiliary tool for the pullback of the pipe in the HDD-method, it was presented for the first time at the Hannover Fair in spring 2006.

The Pipe Thruster embraces the prefabricated and laid out pipeline and pushes it into the ground in strokes of 5 m each. The requisite bore hole is excavated by a slurry microtunnelling machine (AVN) which is mounted at the front of the pipeline.

This method allows the excavation of the borehole and the simultaneous trenchless installation of a prefabricated and tested pipeline in one single continuous step. Similar to pipe jacking, the soil is excavated with a microtunnelling machine. The position along the specified tunnel route is monitored by state-of-the-art techniques of controlled pipe jacking. The excavated material is removed through the slurry circuit placed in the prefabricated pipeline. The forces which are necessary for pushing the pipeline ahead are exerted by a novel push mechanism known as the Pipe Thruster, an innovation of Herrenknecht AG. The Pipe Thruster operates like a jacking frame used for jacking concrete pipes and transmits the push forces to the cutter head via the pipeline.

Just like the microtunnelling method, prior to launching, the machine is positioned at the requisite access angle on a launch rail in front of the launch seal. The pipeline is welded to the conical rear section of the machine and mounted on rollers behind the launch pit. The clamping unit of the Pipe Thruster grips the pipeline and thrusts it into the ground along with the machine. The current maximum pipeline diameter which can be clamped is 60″ (OD = 1524 mm). The forces to be anchored depend on the pipeline access angle and the maximum thrust force to be applied.

This technology can be used also for the application of sea outfalls or intakes. Due to the pullback option of the whole pipeline in case of non-destructible obstacles, the geological risk is minimised (Fig. 6.14).

6.3.4.1 Technical Advantages of Direct Pipe®

- Single-phase mode of operation.
- Permanent borehole support.
- Borehole collapse breaking the surface is thus not possible.
- The cutting wheel can be equipped for any geological condition.
- Excavated material is transported safely out of the borehole by conveyor lines; sedimentation in the borehole is not possible.
- Curved routes can be tracked, precise to the centimeter, thereby avoiding excessive bending stress on the pipeline and excessive laying forces.
- Operations at the target pit only to dismantle the cutting head.

Fig. 6.14 Direct Pipe® view on shaft with Pipe Thruster

6.3.4.2 Economic Advantages of Direct Pipe®

- Limited work area required at the launch pit and target pit.
- A minimum volume of slurry is required.
- Due to a very small overcut, a minimum borehole diameter can be realized, thus minimizing the amount of excavated material.
- Boring time can be reduced, no coupling times are necessary and the pipeline can be laid continuously in one single step.
- No permanent protective piping/casings are necessary

6.3.4.3 Case Study 4 Direct Pipe: Brine outlet Rysumer Nacken

Brine outlet Rysumer Nacken, Emden, Germany	
Use of River outlet	Sewage line
No. of pipelines	1 × Ø 48″–283 m length
Machine type	AVN 1000 XC + HK 750 PT
Method of installation	Direct Pipe
Geology	sand, silt, clay, stones, wood
Ground water level max.	15 m

Europe, and Germany in particular, is among the largest natural gas importers in the world. Natural gas storage facilities are necessary to counter supply security concerns and deal with large fluctuations in demand. Deep below the surface of the earth, gas can be stored safely in salt formations, whose walls provide particularly tight seals. Underground salt mines are flushed out with water to produce huge cavities which are then used as storage caverns. In order to drain the resulting brine, a 42-kilometer-long pipeline had to be laid in the Ems River near the city of Emden.

The final 283 m of the outflow pipe at the Rysumer Nacken artificial dune field were built using the new single-step Direct Pipe® method. The AVN1000XB Direct Pipe® machine worked its way through mixed ground of sand, silt, clay, gravel, wood and stones from the launch pit on the shore towards the target pit in the Ems, with a maximum Pipe Thruster power of 750 t. For this single-step pipe laying method, the microtunnelling machine, with a drilling diameter of 1325 mm, proceeds with a 48-inches (OD = 1220 mm), PE-coated pipeline in tow. Because of the tight space constraints at the construction site, the pipeline could not be laid in one length. For this reason, individual, 36-meter pipe lengths were prepared and joined onto the pipe string. The TBM reached its target with great precision; its deviation from the ideal line was a minimal 3 cm. After reaching the target pit the TBM was detached from the installed pipeline and recovered. At this stage, the sea outfall was complete.

The team proceeded very quickly with performances of up to 9 m/h and reached the target structure in the Ems River only 7 tunnelling days later. The PE-coated 48-inch pipeline was ready for use and put into operation after the completion of the entire 42 km-long pipeline. The result was convincing in terms of both economic efficiency and environmental protection.

After final completion of the construction work in 2016, the storage facility will have a working gas volume of 1.2 billion m^3 and a withdrawal capacity of 1.2 million m^3/h.

6.4 General Decision Criteria

In the following table, the suitability of the four methods for different site conditions is compared. Please note that the final machine choice depends on the specific project parameters and cannot be generalized (Table 6.1).

6.5 Conclusion

As described above, tunnelled sea outfalls offer many advantages over conventional open-trench installations. With minimized impact of installation work on the surrounding environment and existing infrastructure, trenchless technology is often the only way to install outfall lines safely underneath the seabed. Time and cost factors, as well as the extended pipeline life cycle, have to be considered as very important advantages of trenchless installation methods. The application of

Table 6.1 Suitability of trenchless installation methods for different site conditions

		Pipe jacking	Segment lining	HDD	Direct pipe
Geology	Cohesive	+	+	+	+
	Clay, silt, sand	+	+	+	+
	Gravel	+	+	–	+
	Boulders	+	+	–	0
	Mixed	+	+	+	+
	Rock	+	+	+	0
Pipe material	PEHD	–	–	+	–
	Steel	0	0	+	+
	GRP	+	+	+	–
	Concrete	+	+	–	–
Pipe diameter	≤DN 1000	0	–	+	0
	≥DN 1200	+	–	0	+
	≥DN 2000	+	+	–	–
	>DN 3500	0	+	–	–
Laying length	≤100 m	+	–	0	+
	100–500 m	+	0	+	+
	500–1200 m	+	+	+	+
	>1200 m	0	+	+	0

+ particularly suitable, 0 conditionally suitable, – not suitable

one of the described installation methods depends on technical, temporal and commercial factors. Geological ground conditions, specified time and costs targets may be decisive criteria for the selection of the most suitable technology.

List of Figures

References

Avanzini C (2012) Large diameter marine pipeline projects, Case stories and contributions collected at MWWD Organization & Conferences

Dipl.-Ing. Lutz zur Linde (2012) Installation of shore approaches and sealines with trenchless methods by, MWWD Conference

Herrenknecht AG (2013) Herrenknecht tunnelling product portfolio

Herrenknecht Homepage www.herrenknecht.com

Herrenknecht Sea Outfall Animation https://www.herrenknecht.com/de/medien/filme-animationen.html

Pfeff D (2013) Method Statement Direct Pipe®

Chapter 7
Photocatalytic Degradation of Divalent Metals under Sunlight Irradiation Using Nanoparticle TiO_2 Modified Concrete Materials (Recycled Glass Cullet)

M.N. Rashed

Abstract Tackling water shortage issues with the desalination of seawater and salty water is common in desert nations. Recently, new methodologies have been developed based on using nanotechnology for water desalination. Updating the field of water treatment by the photochemical technique has led to an important development in oxidative degradation processes applying titanium dioxide (TiO_2) as a catalyst. Searching for low-cost technologies for materials science applications, glass waste was used as an adsorbent and a promoting catalyst for photocatalytic oxidation of pollutants under sunlight irradiation source. In this study the newly-explored nanocomposite TiO_2 catalyst supported with recycled glass cullet, derived from crushed waste beverage bottles (as adsorbent), was used as a promising alternative catalyst material for the removal of Cd, Cu, Pb and Zn from water. The effects of various parameters (adsorbent and catalyst dose, solution pH, contact time and initial metal concentration) on metal-removal efficiency were investigated. The results revealed a higher removal percentage of metals using glass cullet/TiO_2 (GB/TiO_2) than GB as low-cost materials for water treatment.

Keywords Photodegradation • Desalination • Solar irradiation • Pollution • Treatment • Catalyst

7.1 Introduction

Increased demand for water is a global problem and clean water represents an essential necessity of human life. With only 3 % of all the available water on the planet being fresh water, seawater can be another source made available for drinking water with innovations in the development of novel technologies to desalinate water being a promising method (Mona 2010).

M.N. Rashed (✉)
Chemistry Department, Aswan Faculty of Science, Aswan University, 81528 Aswan, Egypt
e-mail: mnrashed@hotmail.com

M. Baawain et al. (eds.), *Recent Progress in Desalination, Environmental and Marine Outfall Systems*, DOI 10.1007/978-3-319-19123-2_7

Exposure to heavy metals, even at trace levels, is believed to be a risk to human health. New methodologies have been developed recently, based on using nanotechnology for water desalination and heavy metal removal such as chemical precipitation, ion exchange, photocatalysis, adsorption, reverse osmosis, ultra- and nano-filtration and electrochemical technologies (Visa and Anca 2013; Ming et al. 2012).

Material science plays an important role in our life and includes heterogenous photocatalysis which is one of the advanced oxidation processes (AOPs) used for water and wastewater treatment. Photocatalysts aided by titanium dioxide (TiO_2), nanoparticles are used widely in removing inorganic and organic pollutants from water and wastewater. However, TiO_2 photocatalysis lacks efficiency due to the high rate of recombination of electrons and holes. In addition, due to its large band gap energy (of 3.0–3.2 eV), it can only be begun by UV light, which accounts for only 3–4 % of the sunlight spectrum (Hoffmann et al. 1995). TiO_2 particles are illuminated with photons with energy greater than the band-gap energy and its electrons are promoted from the valence band to the conduction band, thus creating electron-hole pairs. In an aqueous system, holes can react with OH^- adsorbed on the surface to produce OH° radicals which are the most oxidizing species in this process.

Adsorption offers flexibility in design, high efficiency, and ease of operation (Mishra et al. 1996; Wang et al. 2003; O'Connell et al. 2008). Therefore, the adsorption process is considered as one of the major techniques for heavy metal removal from water/wastewater.

Combining the use of photocatalysts with adsorbent material is an important development in the field of photocatalytic degradation of inorganic and organic pollutants. Regarding the historical buildup of technology for the use of nanomaterials for environmental applications, photocatalysis has evolved as an important area. The combination of destructive adsorption properties with photoactivity allows a synergistic "photochemical boosting" in which it increases the capacity for the destruction of pollutants. Much research has focused on the preparation as well as on the modification of TiO_2 (Kim et al. 2005), on the composites of SiO_2/TiO_2 and on those of glass/TiO_2 (Holgado et al. 2000; Hilmi et al. 1999), on carbon-coated TiO_2 and on TiO_2-coated exfoliated graphite (Tsumura et al. 2002; Inagaki et al. 2003). Zainal et al. (2009) studied the combined effect of photodegradation-adsorption using a prepared thin film of TiO_2 and chitosan supported on glass (TiO_2-Chitosan/Glass) under the illumination of visible light. Recycled glass cullet, derived from crushed waste beverage bottles, was used as an adsorbent material besides the catalyst TiO_2 (Chen and Poon 2009). Recycled waste porous glass, from municipal solid waste sorting operations, was used as a sorbent for lead ion removal from industrial wastewaters (Petrella et al. 2010).

The present work deals with photocatalytic degradation and adsorption of the heavy metals Cu, Cd, Pb and Zn using a TiO_2 catalyst promoter with glass waste 6 Photocatalytic Degradation of Divalent Metals (GB/TiO_2), as well as with glass waste nanopowder GB as adsorbent. Sunlight was used as the irradiation source for GB/TiO_2. The dependence of metal photo-oxidation and adsorption rates on different parameters (initial dye concentrations, irradiation time, pH, and catalyst dosage) was investigated. It will be very interesting to investigate the combined effect of photodegradation-adsorption mediated by GB/TiO_2.

7.2 Materials and Methods

The catalyst used in this work was non-ionized TiO_2, which was supplied via Degussa " P25, ca. 80 % anatase, 20 % rutile; BET surface area ~ 50 m^2/g, mean diameter ca. 30 nm". This catalyst was used without further treatment.

The waste glass (laboratory glassware) was collected from the waste of our chemical laboratories. All the glass was made of silicate glass, which is a type of glass with silica as the main glass-forming constituent (SiO_2). Typically, the resulting glass (GB) composition is about 57.88 % SiO_2, 3.63 % A_2O_3, 13.97 % Na_2O, 7.74 % K_2O, and 16.96 % CaO. This glass was washed with tap water and then with de-ionized water, crushed by a mechanical crusher in the laboratory, ground in an agate mortar and further sieved to powder of particle sizes 63 μm (GB).

7.2.1 Reagents and Standard Solutions

All reagents used in this study were of analytical grade (BDH, UK). Stock standard solutions of Cd, Cu, Pb and Zn ions (1000 mg/L) were prepared by dissolving the accurately weighed amounts of their nitrate or sulphate salts (BDH, UK or Merck, Darmstadt, Germany) in 1000 ml deionized water. Working standard solutions were prepared daily by appropriate dilution with de-ionized water.

7.2.2 Preparation of GB/TiO_2

Nano-TiO_2 used in the study was supplied by Sinopharm Chemical Reagent Co. Deionized water was used throughout the study.

A composite GB/TiO_2 sample was prepared by ball-milling of nano-TiO_2 and GB powders in a ball miller (QM-1 F, made in Nanjing University, China). 5.0 g TiO_2 powder and three different sizes of zirconia balls were mixed in the miller tank. GB (0.1–5.0 wt %) and H_2O (5 ml) were added. After being milled at the speed of 280 rpm for 120 min, the wet powder was collected and dried at 110 °C in an oven. The chemical composition of GB/TiO_2 is 95.76%TiO_2, 2.96 % SiO_2, 0.76 % Al_2O_3, and 0.65 % Na_2O.

7.2.3 *Instruments and Equipment*

A flame atomic absorption spectrophotometer (Model 3110, Perkin Elmer, USA), equipped with a digital and direct readout of concentration, as well as an air-acetylene burner, were used. Single-element hollow cathode lamps of Cd, Pb, Cu and Zn were used for each element and apH meter (Orion Research, USA), oven furnace (Heraeus Instruments, UK) and a shaker (Burrell Corporation Scientific Instruments, USA) were used.

7.2.4 *Analytical Techniques*

The concentration of the heavy metals Cd, Pb, Cu and Zn were measured by the atomic absorption spectrophotometer, using hollow cathode lamps of Cd, Pb, Cu and Zn. Air- acetylene gas was used as fuel gas with a flow rate that ranged from 1.8 L to 2 L/min. The amount of metal adsorbed was evaluated by the difference between the initial and remaining concentration in the sample.

7.2.5 *Characterization of the Prepared Catalyst*

A scanning electron microscopy dispersive (SEM-EDX) study was carried out on a JEOL, JSM-5500LV electron microscopy instrument (which operates at five kV) in which a thin layer of gold or carbon had been evaporated. The elemental composition and morphology of GB/TiO_2 and GB nanocrystals were investigated. The crystalline phase, particle size, elemental composition and morphology of GB/TiO_2 and GB nanocrystals were investigated. The surface characteristics, both for GB/TiO_2 and GB, were evaluated using a BET surface analyser (Tri Star II 3020 – Micromeritics). The specific surface area of the GB reached 140 m^2/g. The total TiO_2 (GB/TiO_2) surface area developed on the glass spherules was about 60 cm^2/g.

7.2.6 *Adsorption and Photocatalysis of Metals*

7.2.6.1 Photocatalysis of Metals by GB/TiO_2 Catalyst

The photocatalysis of each metal (Cd, Pb, Cu and Zn) was carried out using 20 ml of a 100 mg/L metal solution in a 100 ml closed Pyrex flask over 0.5 g of catalyst (GB/TiO_2) with continuous shaking. The pH of the solution was adjusted to 8 (alkaline media). The solutions were exposed to sunlight with constant stirring. The average total daily short wave radiation for this period was 734 W/m^2, with a

10 h mean sunshine duration from sunrise to sunset. The mean daily temperature was 25 °C. After irradiation, the suspension was centrifuged, filtered through a 0.45 μm membrane filter, and the metal content was analyzed using an AAS spectrophotometer.

The metal removal efficiency was calculated by Eq. (7.1)

$$Removal\, of\, efficiency\,\% = \frac{C_o - C_e}{C_o} x\,100 \tag{7.1}$$

where C_0 and C_e (mg/L) are initial metal concentration and equilibrium concentration, respectively.

7.2.7 Metals Adsorption on GB

Adsorption experiments were carried out in 100 ml polypropylene tubes with 50 ml of 100 mg/L of each metal (Cd, Pb, Cu and Zn) solution, and 2 g GB adsorbent. The suspensions were shaken at room temperature (25 °C) for 4 h. The pH was adjusted at 8 by the addition of 0.1 N NaOH or 0.1 N HCl solutions. On completion of the adsorption procedure, the solutions were filtered through a 0.22 μm membrane. Metal concentrations (Cd, Pb, Cu and Zn) were determined in filtrate by using an atomic absorption spectrometer. The amount of metal ions adsorbed was determined by the difference between the initial and final concentrations.

The adsorption efficiency (%) and amounts of adsorbed metal (q_e) by GB were calculated using Eqs. (7.2) and (7.3), respectively:

$$Adsorption\, efficiency = \% = \frac{C_o - C_e}{C_o} x\,100 \tag{7.2}$$

$$q_e = \frac{(C_o - C_e)V}{m} x\,100 \tag{7.3}$$

where C_0 and C_e (mg/L) are initial metal concentration and equilibrium concentration, respectively. V is the volume of the solution (l), m is the mass of adsorbent (g) and qe (mg/g) is the amount of adsorbed metal at equilibrium.

7.2.7.1 Effect of Initial Metal Concentration

To evaluate the effect of initial metal concentration on GB/TiO_2 catalyst behavior, studies in the previous batch experiment were conducted with initial concentrations of each metal (Cd, Pb, Cu and Zn) of 10, 50, 100, 150 mg/L respectively with a fixed adsorbent or catalyst dose (0.5 g). The amount of metal ions adsorbed was determined by the difference between the initial and final concentrations.

7.2.7.2 Effect of Irradiation Time on Heavy Metal Removal

To determine the GB/TiO_2 catalyst contact time required for the equilibrium of the metal solution, 0.5 g of the catalyst was treated with 50 ml of 100 mg/L metal solution, shaken several times and the final concentration of each metal was calculated after 20, 30, 60, 90 and 120 min, and the previous experimental conditions were followed.

7.2.7.3 Effect of pH on Heavy Metal Removal

0.5 g of the GB/TiO_2 catalyst was treated with 50 ml 100 mg/L metal solution. The pH of the mixture was adjusted to various pH values (3, 5, 7 and 9) by using 0.5 M NaOH or 0.5 M HCl, shaken for one hour, after which the final concentration of each metal was calculated according to the previous experimental conditions.

7.2.7.4 Effect of Dosage on Heavy Metal Removal

Amounts of 0.25, 0.5, 1, 1.5 and 2 g GB/TiO_2 catalyst was treated with 50 ml of 100 mg/L metal solution. The final concentration of the metal after 1 h was determined according to the previous experimental conditions.

7.3 Results and Discussion

7.3.1 Effect of Adsorption (GB) and Photocatalysis (GB/TiO_2) on Metal Removal

The use of TiO_2 alone as a catalyst usually forms a suspension in aqueous solutions, and so mixing it with SiO_2 (or material rich in SiO_2) is the proper way to separate and recover the catalyst. Also, SiO_2 is used as a support to increase the specific surface area. Accordingly, using a mixture of glass powder, as a main constituent of SiO_2, with TiO_2 often leads to an increase in the surface area and could improve the photocatalytic activity of the final produced catalyst (Zhang et al. 1998; Strauss et al. 2011; Mahltig et al. 2011).

Results in Table 7.1 revealed that high removal efficiency of all heavy metals (Pb, Cd, Cu and Zn) was observed when using the modified catalyst (GB/TiO_2), while the GB adsorbent showed less removal efficiency of the metals. This can be explained by the following facts:

Each GB/TiO_2 particle may possess three different reactions: (i) A photoactive zone, which is the outer surface of TiO_2 coating receiving UV illumination; (ii) An

Table 7.1 Effect of GB adsorbent and GB/TiO_2 catalyst on the removal efficiency of the studied heavy metals

Metals	Removal % using GB	Removal % using GB/TiO_2
Pb	86.91	96.16
Cd	85.71	91.02
Cu	88.2	97
Zn	92	99

adsorption zone, which is bulk of GB underneath the photoactive zone and, (iii) A GB /TiO_2 interface.

The mechanism of metal removal efficiency can be explained by: As the main constituent of GB is SiO_2, alkaline media on GB surface new active sites ($\equiv SiO^-$) can develop, allowing metals to form complexes (Eq. 7.4) (Chaizasith et al. 2006)

$$2(\equiv SiO^-) + M^{2+} \rightarrow (\equiv SiO)_2M \quad (7.4)$$

The GB/TiO_2 catalyst exploits the high adsorbent capacity of SiO_2 and the photocatalytic activity of TiO_2 toward the metals, resulting in a higher removal capacity of heavy metal ions (Pb, Cd, Cu and Zn) than that of GB alone.

According to the "site binding" theory, the surface charge on the metal oxides is created by the electrolyte ions adsorption. These processes of TiO_2 can be described by these Eqs. (7.5, 7.6, 7.7) (Janusz and Matysek 2006):

$$\equiv Ti - OH + H+ \rightarrow \equiv Ti - OH_2^+ \quad (7.5)$$

$$\equiv Ti - OH + HO^- \rightarrow \equiv Ti - O^- + H_2O \quad (7.6)$$

$$\equiv Ti - OH + M^{2+} \rightarrow (\equiv Ti - o^-)_2 + 2H \quad (7.7)$$

The photocatalytic processes on TiO_2 involves the formation of active oxidant species as HO^-. Considering the experimental conditions of the GB/TiO_2, the titanium dioxide component will be slightly negatively charged, while the GB matrix preserves an overall negative surface charge. Under these conditions, cationic heavy metals are supposed to be absorbed by electrostatic attractive forces. At the same time, the positive charges on the TiO_2 surfaces may support an increased amount of HO^-.

Harraz et al. (2013) studied the removal of heavy metals by a titania-silica nanoparticles photocatalyst. The adsorption capacity reached a maximum value at 1000 mg TiO_2–SiO_2 catalyst and the maximum removal percentage was 98.6 % for Pb.

It was reported earlier that the TiO_2–SiO_2 catalyst has a relatively high adsorption capacity for such heavy metals. Cr and Pb were removed by adsorption of the modification of the glass beads with glutaraldehyde from an aqueous solution (Ozmen et al. 2009).

Adsorption of heavy metals on fly ash, TiO_2 and their mixtures is studied as a sustainable low-cost technique (Visa et al. 2009). Heavy metals ions (Cr (III), Co (II) and Pb (II)) were removed from their aqueous solutions by the photocatalytic degradation ofthe titania-silica (TiO_2–SiO_2) catalyst prepared by a modified sol-gel technique (Harraz et al. 2013).

7.3.2 *Effect of Catalyst Dosage on Metal Ion Removal*

The effect of catalyst dosage on the removal efficiency of Cd, Pb, Cu and Zn by GB/TiO_2 is shown in Fig. 7.1. The effect of the catalyst dosage on metal-removal efficiency is also represented in Fig. 7.1. The results show that the removal percentage of Cd, Pb, Cu and Zn increased with an increased catalyst dose of up to 1–2 g. This is due to the availability of more surface functional groups and surface area at a higher adsorbent dose (Zhai et al. 2004; Kadirvelu et al. 2000). The optimum dosage of GB/$TiO_2$for the removal of Cd, Pb, Cu and Zn in experimental conditions is 1.5 g and the metal-removal percentage is 98, 98, 87.4, and 90.6 %, respectively.

Visa et al. (2009) proved that good Cd adsorption occurs at a dose of 4 g of fly ash adsorbent and its mixture with TiO_2. Harraz et al. (2013) examined the effect of using different TiO_2_SiO_2 photocatalyst weights (500, 700 and 1000 mg) on the removal efficiency of Cr, Co and Pb, and found that as the catalyst weight was doubled from 500 to 1000 mg, the removal efficiency was increased consequently to 90.0, 91.9 and 98.6 % for Cr, Co and Pb, respectively.

7.3.3 *Effect of Initial Metal Concentration on Metal Ion Removal*

Initial metal concentration affected the removal efficiency of metal under optimum conditions. The adsorption data for the removal of Cd, Pb, Cu and Zn by GB/TiO_2 at different initial metal ion concentrations (10, 30, 50 and 100 mg/L) is shown in Fig. 7.2.

The data in Fig. 7.2 reveals that with the increase of the initial metal concentration (Cd, Pb, Cu and Cu) from 10 to 75 mg/L, the removal percentage of Cd, Pb, Cu and Zn decreases. The high removal efficiency of metals were at a concentration 10 mg l^{-1} (95, 98, 97 and 98 %, respectively), but there was a small decrease in metal removal efficiency observed at 30 and 50 mg/L. These data indicate that the highest metal removal affinity occurred at a lower metal concentration of 10 mg/L.

Rashidi et al. (2010) reported that with the nano-TiO_2 catalyst, the increase of initial Pb (II) and Cu (II) ion concentrations, the percentage adsorption decreased from 99.855 to 94.613 % for Pb (II) and 90.071 to 71.918 % for Cu (II).

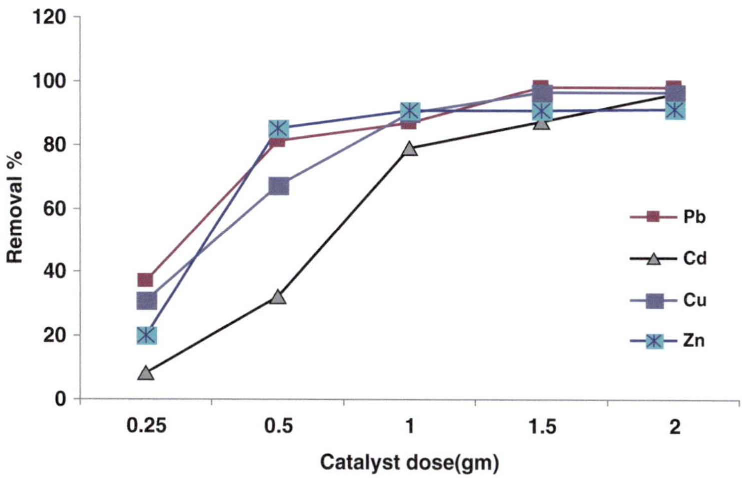

Fig. 7.1 Effect of GB/TiO_2 dosage on the removal of Cd, Pb, Cu, and Zn

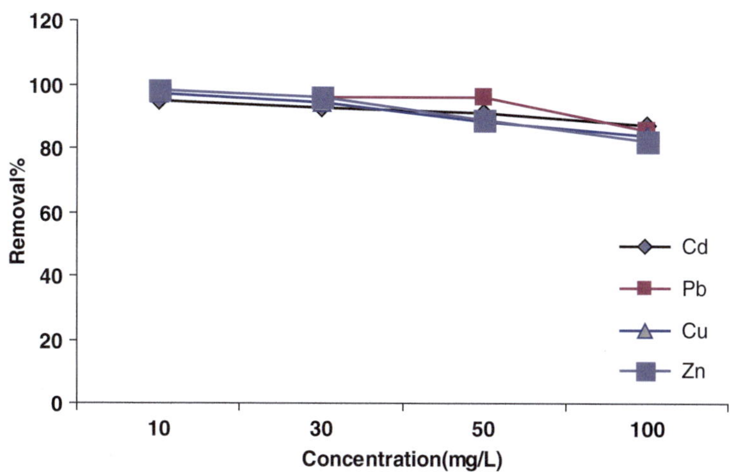

Fig. 7.2 Effect of initial metal concentration on the removal of Cd, Pb, Cu and Zn

7.3.4 Effect of Initial pH on Metal Ion Removal

The effect of initial pH on the removal efficiency of metals is represented in Fig. 7.3. It shows that the removal capacity of metal ions was very low at lower pH, which is attributed to the fact that the higher concentration and higher mobility of H^+ ions present in solution favored adsorption of hydrogen ions to the surface over Pb, Cd, Cu and Zn ions. The removal capacity of GB/TiO_2 increased with increasing pH from 2 to 8 for Pb, Cd, Cu and Zn, with a removal percentage of 98, 95, 98 and 87 %, respectively. The variation in metal removal efficiency in

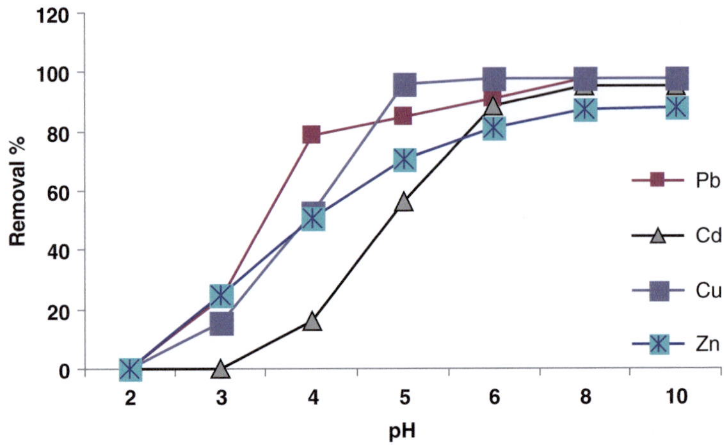

Fig. 7.3 Effect of pH on removal efficiencies of metals

equilibrium pH is largely due to the influence of pH on the adsorption processes of the GB/TiO_2, which indicates that, the adsorption capacity of the catalyst is clearly pH dependent. The results show that the affinity for adsorption of Pb, Cd and Cu is generally higher than for Zn.

The studies of the Pb, Cd, Cu and Zn metal ion uptake on to GB/TiO_2 were dependent on the initial solution pH because the pH affects the surface charge of the catalyst, the degree of ionization of the surface groups and the nature of the adsorbing ions. The removal of Cd, Pb, Cu and Zn metals ions by GB/TiO_2 are explained by electrostatic attraction, surface complex formation and the precipitation mechanism (Deng and Ting 2005; Hawari and Mulligan 2006).

Our result is consistent with the result obtained by Liang et al. (2003) in which nano-TiO_2/g showed adsorptive capacity to Zn and Cd as 15.3 and 7.9 mg/g, respectively, at pH 9.

Nanosized titanium oxides were able to simultaneously remove multiple metals (Zn, Cd, Pb, Ni Cu) from a solution of pH = 8. Rashidi et al. (2010) studied the removal of Pb and Cu ions by nano-TiO_2 at solution pH 6.

Cho et al. (2005) reported that the Cd removal increased proportionally with increasing pH from 0 % at pH 3 to 95 % at pH 8 by fly ash adsorbent. Bouzid et al. (2008) showed that the Cu adsorption by sewage sludge ash increased as pH increased and the maximum adsorption (99 %) was at pH 7.2.

7.3.5 *Effect of Contact Time on Metal Removal*

Adsorption and photocatalysis of metal removal from an aqueous solution are a time-dependent process, so it is of importance to study the effects of equilibrium contact time with the GB/TiO_2 catalyst on heavy metal ion removal. Figure 7.4

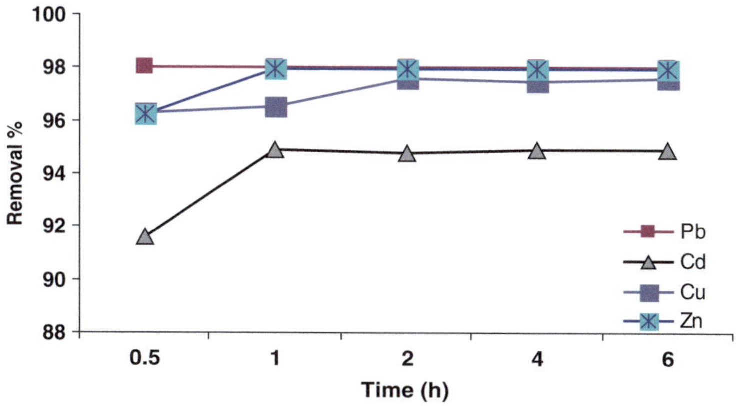

Fig. 7.4 Effect of contact time on the removal of Cd, Pb, Cu and Zn

shows the effect of the contact time on the removal efficiency of the studied metals. It was found that the amount of adsorbed ions increased rapidly from 0.5 to 1 h, which indicates that there were enough adsorption sites for the ions to be accommodated.

Subsequently, the process slowed to a constant level as the adsorption and active sites gradually became saturated. Also, the adsorption of metal ions increased with agitation time, perhaps due to a decrease in the boundary layer resistance to mass transfer in the bulk solution and an increase in the kinetic energy of the hydrated ions.

The result means that the equilibrium of metal adsorption occurred after 1 h. The fast adsorption at the initial stage may be due to the higher driving force making fast transfer of metal ions to the surface of adsorbent particles and the availability of the uncovered surface area and active sites on the adsorbent. With increasing time the availability of the uncovered surface area and the remaining active sites decreased (Wu et al. 2008; Aroua et al. 2008).

The studies on the Cd^{2+} and Cu^{2+} removal efficiency on TiO_2/fly ash indicated that the process was relatively fast and the maximum adsorption occurred after 30 min. A contact time of 30 min is technologically feasible (Visa et al. 2009).

The effect of the contact time for the adsorption of Cr (VI) on TiO-MCM-41 was studied by Parida et al. (2012) who concluded that the equilibrium time for the adsorption of Cr (VI) was considerable until 80 min had elapsed. Fonseca et al. (2006) reported that the equilibrium times for Mn and Cd by a clay mineral are 48 and 72 h respectively which is a long time to achieve adsorption.

7.4 Characterization of the Prepared Catalyst (GB/TiO_2)

7.4.1 Scanning Electron Microscopy (SEM)

The morphology of the adsorbent (GB) and the catalyst (GB/TiO_2) samples were examined by SEM. Figure 7.5 shows the SEM micrographs of the two samples, glass waste powder (GB) and after mixing with TiO_2, GB/TiO_2 whereas the latter sample is constituted by well-defined micro-size particles of glass particles with titania. Images indicate changes in the surface morphology after mixing with TiO_2. The TiO_2 particle size in GB/TiO_2 measured was not more than 10 μm, while the other glass powder (GB) particle consists of agglomerates of particles in nanometer size. The SEM-overview image in Fig. 7.5 reveals that the sample consists of two kinds of particles: scale-shaped GB, and TiO_2 particles with no characteristic morphology. The TiO_2 nanoparticles are homogeneously dispersed on the surface of GB.

7.4.2 Equilibrium Metal Adsorption Measurements and Modeling

7.4.2.1 Adsorption Isotherms

Langmuir and Freundlich isotherms were applied to describe the adsorption equilibrium of metal ions on the adsorbent. The Freundlich expression is an empirical equation based on a heterogeneous surface, which relates the concentration of a solute on the surface of an adsorbent to the concentration of the solute in the liquid. The Freundlich Eq. (7.8) is given as follows:

$$q_e = K_{fCe}^{1/n} \tag{7.8}$$

Equation (7.8) can also be expressed in linearized logarithmic form

$$Log\, q_e = Log\, k_f + (1/n) Log\, C_e \tag{7.9}$$

where q_e is the amount adsorbed (mg/g), C_e is the equilibrium concentration of the adsorbate (mg/L), and k_f and n are the Freundlich constants related to adsorption capacity and adsorption intensity, respectively. When Log q_e is plotted versus Log $C_{e,}$ the slope is equal to 1/n and the intercept is equal to Log $k_{f.}$. The high value of R^2 (the correlation coefficient of the line) indicates that the adsorption follows the Freundlich isotherm model perfectly (Zhang et al. 2005). When the value of 1/n is between 0 and 1, this indicates the heterogeneity of the sorbent (Mishra et al. 1996). Furthermore, the smaller 1/n and larger K_f values for the adsorbent indicate that the adsorbent has high adsorption capacity, intensity and affinity for metal ions (Zhang and Itoh 2003).

Fig. 7.5 SEM Micrographs of GB and GB/TiO_2 samples

The Langmuir equation is expressed as follows

$$C_e/q_e = C_e/Q_o + 1/Q_o b \tag{7.10}$$

where C_e (mg/L) is the metal equilibrium concentration, q_e (mg/g) the amount adsorbed at equilibrium, b (mg/L) is the Langmuir constant which is related to the

Table 7.2 Langmuir and Freundlich constants for the adsorption of Cd, Pb, Cu and Zn by GB/TiO_2

Freundlich constants			Langmuir constants			
1/n	K_f	R^2	Q^o	b	R^2	Metal ion
0.618	5.59	0.998	31.44	0.219	0.958	Cd
0.731	10.38	0.997	46.72	0.314	0.859	Pb
0.442	6.511	0.966	21.97	0.489	0.993	Cu
0.760	3.477	0.988	44.44	0.079	0.8562	Zn

affinity of the binding site, and Q_o (mg/g) the capacity parameter. When C_e/q_e is plotted versus C_e, the slope is equal to $1/Q_o$ and the intercept is equal to$1/Q_o$ b.

The results obtained on the adsorption of Cd, Pb, Cu and Zn by GB/TiO_2 at different initial metal ion concentrations (10, 30, 50 and 75 mg/L) and analyzed by the Freundlich and Langmuir models are presented in Table 7.2.

From the Freundlich constants, the values of 1/n for adsorption on GB/TiO_2 were 0.618 for Cd, 0.731 for Pb, 0.442 for Cu and 0.760 for Zn. The maximum adsorption from the Langmuir parameters of Cd, Pb, Cu and Zn on GB/TiO_2 are 31.4, 46.7, 21.9, and 44.4 mg/g, respectively. This indicated that Pb and Zn are highly adsorbed on GB/TiO_2. These values indicated that the order of metal ions according to their affinity to adsorption on GB/TiO_2 is Pb $>$ Zn $>$ Cd $>$ Cu.

7.5 Conclusion

The GB/TiO_2 catalyst was successively synthesized via nano-mechanical synthesis. TEM analysis confirmed the presence of the crystalline anatase phase with an almost uniform diameter and a narrow range distribution of spherical nanoparticles in the GB/TiO_2 catalyst. The synthesized catalyst exhibited good photocatalytic activity for heavy metals (Cd, Cu, Pb and Zn) removal under sunlight irradiation and mild conditions in aqueous solutions. The photocatalytic performance in both cases was markedly dependent on catalyst dose, metal concentrations and contact time.

List of Figures

References

Aroua MK, Leong SPP, Teo LY, Yin CY, Daud WMAW (2008) Real-time determination of kinetics of adsorption of lead (II) onto palm shell-based activated carbon using ion selective electrode. Bioresour Technol 99:5786–5792

Bouzid J, Elouear Z, Ksibi M, Feki A, Montiel A (2008) A study on removal characteristics of copper from aqueous solution by sewage sludge and pomace ashes. J Hazard Mater 152 (2):838–845

Chaizasith S, Chaizasith P, Septhum C (2006) Removal of cadmium and nickel from aqueous solutions by adsorption onto treated fly ash from Thailand. J Sci Technol 11:13–19

Chen J, Poon CS (2009) Photocatalytic activity of titanium dioxide modified concrete materials – influence of utilizing recycled glass cullets as aggregate. J Environ Manage 90:3436–3442

Cho M, Chung H, Choi W, Yoon J (2005) Different inactivation behaviors of MS-2 phage and Escherichia coli in TiO2 photocatalytic disinfection. Appl Environ Microbiol 71:270–275

Deng S, Ting YP (2005) Characterization of PEI-modified biomass and biosorption of Cu(II), Pb (II) and Ni(II). Water Res 39(10):2167–2177

Fonseca MG, de Oliveira MM, Arakaki LNH (2006) Removal of cadmium, zinc, manganese and chromium cations from aqueous solution by a clay mineral. J Hazard Mater B137:288–292

Harraz FA, Abdel-Salam OE, Mostafa AA, Mohamed RM, Hanafy M (2013) Rapid synthesis of titania–silica nanoparticles photocatalyst by a modified sol-gel method for cyanide degradation and heavy metals removal. J Alloys Compd 551:1–7

Hawari AH, Mulligan CN (2006) Biosorption of lead(II), cadmium(II), copper(II) and nickel(II) by anaerobic granular biomass. Bioresour Technol 97(4):692–700

Hilmi A, Luong JHT, Nguyen AL (1999) Utilization of TiO_2 deposited on glass plates for removal of metals from aqueous wastes. Chemosphere 38(4):865–874

Hoffmann MR, Martin ST, Choi W, Bahnemann DW (1995) Environmental applications of semiconductor photocatalysis. Chem Rev 95:69–96

Holgado M, Cintas A, Ibisate M, Serna CJ, Lopez C, Meseguer F (2000) Three-dimensional arrays formed by monodisperse TiO2 coated on SiO2 spheres. J Colloid Interface Sci 229(1):6–11

Inagaki M, Hirose Y, Matsunaga T, Tsumura T (2003) Carbon coating of anatase-type TiO_2 through their precipitation in PVA aqueous solution. Carbon 41:2619–2624

Janusz W, Matysek M (2006) Coadsorption of Cd (II) and oxalate ions at the TiO_2 electrolyte solution interface. J Colloid Interface Sci 296:22–29

Kadirvelu K, Faur-Brasquet C, Le Cloirec P (2000) Removal of Cu (II), Pb (II), and Ni (II) by adsorption onto activated carbon cloths. J Langmuir 16(22):8404–8409

Kim TK, Lee MN, Lee SH, Park YC, Jung CK (2005) Development of surface coating technology of TiO_2 powder and improvement of photocatalytic activity by surface modification. Thin Solid Films 475:171–177

Liang MJ, Kong CH, Wang C (2003) Characterization and electrophoretic deposition of apatite on titanium substrate. J Mater Sci Mater Med 14:797–801

Mahltig B, Gutmann ED, Meyer C, Solvo D (2011) Thermal preparation of nanocrystalline anatase containing TiO_2 and TiO_2/SiO_2 coating agents for application of photocatalytic treatments. Mater Chem Phys 127(1-2):285–291

Ming H, Shujuan Z, Pan B, Zhang W, Lu L, Zhang Q (2012) A review – heavy metal removal from water/wastewater by nanosized metal oxides. J Hazard Mater 211/212:317–331

Mishra SP, Singh VK, Tiwari D (1996) Radiotracer technique in adsorption study. Efficient removal of mercury from aqueous solutions by hydrous zirconium oxide. Appl Radiat Isot 47:15–21

Mona BM (2010) Low cost nanomaterials for water desalination and purification desalination. Final UNESCO Report, Contract No. 4500103693, Nano Tech 28 pp

O'Connell DW, Birkinshaw C, O'Dwyer TF (2008) Heavy metal adsorbents prepared from the modification of cellulose: a review. Bioresour Technol 99:6709–6724

Ozmen M, Keziban C, Ilker A, Arslan G, Cengeloglu Y, Ersoz M, Ali T (2009) Surface modification of glass beads with glutaraldehyde: characterization and their adsorption property for metal ions. J Hazard Mater 171:594–600

Parida K, Mishra KG, Dash SK (2012) Adsorption of toxic metal ion Cr (VI) from aqueous state by TiO_2-MCM-41: equilibrium and kinetic studies. J Hazard Mater 241–242:395–403

Petrella A, Petruzzelli V, Basile T, Petrella M, Boghetich G, Petruzzelli DA (2010) Recycled porous glass from municipal/industrial solid wastes sorting operations as a lead ion sorbent from wastewaters. React Funct Polym 70:203–209

Rashidi F, Sarabi RS, Ghasemi Z, Seif A (2010) Kinetic, equilibrium and thermodynamic studies for the removal of lead (II) and copper (II) ions from aqueous solutions by nanocrystalline TiO_2. Superlattice Microstruct 48:577–591

Strauss M, Maroneze CM, Souza D, Silva JM, Sigoli FA, Gushikem Y, Mazali MIO (2011) Annealing temperature effects on sol–gel nanostructured mesoporous TiO_2/SiO_2 and its photocatalytic activity. Mater Chem Phys 126(1-2):188–194

Tsumura T, Kojitani N, Umemura H, Toyoda M, Inagaki M (2002) Composites between photoactive anatase-type TiO_2 and adsorptive carbon. Appl Surf Sci 196:429–436

Visa M, Anca D (2013) TiO2/fly ash novel substrate for simultaneous removal of heavy metals and surfactants. Chem Eng J 223:860–868

Visa M, Carcel RA, Andronic L, Duta A (2009) Advanced treatment of wastewater with methyl orange and heavy metals on TiO2, fly ash and their mixtures. Catal Today 144:137–142

Wang YH, Lin SH, Juang RS (2003) Removal of heavy metal ions from aqueous solutions using various low-cost adsorbents. J Hazard Mater 102:291–302

Wang T, Liu W, Xu N, Ni J (2013) Adsorption and desorption of Cd(II) onto titanate nanotubes efficient regeneration of tubular structures. J Hazard Mater 250–251:379–386

Wu Yun, Shuzhen Zhang, Xueyan Guo, Honglin Huang (2008) Adsorption of chromium(III) on lignin. Bioresour Technol 99(16):7709–7715

Zainal Z, Hui LK, Hussein MZ, Abdullah AH, Khair IM, Hamadneh R (2009) Characterization of TiO_2–Chitosan/Glass photocatalyst for the removal of a monoazo dye via photodegradation–adsorption process. J Hazard Mater 14:138–145

Zhai YB, Wei XX, Zheng GM, Zhang DJ, Cfu KF (2004) Study of adsorbent derived from sewage sludge for the removal of Cd2+, Ni2+ in aqueous solutions. Sep Purif Technol 38:191–196

Zhang FS, Itoh H (2003) Adsorbents made from waste ashes and post-consumer PET and their potential utilization in wastewater treatment. J Hazard Mater B101:323–337

Zhang Z, Wang C, Zakaria R, Ying J (1998) Role of particle size in nanocrystalline TiO_2-based photocatalysts. J Phys Chem B102(52):10871–10878

Zhang K, Cheung WH, Valix M (2005) Roles of physical and chemical properties of activated carbon in the adsorption of lead ions. Chemosphere 60:1129–1140

Chapter 8
Cost Effective Fouling Control in Cooling Water Intake Systems with Environmental and Operational Benefits

M.C.M. Bruijs

Abstract Colonization of intake cooling water systems by fouling organisms is a major concern for industries, power and desalination plants. Biofouling can result in increased risk of operational problems, resulting in a reduction of output or even an unplanned shutdown which would have a high cost impact. Biofouling results in an increased wall roughness and reduction of the inner pipe diameter which leads to a significant head loss in the intake structure which then requires additional pump capacity. To prevent settlement and growth of fouling species, an effective anti-fouling treatment is required. Worldwide, the standard industrial practice in coastal areas is the application of an oxidative biocide, e.g. hypochlorite, which is dosed at the water intake. Opportunities exist for science-based decisions to optimize site-specific biocide dosing regimes which enable cost-efficient and reliable fouling control while complying with stringent regulatory discharge limits. Research and development in this work field has resulted in the dosing technology Pulse-Chlorination. This applies a timed-based on/off low-level chlorine dosing regime based on the reactional patterns of local bivalve fouling species. The adoption of this dosing technology has resulted in significant reductions of chlorine use (up to 50 %) and has proved to be more effective in preventing biofouling settlement in intake and cooling water systems compared to continuous and shock dosing regimes. It has resulted in major cost savings and consequently reduced environmental impact significantly.

Keywords Biofouling mitigation • Chlorination • Environmental impact • Pulse-Chlorination

M.C.M. Bruijs (✉)
Department Process & Cooling Water, DNV GL Energy, Arnhem, The Netherlands
e-mail: maarten.bruijs@dnvgl.com

M. Baawain et al. (eds.), *Recent Progress in Desalination, Environmental and Marine Outfall Systems*, DOI 10.1007/978-3-319-19123-2_8

8.1 Introduction

Industries worldwide extract enormous volumes of surface waters to cool their operation processes, e.g. power plants, (petro) chemical installations, waste incinerators, etc. In addition, desalination plants utilise seawater as a source to produce potable water or process water. Larger facilities are mainly located in coastal areas using seawater for cooling or make-up water. The intake facilities can either be open, directly located on the sea shore, or consist of a submerged intake pipe with an intake head located below sea level. With seawater, a variety of marine fouling organisms enter the intake system, such as mussels, oysters, barnacles, etc. This type of fouling can result in major operational problems due to blockage of the cooling water system, e.g. heat exchangers. To guarantee operational reliability, an effective fouling mitigation method needs to be applied, usually dosing of a biocide. To reduce the environmental impact, a balanced dosing regime of biocide needs to be applied to maintain reliable plant operation. Pulse-Chlorination is a dosing method which reduces the amount of biocide dosed by up to 50 % and secures safe plant operation. This has resulted in major operational cost savings and has proven to be a reliable dosing technique which can be applied worldwide.

8.2 Biofouling Settlement

Intake structures and cooling water conduits are generally in ideal environments, providing optimal conditions for settlement and growth of foulers. The continuous flow of seawater provides sufficient oxygen and nutrients, the water flow is turbulent, it is dark and there are no predators.

The process of settlement starts with chemically conditioning the surfaces to create optimal conditions for colonization by fouling organisms in a reasonably standard pattern. Firstly, organic molecules are deposited, followed by colonisation by microorganisms which in their sessile phase produce 'slime' (xPS), creating a so-called biofilm. Thereafter, colonization of the surfaces by other organisms becomes possible. Both the microfouling and the macrofouling species constitute the overall biofouling community. Clearly, the types of fouling species and growth patterns are dependent on the geographical location, climatic conditions, and local water conditions such as salinity and water quality and any seasonal changes.

Macrofouling organisms enter the intake system as larvae, which settle on the surfaces and develop into adults if conditions are suitable. There is a wide range of sessile species which can cause macrofouling problems, such as bivalves, barnacles, hydroids, tube worms, tube building amphipods, bryozoans and ascidians. This is especially the case with fouling organisms that cement themselves (barnacles and some oyster species) which results in an irreversible increase in wall roughness. This because after dying, part of the animal remains on the surface. Therefore, it is very important to prevent settlement of macrofouling larvae in the intake and cooling water system from the start of operation. Fouling organisms will settle on the surface of intake pipes and may, in competition for substrate, grow on top of each other forming thick layers.

Of all biofouling species, bivalves (mussels, oysters and clams) and barnacles are known to cause serious operational problems to industrial cooling water systems. Hydroids can also cause operational problems especially when they foul filter systems or form hard layers in water with high concentrations of sediment. Hydroids can act as a web structure and in this way filter sediment particles to form obstructions. In addition, hydroids are known to foul sieves of small mesh sizes. Measures to control bivalve species will also control the other sessile biofouling species but this depends on their tolerance to the treatment. The growth rate depends on the species, water temperature and availability of nutrients.

8.2.1 Impact of Biofouling

A coastal industry that uses seawater for cooling purpose can suffer from biofouling build-up of a potential biomass of up to hundreds of tons within 2 years (Polman et al. 2012). As a consequence, the cooling water flow is interfered with due to the decreased diameter size of the pipe and the increased wall roughness. This results in an increased head loss and decreased efficiency for the pumping station (Woods Hole Oceanographic Institution 1952; Hall et al. 1981). The head loss is mainly caused by the riser head, intake pipelines (including the chlorination system) and screens. In other parts of the cooling water system, hydraulic conditions within the system determine whether the location is prone to becoming fouled (Bruijs and Jenner 2012).

Both the diameter of the pipeline and the wall roughness are affected by biofouling. The head loss over a pipeline is determined by the flow velocities, wall roughness and local losses. The local losses are small compared with the losses due to the wall roughness and are therefore neglected. In addition, local losses are too specific for any system to take into consideration.

Besides head loss, there is a continuous risk of blockage of condenser tubes, valves, orifices and other constricted places by organisms that become detached. In Europe alone, companies lose millions of Euros due to biofouling, often not quantifiable due to lack of information. An example of fouling in intake systems and condensers is presented in Fig. 8.1.

8.2.2 Biofouling Mitigation

To prevent settlement of biofouling there are numerous options, e.g. filtration, coatings or biocides. However, most options do not protect the cooling water system completely from settlement of macrofouling larvae. In addition, most

Fig. 8.1 Macrofouling settlements in condenser A (1996) and breech at the intake of the Maasvlakte Power Station, Unit EFM 1. Cooling water intake was last manually cleaned at September 1994. Walls and floor intake channel were covered with mussels in a layer of a maximum 30 cm (Polman et al. 2010)

treatment options are not capable of completely removing any settled organism. Biocides have proven to be a reliable and effective option to prevent the settlement and growth of macrofouling species. In addition, biocides are usually capable of mitigating the numbers of any settled organisms. Worldwide, the typical industry practice in coastal areas includes continuous chlorination of the seawater with periodic shock-dosing. Chlorination of (sea) water results in the formation of halogenated compounds (Chlorination By-Products, CBPs). In previous studies carried out in the 1990s, no clear negative ecological effect from CBP's was found either near the outfall of power stations or in the far-field (Jenner et al. 1997, 1998; Taylor 2006). Although the study did not show clear evidence CBP's resulting in negative impacts, water authorities urged alternative methods to reduce the amount of chlorine dosed. The typical industry practise of dosing chlorine (a combination of continuous and shock dosing) is not based on ecotoxicological data of the targeted species, but is generally based on a post-hoc observation of antifouling efficiency or performed as an attempt to meet the discharge limits of residual biocide concentrations. Shock dosing is applied in the erroneous notion that it prevents fouling species from adapting to continuous chlorination. Such typical dosing procedures are practiced also in the Arabian Gulf. Therefore, opportunities exist for science-based decisions to optimize site-specific biocide dosing regimes which enable cost-efficient and reliable fouling control while complying with stringent regulatory discharge limits.

8.2.3 Chlorination Chemistry in Seawater

In chlorination chemistry a distinction is normally made between free (active/available) chlorine and combined chlorine. Free oxidant (FO) is present as an equilibrium mixture $HOCl \rightarrow OCl^- + H^+$ (hypochlorous acid and hypochlorite). Combined chlorine is available in chloramines or other compounds having oxidising properties. Total residual oxidant (TRO) is defined as the total oxidising capacity (free and combined) which is available after chlorination. Chlorine demand is defined as the difference between the amount of chlorine added and the FO concentration remaining at the end of a specified contact period.

When chlorine is added to seawater, naturally containing 68 mg/L bromide at full salinity, the bromide is oxidized and the hypochlorite is displaced by hypobromous acid (HOBr). This reaction is rapid, with 99 % conversion within 10 s at full salinity and within 15 s even at half salinity. Within the Arabian Gulf, where salinities are above the normal 35‰ and typically vary between 39 and 42‰ depending upon the season in well-mixed deep locations, this would further increase the conversion rate. However, since hypochlorite is produced and stored on-site (in a 10 m^3 tank at approximately 500–2000 mg/L) prior to dosage, there is an opportunity for chlorine- dominated chemistry to produce various by-products.

Hypochlorite and -bromite immediately react with suspended and dissolved organic matter within seawater, especially the N-containing compounds. This process is called "chlorine (bromine) demand". Reactions between N-containing compounds and chlorine produce halogenated amines which are referred to as "bound oxidants". During chlorination in sea- or brackish water, these oxidants provide an extra toxic effect on bivalves by the reaction product bromamines. Bromamines are, in contrast with chloramines, acutely toxic for bivalves. Brominated amines are, more or less, as toxic as hypobromous acid. The production of these chlorinated by-products accounts for using the term TRO in seawater.

In summary, the effective part of the hypochlorite dosing in seawater is the total toxicity of free (bromine) oxidants (FO) and bound (bromine) oxidants (the latter defined as TRO). For this reason the chlorine concentration is generally defined by the amount of FO, when used in freshwater, and TRO, when used for either seawater or brackish water.

8.3 Pulse-Chlorination®

Pulse-Chlorination® (P-C), a method for chlorine dosing through cooling water systems developed by KEMA in 1998, is based on the observation that bivalve biofouling species like oysters, mussels and clams, have a recovery period after exposure to a chlorination period, before fully opening their valves and starting to filter water. Following exposure to chlorine, bivalves normally require time to recover before they can open again fully and restart filtration for oxygen and

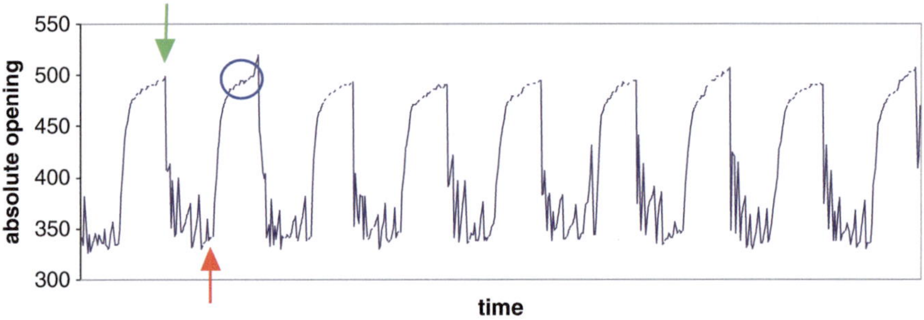

Fig. 8.2 Movement behaviour (opening and closing of the valves) of fouling bivalves detected by a valve movement monitor during a P-C regime. The top arrow (*green*) indicates the start ('on') and the bottom arrow (*red*) shows the timing of the stop ('off') of chlorination. The circle indicates the recovery period during which the bivalve slowly opens. Normal behaviour would reflect a continuously open valve position allowing respiration and feeding to occur

nutrients. P-C enhances a cyclic mode of hypochlorite dosing (on/off dosing regime), based on the behavioural response of the specific bivalve to chlorine dosing, thereby taking advantage of this recovery period to delay the restart of the next dosing pulse. By applying P-C, bivalves are forced to switch their metabolic mode continuously between aerobic (when open) and anaerobic (when closed), without the opportunity to return to normal filtration modus for nutrition and respiration. In this situation, the target organisms rapidly use their own energy reserves (i.e. glycogen and muscles). In adult specimens this leads to physiological exhaustion and subsequently death. Thus, the effect of P-C upon the target organisms is based on the repetitive short recovery period after exposure to short successive periods of chlorination. In addition, other fouling species, such as barnacles, hydroids, ascidians, etc, do not have the means to close and thereby avoid the chlorine treatment. These species are very efficiently mitigated by the Pulse-Chlorination dosing strategy.

P-C results in a more rapid effect, i.e. mortality of bivalves, compared to the conventional continuous chlorination method. For a better understanding of P-C, an example of the valve movement behaviour which shows the reaction pattern of bivalves in general during P-C is given in Fig. 8.2. Typical behaviour of a bivalve in seawater would be represented by valves being fully open most of the time.

The required effective initial dosage of hypochlorite concentration depends on the target organism's behaviour, and the seawater quality parameters at the intake.

In the past 15 years, P-C has proven to result in effective fouling mitigation. Some of these results were published (Jenner et al. 2008; Polman and Jenner 2002; Polman et al. 2008, 2010, 2013; MacDonald et al. 2012). Overall P-C resulted in a reduction of the amount of chlorine dosed to about 50 % compared to previous continuous dosing regimes. The reduced P-C dosing also proved to be more effective in comparison to a continuous dosing using higher levels of FO in the cooling water system. The reason for this increased effect is related to the fact

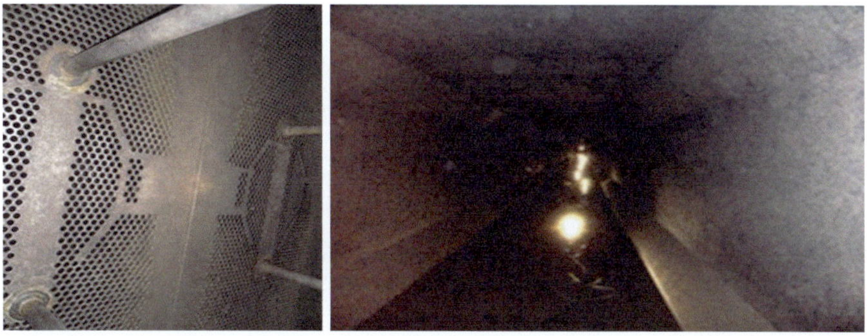

Fig. 8.3 Maasvlakte power station, Unit EFM 1 – return waterbox and intake channel after implementation of Pulse-Chlorination

Fig. 8.4 Qatargas operating company limited – condenser (*left*) and tube (*right*) endplate after implementation of Pulse-Chlorination

chlorine is not used as a biocide, but as a trigger to stimulate the bivalves to become highly active and consume all their energy in a relative short time period. In Figs. 8.3 and 8.4 some examples of cooling water system inspections are presented after P-C dosing.

The method is universally applicable, but needs to be attuned to local conditions. The overall aim is to ensure an optimal and reliable cooling seawater system and condenser/heat exchanger performance. The application of P-C between 1998 to date, has resulted in extremely clean condensers and heat exchangers. This has been observed in all companies that have implemented P-C since 1998. The overall result is better performance of the cooling water system (better heat transfer efficiency of the condenser, expressed as the K-value). In addition, due to the prevention of fouling and thus absence of large quantities of fouling specimens, less maintenance is required. This in turn allows longer intervals between planned outages and brings down the running costs on the basis of about €50,000 per day spread out over 3 years rather than 2 years.

When building new facilities, it is very important to include an effective biofouling control method during the commissioning phase, as soon as a water intake system is started up. Potential biofouling should be taken into account during the design phase of an intake and pumping system (Bruijs and Jenner 2012). This can be done by not only proper design, but also selecting the correct chlorination procedure and applying this from the very beginning of the operation of the facility. Alternatively, sufficient margin could be built into the system design, so that biofouling will not cause operational problems and additional cost shortly after start-up. The easiest way to achieve this is to use intake pipes with relatively large diameters and low flow velocities. The impact of biofouling on operation will be smaller in that case whereas a larger system will have a significant impact on construction costs and lower velocities will enhance quick colonisation by fouling. Therefore biofouling control by means of optimal prevention methods is the most cost-effective way forward.

It is important to prevent biofouling species which cement themselves onto the walls, to have the opportunity to settle and grow. Although these animals can be killed at a later stage with an effective anti-fouling treatment, the attached shell halve of these animals will stay cemented to the surface, even after manual cleaning. This will result in permanent increased wall roughness and could have a significant yearly cost impact.

An additional effect of P-C is the reduction of the amount and concentration of oxidants discharged into the receiving water body. This is in line with (new) stringent regulatory requirements which will lower the operational costs of related equipment (e.g. electro-chlorinators). Since P-C uses a minimum amount of chlorine, it produces a minimum amount of CBPs. Based on its effectiveness and reduced discharge of CBP's, P-C is the best available technique under the terms of the European Integrated Pollution Prevention and Control (IPPC-BREF Industrial Cooling) for macrofouling mitigation in once-through seawater systems. It has also gained worldwide recognition and acceptance as an environmentally benign method of biofouling mitigation and is officially recognized as such by local regulatory bodies in the Middle East.

8.4 Conclusions

For industrial facilities, it is of high importance to control biofouling in cooling water systems since this can result in major risks for operational reliability which can result in unplanned outages. In addition, fouling has an impact on wall roughness and can reduce the pipe diameter of the cooling water system, especially in the case of fouling organisms that cement themselves (barnacles and some oyster species) which results in an irreversible increase in wall roughness because after dying, a part of the animal remains on the surface. Biofouling results in major costs and makes the control of biofouling of great importance.

The choice of effective fouling control methods are limited as most methods have some limitations in efficacy or applicability. Chlorine is still the most commonly applied biocide to control biofouling and has proven to be efficient and applicable in most cooling water systems. To reduce the use of chlorine and minimise the discharge of residuals or CBPs, the dosing technique Pulse-Chlorination has proven to be the best available technique. The adaptation of this dosing technology has resulted in significant reduction of chlorine use (up to 50 %) and proved to be more effective in preventing biofouling settlement in intake and cooling water systems compared to continuous and shock dosing regimes. P-C has resulted in major cost savings and consequently reduced the environmental impact significantly.

List of Figures

Fig. 8.1 Macrofouling settlements in condenser A (1996) and breech at the intake of the Maasvlakte Power Station, Unit EFM 1. Cooling water intake was last manually cleaned at September 1994. Walls and floor intake channel were covered with mussels in a layer of a maximum 30 cm

Fig. 8.2 Movement behaviour (opening and closing of the valves) of fouling bivalves detected by a valve movement monitor during a P-C regime. The top arrow (*green*) indicates the start ('on') and the bottom arrow (*red*) shows the timing of the stop ('off') of chlorination. The circle indicates the recovery period during which the bivalve slowly opens. Normal behaviour would reflect a continuously open valve position allowing respiration and feeding to occur

Fig. 8.3 Maasvlakte power station, Unit EFM 1 – return waterbox and intake channel after implementation of Pulse-Chlorination

Fig. 8.4 Qatargas operating company limited – condenser (*left*) and tube (*right*) endplate after implementation of Pulse-Chlorination

References

Bruijs MCM, Jenner HA (2012) Cooling water system design in relation to fouling pressure. In: Rajagopal S, Jenner Venugopalan VP (eds) Operational and environmental consequences of large industrial cooling water systems. Springer. doi:10.1007/978-1-4614-1698-2_19. Springer Science+Business Media, LLC 2012

Hall LW, Helz GR, Burton DT (1981) Power plant chlorination: a biological and chemical assessment, 1st edn. Ann Arbor Science Publishers, Palo Alto, 237 pp

Idelchik IE (1996) Handbook of hydraulic resistance, 3rd Edition. Begell House Inc., New York, 790 pp

Jenner HA, Taylor CJL, van Donk M, Khalanski M (1997) Chlorination by-products in chlorinated cooling water of some European coastal power stations. Mar Environ Res 43(4):279–293

Jenner HA, Whitehouse JW, Taylor CJL, Khalanski M (1998) Cooling water management in European power stations: biology and control of fouling. Hydroécologie Appliquée 10 (1-2):1–225

Jenner HA, Polman HJG, Bruijs MCM (2008) Optimising cooling seawater antifouling strategy through pulse-chlorination: associated benefits to operating facilities from adopting an environmentally friendly (BAT) technology. In: Proceedings: OPENWAC-2008, Kalpakkam, 15–16 December 2008, pp 37–42

MacDonald IA, Polman HJG, Jenner HA, Quyam SQBM (2012) Industrial cooling seawater antifouling optimisation through the adoption of Pulse-Chlorination®. In: Rajagopal S, Jenner Venugopalan VP (eds) Operational and environmental consequences of large industrial cooling water systems. Springer, New York, 480pp

Polman HJG, Jenner HA (2002) Pulse-Chlorination®, the best available technique in macrofouling mitigation using Chlorine. Power Plant Chem 4(2):93–97

Polman HJG, Bruijs MCM, Calneggia F, Jenner A (2008) Optimisation chlorination strategy cooling water system Verve energy's Cockburn and Kwinana power plants. In: Proceedings of API PowerChem 2008: power station chemistry solutions for the 21st century, Queensland, 25–30 May 2008

Polman HJG, Bruijs MCM, Venhuis LP, van Dijk SA (2010) More than 10 year experience with Pulse-Chlorination dosing regime against macrofouling. In: Müller-Steinhagen H, Zettler HU (eds) Handbook heat exchanger fouling: mitigation and cleaning technologies, 2nd edn. PP-Publico, Essen, pp 240–251

Polman HJG, Verhaart F, Bruijs MCM (2012) Impact of biofouling in intake pipes on the hydraulics and efficiency of pump capacity. Publication in Desalination and Water Treatment 1/20 12, www.deswater.com. 25 June 2012

Polman HJG, Bruijs MCM, Khan F, Baldwin J (2013) Results and benefits of the adoption of Pulse-Chlorination® for industrial cooling seawater antifouling at Qatargas, Ras Laffan Industrial City, Qatar. Download pdf at http://www.utilities-me.com/article-2237-case-study-on-the-pulse/1/print/

Taylor CJL (2006) The effects of biological fouling control at coastal and estuarine power stations. Mar Pollut Bull 53(1–4):30–48

Woods Hole Oceanographic Institution (1952) Marine fouling and its prevention, 1st edn. U.S. Naval Institute, Annapolis, 391 pp

Chapter 9
Chlorination By-Products and Other Quality Parameters in Desalinated Drinking Water System in Muscat

Aliaa Al-Kalbani, Sara Al-Kiyumi, Zainab Ambu Saidi, Salwa Al-Rawahi, and Hamed Al-Hasni

Abstract Disinfection is important in the supply of safe drinking water. Chlorination is the major disinfectant process for drinking water. However, the reaction between the disinfectant and the organic matter could form disinfection by-products like trihalomethanes (THMs). The study was done in order to check the presence of chlorination by-products in the drinking water system in Muscat. This study presented the monitoring results of Muscat with collaboration with the Public Authority of Electricity and Water (PAEW) in the period of June 9, 2013 to August 1, 2013. Seventeen drinking water reservoirs were selected to assess the presence of the chlorination by-products. The chlorination by-products were analyzed or detected by using Gas Chromatography Mass Spectrometry (GCMS). According to the analyses, the concentration of the THMs did not exceed 0.3 ppb. Chloroform, bromodichloromethane and dibromochloromethane concentrations did not exceed 1.6 ppb while the bromoform reached 28 ppb. Therefore, the results showed that the bromoform concentration is the most dominant THM species observed in all the samples. The bromoform and the THMs concentrations did not exceed the maximum value of the Omani standard for drinking water (1 mg/l). The drinking water quality in the Muscat Governate is considered to be of high water quality based on the THM standards. Thus, at present, there is no health concern associated with the presence of THMs in drinking water.

Keywords Chlorination • Disinfection by-products • Organic matter • Trihalomethanes (THM)

A. Al-Kalbani (✉) • S. Al-Kiyumi • Z. Ambu Saidi • S. Al-Rawahi
Department of Soils, Water and Agricultural Engineering, College of Agricultural and Marine Sciences, Sultan Qaboos University, Al-Khoud, Muscat, Oman
e-mail: squ1071991@gmail.com; saraqme91@gmail.com; zainabsultan90@gmail.com; squ-study@hotmail.com

H. Al-Hasni
Public Authority of Electricity and Water, Muscat, Oman
e-mail: hamed.alhasni@paew.gov.om

M. Baawain et al. (eds.), *Recent Progress in Desalination, Environmental and Marine Outfall Systems*, DOI 10.1007/978-3-319-19123-2_9

9.1 Introduction

Water is a key driver of the development of countries worldwide as it plays a central role in sustainable growth and alleviates poverty (Grey and Sadoff 2007). Water is an input for production in the agriculture, industry and health sectors. Arid and semi-arid regions, where drought conditions prevail, suffer from infrequent and sporadic rainfall. Oman is one of these countries with an average rainfall of 100 mm/year (MRMWR 2010). Due to the limited groundwater resources and the absence of surface water resources, the water shortage issue has reached a crisis level. Moreover, water contamination is another threat to the availability of water. Sources of contamination might be chemical and/or biological (Al-Rashed and Sherif 2000). Surface water is more vulnerable to contamination as it is an open water body, but groundwater is also contaminated through many anthropogenic and natural processes (WHO 2008).

Extensive governmental efforts to augment water resources are practiced in Oman to secure water supply for different purposes. One of the most effective ways of water resource augmentation (mainly groundwater) is the harvesting of flashfloods by means of recharge dams and finding non-conventional water supplies such as desalinated water (Abdel-Rahman and Abdel-Magid 1993). The concerned authority in the country, PAEW, is distributing water to different users from both sources. Along with water quantity, the water quality of the distributed water is also of paramount importance as it is of concern to health. The distribution systems of domestic water use consist of storage tanks, pipeline networks and household tanks before the water reaches the end user. The distributed water in Oman originates from three different sources: desalination, blending groundwater with desalinated water and only groundwater from well-fields. Proper analyses (physical, chemical and biological) of water are performed efficiently prior to distribution by the authority to ensure that the quality of water is up to the specified standards. Chlorination is one of the essential steps in the cleaning process (WHO 2008) and is commonly used for disinfection. Globally, it is the chemical most used for drinking water disinfection due to its low cost, ease of use and control. Furthermore, it can also be simply maintained, is safe to use and remains within the system for a long time (Wagenet et al. 1988). Chlorine is used at one or two points in a pre-treatment and/or post-treatment process to maintain a disinfectant residual in the distribution system (Debordea and Guntena 2008) since it helps to remove biological contaminants from water (Prathapar et al. 2006). However, there is risk arising from the potential reaction of disinfectants (chlorine, ozone, chlorine dioxide, or chloramines) with naturally occurring OM, anthropogenic contaminants, bromide, and iodide during the production of drinking water (Richardson et al. 2007) by the formation of by-products in the chlorination process according to factors such as time, pH, temperature, bromide concentration, and the concentration and type of dissolved organic matter (DOM). The general reaction of DOM with chlorine, which produces disinfection by-products (DBPs), is as follows (Taha and Doanh 2000):

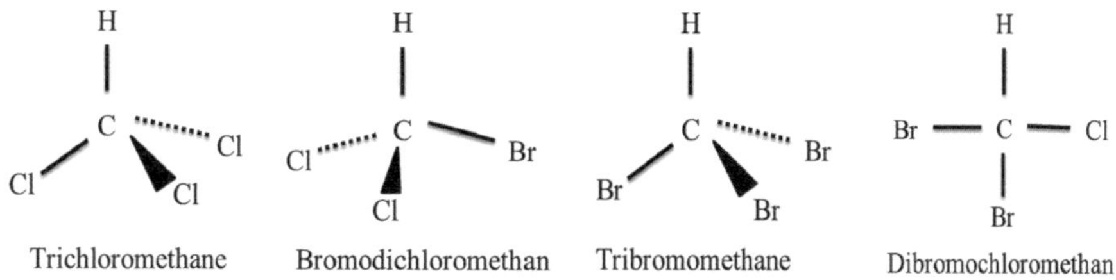

Fig. 9.1 The chemical structure of THMs, BDCM and DBCM (www.mermaidproject.eu, www.en.wikipedia.org, www.clu-in.org, and www.commons.wikimedia.org (16 May 2013))

$$Cl_2 + H_2O \rightarrow HOCl + HCl$$

$$HOCl + organic\ matter \rightarrow By\text{-}products$$

Along with other potential contaminants, OM is of interest for this paper. Natural organic matter (NOM) is a group of carbon-based compounds that are produced by various decomposition and metabolic reactions in the water supply (dspace.cc.tut.fi 2007). They commonly include proteins, polysaccharides, and humic substances (www.env.gov.nl.ca 2011). NOM concentration is usually measured as dissolved organic carbon (DOC) or total organic carbon (TOC) (www.epa.ie 2012). Although NOM does not pose a risk to human health, some NOM compounds react with chlorine to produce disinfection by-products (DBPs), such as trihalomethanes (THMs) and haloacetic acids (HAA) that are thought to be carcinogenic and/or genotoxic. These DBPs have been related to the development of various cancers and birth defects (www.env.gov.nl.ca 2011).

However, special concerns are associated with the trihalomethanes (THMs) because they have been recognized as potentially hazardous and are the major by-products of chlorination (Debordea and Guntena 2008). THMs are a group of organic chemicals that occur in drinking water as disinfection by-products (DBPs) consequent to chlorine treatment. This group is trichloromethane (chloroform), bromodichloromethane (BDCM), dibromochloromethane (DBCM), tribromomethane (bromoform). Figure 9.1 shows the chemical structure of these by-products.

This group is called THM because of the well-known organic chemicals occurring in it in the highest concentrations, while BDCM and DBCM have a very limited amount of these chemicals (Hazen 2006). THM levels tend to increase with pH, temperature, time, and the level of OM and residual chlorine (Pavelica et al. 2005).

Haloacetic acids (HAAs) are monobromoacetic acid, dibromoacetic acid, monochloroacetic acid, dichloroacetic acid, and trichloroacetic acid which and are all similar to THMs. and can be formed at the highest levels with chlorination. These acids induces DNA damage and other health effects such as chromosomal aberrations in mammalian cells. In general, THMs and HAAs are the most common DBPs present at the highest level (Richardson et al. 2007).

THMs have been recognized as potentially hazardous and they are the major by-products of chlorination (Bull et al. 1995). Epidemiological studies have found that inhabitants exposed to chlorination by-products have high rates of bladder, colorectal and brain cancers (Hsu et al. 2000; Cantor et al. 1999). According to a meta-analysis by Morris et al. (1992), approximately 10–40 % excess risk of bladder and colorectal cancers may be related to the high exposure to chlorination by-products in drinking water. In addition, according to Fetter (1992), "Cancer risk relative to drinking a litre (1 L) of chlorinated tapwater a day has a risk of 1.0. The relative risks were determined as an index obtained by dividing the daily lifetime human exposure in mg/kg of body weight by the daily dose rate for rodents in mg/kg of body weight. The dose rate of rodents is the daily dose necessary to give cancer to half the rodents at the end of the standard lifetime."

A study conducted by the Environmental Health Program (Drive 2006) pointed out that pregnant women who drank water containing the disinfection byproduct THM could increase the risk of early-term miscarriage. On other hand, there is a study showing that miscarriage was not related to the THMs (www.hc-sc.gc.ca 2004).

9.2 Objective

The main aim of this paper is to investigate the potential organic carbon and chlorination by-products.

9.3 Methodology

9.3.1 Study Area

Muscat governate reservoirs were targeted in this project. Seventeen (17) reservoirs (where Barka desalination plant and Al-Ghubrah desalination plant are the water supply sources), were selected to investigate chlorination by-products. The reservoirs were distributed in Muscat, Bushar, Muttrah, Al-Amirat, Qurayyat and As-Seeb.

9.3.2 Samples Collection

The samples were collected according to the PAWE sampling method. For chemical and physical analyses, the water was collected in plastic bottles without adding any preserver reagent. Thus, the bottles were washed at least three times in the water of the same source. Also, the bottles were filled completely with the water to prevent the change of its physical and chemical properties. Moreover, for hydrocarbon analyses, the water was collected in sterilized glass bottles with a preserver reagent like sodium thiosulfate ($Na_2S_2O_3$), so that in this case, the bottles were not washed during the sampling and the bottles were filled completely. For the total organic carbon and BOD analyses, unsterilized glass bottles were used and there was no preserver reagent added to the bottles.

In addition, a sampling form was printed for each location (sampling point) and each sample was labeled. The form contain information like the sample source, collecting time and time of arrival at the laboratory, date and the physical and chemical on-site measurements (temperature, EC, pH and free chlorine). After collection, the samples were placed in a cold box to prevent any change of water properties, such as the concentration of the THMs due to increase in temperature which could enhance THM formation.

From each reservoir, two (2) samples were collected; one (1) sample for each month. S1 indicates the samples of the first month while S2 presents the second month samples.

9.3.3 *Sample Analyses*

Each parameter was analyzed by a special instrument as following:

- Electric conductivity (EC), pH and temperature by the multi-analyzer
- Dissolved oxygen by DO-meter
- Turbidity by turbid-meter
- Free chlorine by chlorometer and ultra-violet (UV) spectroscopy
- Concentration of total carbon, total organic carbon and total inorganic carbon by Total Organic Carbon instrument (TOC)
- Concentration of anions and cations that are present in the water samples by Ion Exchange Chromatography (IC)
- Concentration of the chlorination by-products by Gas Chromatography Mass Spectrometry (GCMS)

9.4 Results and Discussion

From Fig. 9.2a, it is clearly seen that the temperature readings was in the range of 34–41 °C for the first sampling. However, the range of the second sampling was from 32 to 40 °C, where most readings were centered between 32 and 37 °C. Therefore, the temperature readings showed that the temperature was higher in the first sampling than in the second one as is shown in Fig. 9.2a.

However, pH readings were relatively the same in that there was no significant difference in the pH readings between both samplings. The pH readings ranged

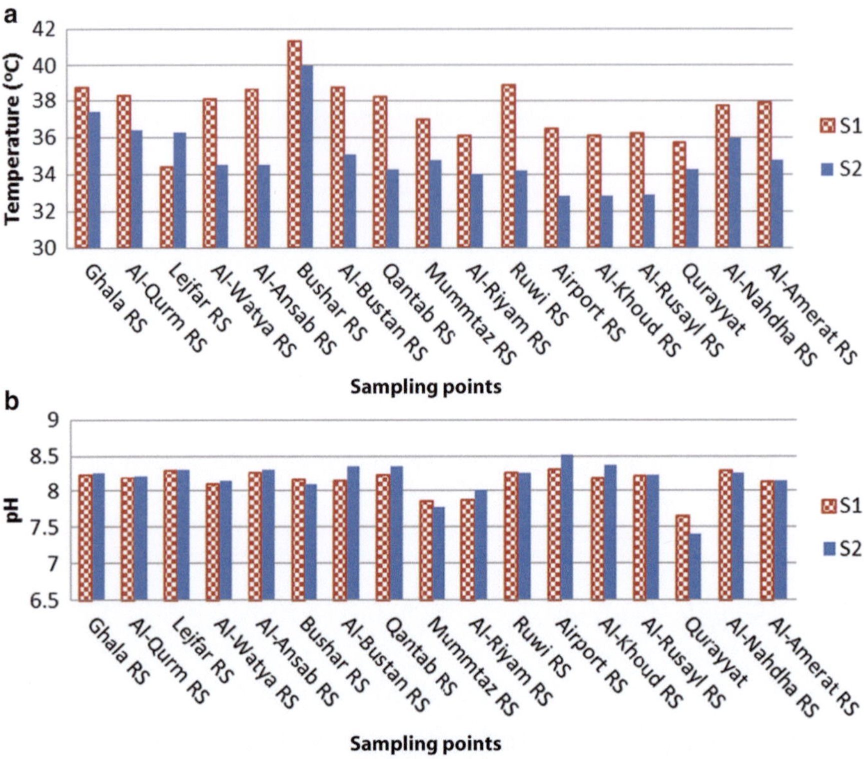

Fig. 9.2 (**a**) Temperature (°C). (**b**) pH readings of S1 and S2

from 7.65 to 8.31 for the first sampling and from 7.4 to 8.52 for the second sampling. These results were acceptable according to the Omani standards for drinking water where the pH value must be from 6.5 to 8 with the maximum value as 9.

The relationship between the temperature and the pH is an inverse relationship (chemwiki.ucdavis.edu). This relation was identified by the project samples results. However, there were unexpected results in some samples like Lejfar, Bushar and Mummtaz. This abnormality could have been due to the effect of other physical or chemical parameters or an instrumental error. For example, debris and contaminants which could be present in such reservoirs could have affected the pH. Also, gaseous chlorine (disinfectant) affects pH.

For the THM, the temperature and the pH readings cannot be used to identify the level of the THM in the project samples. However, according to previous studies, the formation of the THMs depends on the temperature and the pH. The formation of the THMs increases as the pH of the source water decreases (an inverse relationship) (Kutty et al. 1995). However, according to Hassani et al. (2010), the relationship between the formation of chlorination by-products (THMs) and the pH is a proportional relation. Thus, it could be that there is another factor or parameter

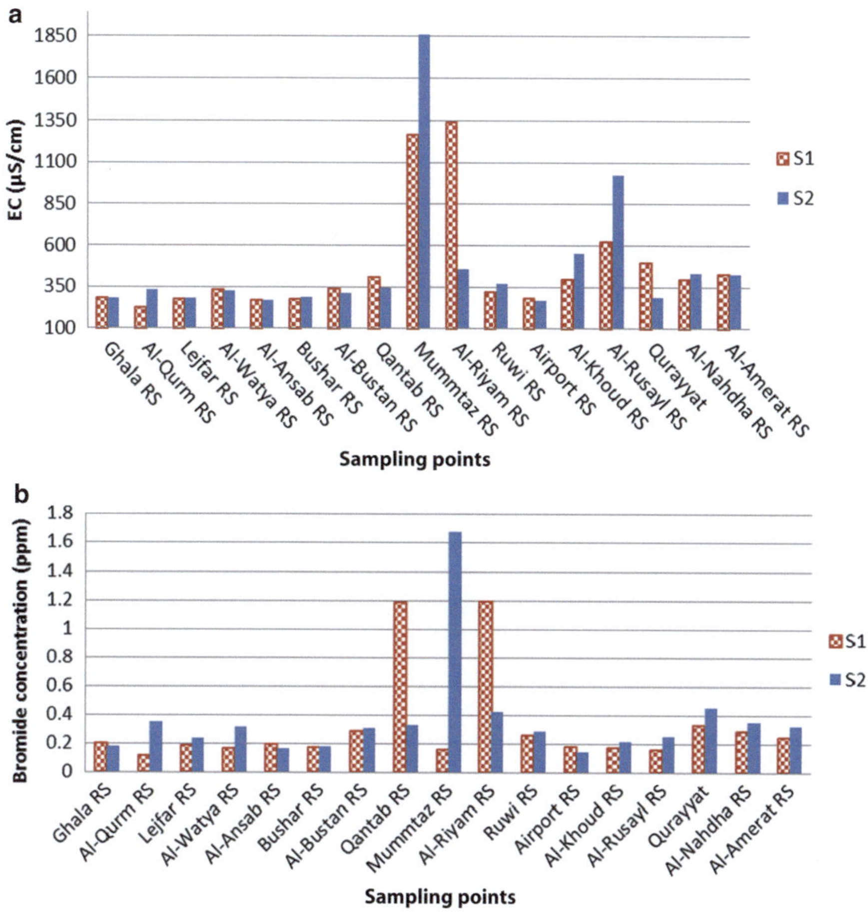

Fig. 9.3 (**a**) EC (µS/cm). (**b**) Bromide concentration (mg/l)

that effects the pH value which leads to affect the THMs formation as well. Moreover, there is a strong effect of temperature on the THMs formation. The reaction is faster and higher chlorine doses are required for disinfection as the temperature increases. Consequently, the THMs will be expected to be higher as the temperature increases (Hassani et al. 2010).

EC is the ability of the water to pass an electrical current and it is a function of the total dissolved solids (TDS). From Fig. 9.3a, it is clearly seen that the EC values ranged between 200 and 1900 µS/cm (1216 mg/l of TDS) in both samplings. According to Omani standards, the TDS must be between 200 and 600 mg/l, with a maximum allowable limit of 1200 mg/l. Thus, some of the values of the electrical conductivity of the Muscat reservoirs are not acceptable. The following is proof of this rejection.

In general, water is healthier when the EC is low due to the low salt concentrations present in the water. Increasing salts in the water can affect human health

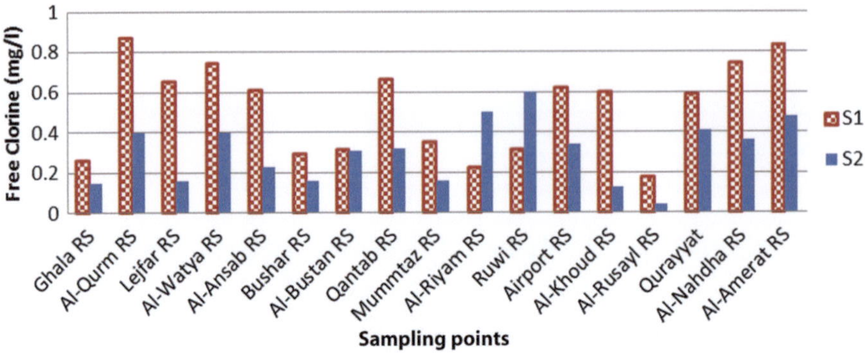

Fig. 9.4 Free chlorine concentration in mg/l for S1 and S2

because an over-intake of these nutrients can result in adverse health consequences. For instance, the presence of calcium and magnesium in high concentrations leads to an increase in water hardness which will affect human health. In fact, exposure to hard water has been suggested as a risk factor that could exacerbate eczema (WHO 2011). The explanation relative to hard water is that increased soap usage in hard water results in metal or soap salt residues on the skin or on clothes which are not easily rinsed off which leads to contact irritation (WHO 2011). Moreover, increasing the calcium concentration in drinking water is associated with increasing the risk of such diseases as nephrolithiasis (kidney stones), colorectal cancer, hypertension and others (WHO 2011).

Electrical conductivity is also affected by temperature so that the warmer the water, the higher the conductivity. The electrical conductivity of water increases by 2–3 % for an increase of 1 °C of water temperature (www.smart-fertilizer.com 2013). However, this relation was not found from the results in this project. The reason for that could be due to the effect of other parameters.

Moreover, as a relationship, the presence of bromide ions in a high concentration, where the EC is high, is a normal condition because bromide ions are one of the total dissolved solids (TDS). However, the presence of bromide ions in a high concentration is not indicative of the increase in the EC value because EC is not a function of bromide ions alone. Therefore, it is normal that electric conductivity is low when bromide ion concentration is high and vice versa, as is clear for Qantab SR in the first sampling.

The bromide ion is a catalyst ion for the formation of the TTHM. Thus, when the presence of the bromide ion is in the appropriate condition, chlorination by-products will form.

The free chlorine concentration in Fig. 9.4 shows the concentration of the chlorine (Cl_2) in drinking water. As is clear from the readings, the concentration of free chlorine in the water samples ranged between 0.1 and 0.83 mg/l for the first sampling, while the Cl_2 concentrations ranged from 0.1 to 0.6 mg/l for the second sampling. Therefore, the second sampling readings are more acceptable than the

Table 9.1 Comparison of the BOD results obtained by using the DO-meter and titration method

	BOD (mg/l)	
Sample source	DO-meter	Titration method (Winkler)
Ruwi SR	0.26	0.27
Al-Wadi Al-Kabir SR	0.51	0.56
Al-Wadi Al-Kabir house 2	0.19	0.17
Al-Riyam	0.03	0.05

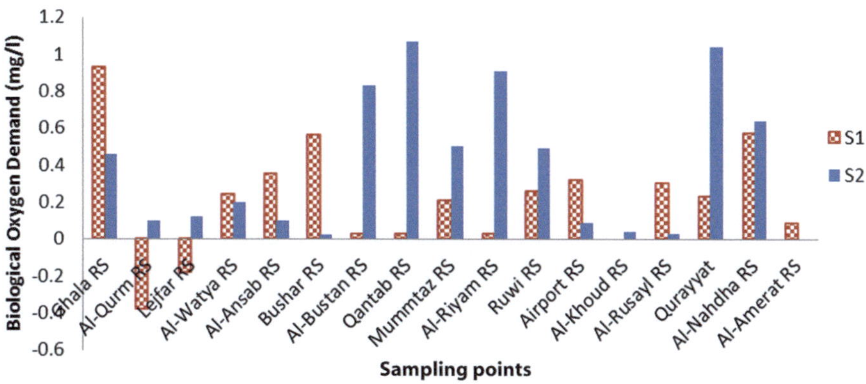

Fig. 9.5 BOD results mg/l for S1 and S2

first one according to the Omani standards for drinking water. This shows that the concentration of free chlorine must fall within the range of 0.2–0.5 mg/l as Cl_2. The highest concentration of free chlorine can enhance the formation of such by-products like THMs with the presence of other formation factors because the reaction rates of chlorine are dependent on several factors such as source characteristics, treatment type, contact time and the characteristics of the distribution system (Dumitru et al. 2008).

The biological oxygen demand (BOD) is an important parameter which is a measure of the quantity of oxygen used by microorganisms (e.g., aerobic bacteria) in the oxidation of organic matter. According to basic water quality experiments, the BOD parameter was measured by using the Winkler principle. A DO-meter was used to measure the dissolved oxygen for day one and day five. However, to ensure the accuracy of the DO-meter, the titration method was used for four (4) samples. The results of the four samples were relatively the same as the DO-meter results, as show in Table 9.1.

From Fig. 9.5, it is clearly seen that the BOD values exceeded 1 mg/l in some sample points which indicates the presence of organic matter in high concentration. Thus, in general, as the BOD is less, the water is more potable because a low number generally means little pollution and/or little aerobic activity. A high BOD means the opposite. The figure shows two samples with negative values (Al-Qurm RS and Lejfar RS) indicating an increase of the dissolved oxygen after 5 days at these sample points. This could be due to the presence of oxygen in the samples during the 5 days if the bottle was not completely full of water. This could have resulted in microorganisms, e.g. aerobic bacteria, consuming the dissolved oxygen, and the reduction could have been replaced by the available oxygen dissolving in the bottle. However, in case there were no microorganisms present in the sample, the available oxygen dissolved in water over time.

However, the BOD measurement is not important that much in this project because the tested water is for drinking purposes.

Turbidity is the amount of particulate matter that is suspended in water which makes the water cloudy or opaque. Material that causes water to be turbid includes clay, silt, finely divided organic and inorganic matter, soluble colored organic compounds, etc. From Fig. 9.6a, it is clearly seen that the turbidity values ranged from 0.04 to 0.56 NTU for S1 and exceeded 5.16 NTU in the airport reservoirs sample for S2. Thus, the turbidity readings are acceptable except for the airport RS because, according to Omani standards, the turbidity level must not exceed 5 NTU.

Moreover, the total organic carbon (TOC) is another parameter that can help to indicate the turbidity level but not vice-versa because the turbidity may indicate organic carbon, inorganic or both. From Fig. 9.6b, it is obvious that the concentration of TOC is relatively low. It does not exceed 400 ppb (μg/l) which is equal to 0.4 ppm (mg/l). Some researchers have found reasonably strong correlations between TOC and DBP formation while others have found these relationships to be more tenuous (www.env.gov.nl.ca 2011). According to Phil et al. (2004), there is no universal relationship that can be used to predict the formation of the THMs based on the TOC values in the water system.

Trihalomethane is a chlorination by-product that forms according to many factors such as pH, time, temperature, bromide concentration and others. From Fig. 9.7, it is clearly seen that the concentration of THMs does not exceed 0.3 μg/l (ppb) for both samples. Thus, these results are acceptable according to Omani standards for drinking water which stipulate that the sum of the ratio of the concentration of each value to its respective guideline value should not exceed 1 mg/l.

In addition, as can be seen in Table 9.2 the bromoform concentration is the most common trihalomethanes species compared to chloroform, bromodichloromethane (BDCM) and dibromochloromethane (DBCM) concentrations. The latter three THM types did not exceed 1.6 μg/l while bromoform reached 28 μg/l. However, the results of each species are still acceptable because in all the sampling points, the bromoform and the total trihalomethane concentration did not exceed the maximum permissible value. According to Omani standards, the maximum limit of bromoform is 0.1 mg/l, bromodichloromethane (BDCM) 0.06 mg/l, chloroform 0.2 mg/l and dibromochloromethane (DBCM) should not exceed 0.1 mg/l.

On other hand, the THM results of the first sampling may not be as accurate as the second sampling because the holding time for the trihalomethanes is 14 days (www.dshs.state.tx.us 2010). However, some samples of the first sampling were analyzed after more than 14 days. In fact, some samples were analyzed after 2 months after they were taken.

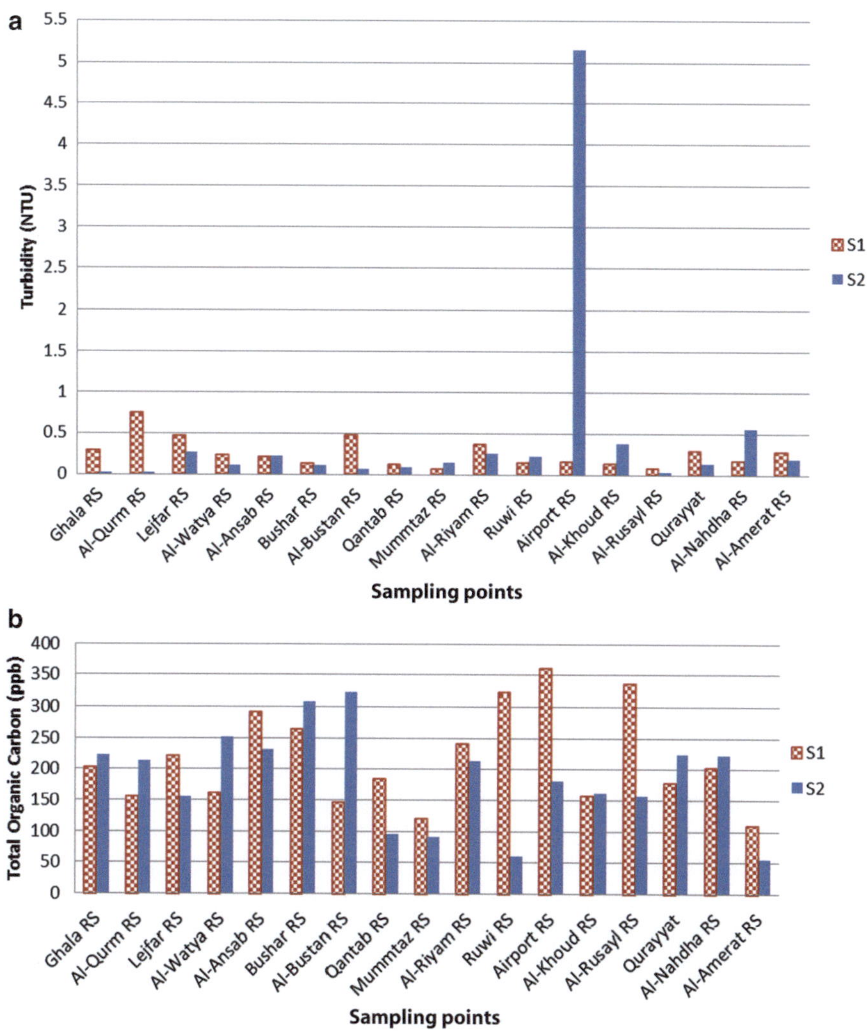

Fig. 9.6 (**a**) Turbidity measurements for S1 and S2 (NTU). (**b**) Total organic carbon (ppb)

Moreover, the THM level in drinking water depends on temperature, organic matter nature and concentration, chlorine dose, pH, reaction time and inorganic ions like bromide (Dumitru et al. 2008). Generally, the longer the contact time between chlorine and natural organic matter (NOM), the greater the amount of THM that can be formed. In addition, high THM values usually occur at points in the distribution system with the longest residence time or water age, such as reservoirs, oversized pipes and network dead ends (www.epa.ie 2012).

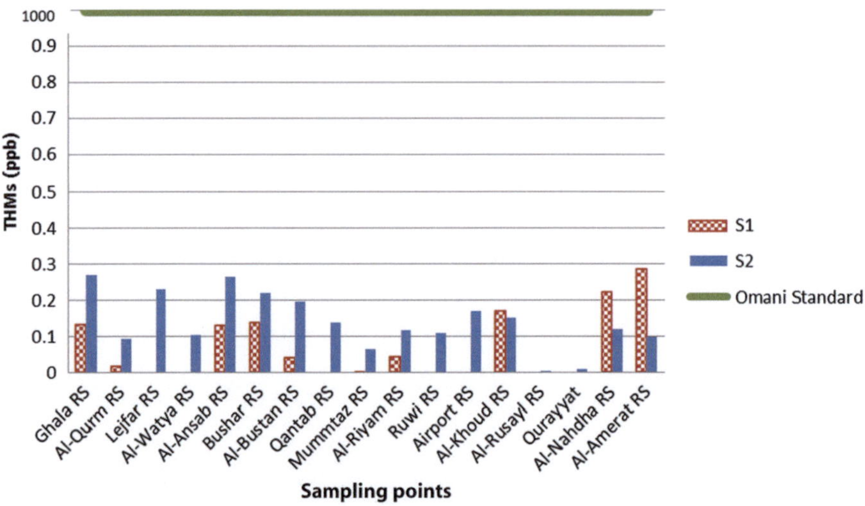

Fig. 9.7 THMs results in ppb for S1 and S2

9.4.1 Standards Comparison

There is a worldwide effort to try and control the formation of chlorination by-products by promulgating regulations and guidelines for drinking water because of the dangerous effects of THM. Based on these guidelines, people can know the maximum and the acceptable levels of such chemicals and parameters. However, these guidelines vary between governmental health bodies from country to country as is shown in Table 9.3. Table 9.3 shows that Omani standards are about the same as WHO standards but varies from other standards. The variation in physical phenomena (climate, topography, source of water, etc) is a reason for this difference in drinking water standards.

According to Table 9.4, for the first sampling, statistically, there is a positive correlation between some parameters such as Temperature and pH, Temperature and TOC, Temperature and THM, EC and Br, Free Chlorine and pH, Free Chlorine and Turbidity, Free Chlorine and THM, pH and TOC, and pH and THM. According to the said table, the correlation between EC and Bromide Ions (Br) is the strongest proportional relation with a magnitude of 0.4495. In addition, the correlation between free chlorine and TOC has the strongest inverse relationship with a value of −0.3863.

Table 9.5 shows that there is a positive relationship between Temperature and TOC, Ttemperature and THM, EC and Br, pH and Turbidity, pH and THM, TOC and THM and, Turbidity and THM. The strongest relationship is between Electric Conductivity and Bromide Ions (Br) which has a magnitude of 0.86877. The strongest negative relationship has a magnitude of −0.5334 which is between pH and Bromide Ions (Br).

Table 9.2 The GCMS results of the project samples

Sampling Source	Chloroform (ppb)	BDCM (ppb)	DBCM (ppb)	Bromoform (ppb)	THM (ppb)
First sampling (S1)					
Ghala RS	0.02	0.07	0.68	12.41	0.13217
Al-Qurm RS	0.01	0.01	0.07	1.37	0.01462
Lejfar RS	0	0	0	0	0
Al-Watya RS	0	0	0	0	0
Al-Ansab RS	0.03	0.08	0.72	12.04	0.12908
Bushar RS	0.03	0.11	0.84	12.7	0.13738
Al-Bustan RS	0	0.01	0.13	3.74	0.03887
Qantab RS	0	0	0	0	0
Al-Mumtaz RS	0	0	0	0.61	0.0061
Al-Riyam RS	0.01	0	0.06	4.17	0.04235
Ruwi	0	0	0	0	0
Airport	0	0	0	0	0
Al-Khoud	0.07	0.22	1.17	15.37	0.16942
Al-Rusyl RS	0	0	0	0	0
Qurayyat	0	0	0	0	0
Al-Nahdha	0.01	0.02	0.45	21.63	0.22118
Al-Amerat	0.01	0.03	0.78	27.68	0.28515
Second sampling (S2)					
Ghala RS	0.09	0.24	1.47	25.01	0.26925
Al-Qurm RS	0.05	0.05	0.52	8.58	0.09208
Lejfar RS	0.06	0.2	1.38	21.29	0.23033
Al-Watya RS	0.1	0.11	0.53	9.53	0.10293
Al-Ansab RS	0.09	0.24	1.53	24.34	0.26315
Bushar RS	0.07	0.21	1.34	20.05	0.21775
Al-Bustan RS	0.3	0.35	1.56	17.2	0.19493
Qantab RS	0.02	0.08	0.96	12.66	0.13763
Mummtaz RS	0.04	0.04	0.3	5.84	0.06227
Al-Riyam RS	0.07	0.1	0.76	10.6	0.11562
Ruwi RS	0.04	0.08	0.7	10.02	0.10873
Airport RS	0.07	0.23	1.17	15.17	0.16758
Al-Khoud RS	0.05	0.15	0.93	13.81	0.15015
Al-Rusayl RS	0	0	0	0.3	0.003
Qurayyat	0.01	0	0.08	0.81	0.00895
Al-Nahdha RS	0.02	0.07	0.65	11.15	0.11927
Al-Amerat RS	0.04	0.08	0.58	8.99	0.09723

THM has a directly proportional relationship with Temperature, pH, TOC and turbidity, while it has an inverse relationship with EC, Free Chlorine and Bromide Ions. The strongest correlation of THM is with pH with a magnitude of 0.5515 (Table 9.5).

Table 9.3 Standards/Guidelines related to THMs (mg/l) in various jurisdictions of the world

Compounds	Oman (2006)	Oman (2012)	WHO (2006)	WHO (2008)	USEPA (2012)	Canada (2012)
$CHCl_3$ (mg/l)	0.2	0.2	0.3	0.3	0.07	–
$CHCl_2Br$ (mg/l)	0.06	0.06	0.06	0.06	–	–
$CHClBr_2$ (mg/l)	0.1	0.1	0.1	0.1	0.06	0.05
$CHBr_3$ (mg/l)	0.1	0.1	0.1	0.1	–	–
TTHMs (mg/l)	≤ 1	≤ 1	≤ 1	≤ 1	0.08	0.1

9.5 Conclusion

To conclude, trihalomethane (THM) is a chlorination by-product that is formed as a result of the chlorination of organic matter present naturally in water. THM concentration depends on the applied initial chlorine concentration, water pH, reaction time, water temperature, bromide concentration and others. According to the project results, bromoform concentration is the most common trihalomethanes species observed in all the samples. Fortunately, the bromoform and the THM concentrations did not exceed the maximum permissible value of Omani standards for drinking water (1 mg/l).

Table 9.4 Correlation between the measured parameters for the first sampling

	Temperature (C)	EC (μS/cm)	Free Chlorine (mg/l)	pH	Br	TOC (ppb)	Turbidity (NTU)	THM
Temperature (C)	1							
EC (μS/cm)	−0.3407	1						
Free Chlorine (mg/l)	−0.1299	−0.4289	1					
pH	0.2623	−0.6395	0.1703	1				
Br	−0.0985	0.4496	−0.1326	−0.2287	1			
TOC (ppb)	0.0080	−0.1482	−0.3863	0.4011	−0.0309	1		
Turbidity (NTU)	−0.0772	−0.1931	0.3091	−0.0262	−0.0341	−0.3248	1	
THM	0.2875	−0.1727	0.2422	0.2032	−0.1603	−0.2714	−0.1360	1

Table 9.5 Correlation between the measured parameters for the second sampling

	Temperature (C)	EC (μS/cm)	Free Chlorine (mg/l)	pH	Br	TOC (ppb)	Turbidity (NTU)	THM
Temperature (C)	1							
EC (μS/cm)	−0.2200	1						
Free Chlorine (mg/l)	−0.1699	−0.3648	1					
pH	−0.0806	−0.3303	−0.1534	1				
Br	−0.0688	0.8688	−0.0811	−0.5335	1			
TOC (ppb)	0.4257	−0.3886	−0.1930	−0.0159	−0.3185	1		
Turbidity (NTU)	−0.3289	−0.1392	0.0781	0.3469	−0.1641	−0.0289	1	
THM	0.4644	−0.4670	−0.2623	0.5515	−0.4031	0.3556	0.1025	1

Acknowledgements We appreciate assistance given by SQU and PAEW. We would also like to express our gratitude to our academic advisors Dr. Ali Al-Maktomi and Dr. Mushtaque Ahmed. Thanks also are extended to the Soils, Water and Agricultural Engineering Department staff as well as the Water Quality Department staff at PAEW.

Finally, we would like to give a big thank you to Mr. Yaqoob Al-Rabaani (HSSE Section Head in Al-Ghubrah Power and Desalination Company) for his help.

List of Figures

References

Abdel-Rahman HA, Abdel-Magid IM (1993) Water Conservation in Oman 18(2):95–102

AL-Rashed MF, Sherif MM (2000) Water resources in the GCC countries. Kluwer Academic Publishers, Dordrecht, 14(1):59–75

Bull RJ, Birnbaum LS, Cantor KP, Rose JB, Butterworth BE, Pegram R, Tuomisto J (1995) Water chlorination: essential process or cancer hazard. Fundam Appl Toxicol 28:155–166

Cantor KP, Lynch CF, Hildesheim ME, Dosemeci M, Lubin J, Alavanja M, Craun G (1999) Drinking water source and chlorination byproducts in Iowa. III. Risk of brain cancer. Am J Epidemiol 150:552–560

Debordea M, Guntena UV (2008) Reactions of chlorine with inorganic and organic compounds during water treatment – kinetics and mechanisms: a critical review. Water Res 42:13–51

Drinking water chlorination. Health Canada's water quality activities web site at www.hc-sc.gc.ca. Accessed 16 May 2013

Drive H (2006) Trihalomethanes: health information summary. Department of Environmental Services, Environmental Health Program

Dumitru R, Melinda-Haydee K, Iovance H, Sidonia V (2008) Disinfection efficiency – trihalomethanes (Thms) formation after chlorination process. pp 150–151.

Environmental Tests – Drinking Water- SVOC (2010). Texas department of state health services. Last updated 14 September 2010. http://www.dshs.state.tx.us/lab/mrs_en_test_dw_svoc.htm. Accessed 1 Sep 2013

Fetter CW (1992) Contaminant hydrology. United State of America

Grey D, Sadoff CW (2007) Sink or swim? Water security for growth and development pp. Water Policy 7:545–571

Hassani AH, Jafari MA, Torabifar B (2010) Trihalomethanes concentration in different components of water treatment plant and water distribution system in the North of Iran. pp 887–892

Hazen D (2006). Trihalomethanes: health information summary. http://des.nh.gov/organization/commissioner/pip/factsheets/ard/documents/ard-ehp-13.pdf. Accessed 16 May 2013

Hsu CH, Jeng WL, Chang RM, Chien LC, Han BC (2000) Estimation of potential lifetime cancer risks for Trihalomethanes from consuming chlorinated drinking water in Taiwan. Water Conservation in Oman 85:77–82

Kutty PCM, Nomani AA, Thankachan TS, Al-Rasheed R (1995) Studies on Thms formation by various disinfectants in seawater desalination plants. Research and Development Center, Saline Water Conversion Corporation, Kingdom of Saudi Arabia. http://www.swcc.gov.sa/files/assets/Research/Technical%20Papers/Chemstry/STUDIES%20ON%20THMs%20FORMATION%20BY%20VARIOUSDISINFECTANTS%20IN%20SEAWATE.pdf. Accessed 23 September 2013

Ministry of Regional Municipalities and Water Resources (MRMWR) (2010) Water resources sector annual report 2010 (Arabic Language). http://www.mrmwr.gov.om/en/Page.aspx?id=144&li=7&Type=&Slide=true. Accessed 5 May 2013

Morris RD, Audet AM, Angelillo IF, Chalmers TC, Mosteller F (1992) Chlorination by-products and cancer: a meta-analysis. Am J Public Health 82:955–962

Pavelica P, Nicholson BC, Dillona PJ, Barrya KE (2005) Fate of disinfection by-products in groundwater during aquifer storage and recovery with reclaimed water. J Contam Hydrol 77:119–141

Phil JC, Peter JL, Lusardi Rita K, Rhonda LM (2004) Total organic carbon: a reliable indicator of TTHM and HAA5 formation? Department of Environmental Protection; Bureau of Water Supply and Wastewater Management. American Water Works Association; WQTC Conference

Prathapar SA, Ahmed M, AL-Adawi S, AL-Sidiari S (2006) Design, construction and evaluation of an ablution water treatment unit in Oman: a case study. Int J Environ Stud 63:283–292

Removal of the natural organic matter in the different stages of the drinking water treatment process (2007) http://dspace.cc.tut.fi/dpub/bitstream/handle/123456789/35/matilainen.pdf?sequence=1. Accessed 15 Sep 2013

Richardson SD, Plewa MJ, Wagner ED, Schoeny R, DeMarini DM (2007) Occurrence, genotoxicity, and carcinogenicity of regulated and emerging disinfection by-products in drinking water: a review and roadmap for research. Mutat Res 9:178–242

Study on characteristics and removal of natural organic matter in drinking water systems in Newfoundland and Labrador. http://www.env.gov.nl.ca/env/waterres/reports/drinking_water/NOM_Study_August_19_2011.pdf. Accessed 15 Sep 2013

Taha FM, Doanh V (2000) The variation of mass and disinfection by-product formation potential of dissolved organic matter fractions along a conventional surface water treatment plant. J Hazard Mater 74:133–147

Temperature dependent of the pH of pure water. http://chemwiki.ucdavis.edu/Physical_Chemistry/Acids_and_Bases/Aqueous_Solutions/The_pH_Scale/Temperature_Dependent_of_the_pH_of_pure_Water. Accessed 26 Jan 2014

The electrical conductivity of water. http://www.smart-fertilizer.com/articles/electrical-conductivity. Accessed 15 Sep 2013

The Environmental Protection Agency (EPA) (2012) Drinking water guidance on disinfection by-products advice note No. 4. Version 2. http://www.epa.ie/pubs/advice/drinkingwater/DrinkingWaterGuide4_v8.pdf. Accessed 15 Sep 2013

Wagenet L, Heidekamp A, Lemley A (1988) Chlorination of drinking water. Cornell Cooperative Extension, College of Human Ecology, Ithaca. Kluwer Academic, Dordrecht, pp 1–6

World Health Organization (WHO) (2008) Guidelines for drinking-water quality, vol 1, 3rd edn, Recommendations. WHO, Geneva

WHO (2011) Hardness in drinking-water: background document for development of WHO guidelines for drinking-water quality. http://www.who.int/water_sanitation_health/dwq/chemicals/hardness.pdf. Accessed 23 Sep 2013

Links

Bromodichloromethane. Last modified on 25 February 2013. http://en.wikipedia.org/wiki/Bromodichloromethane. Accessed 16 May 2013

Bromoform-2D.png. Last modified on 20 March 2011. http://commons.wikimedia.org/wiki/File:Bromoform-2D.png. Accessed 16 May 2013

Trichloromethane. Last modified on 19 March 2013. http://www.mermaidproject.eu/wiki/Trichloromethane. Accessed 16 May 2013

Chapter 10
The Viability of Renewable Energy and Energy Storage as the Power Source for Municipal Scale Reverse Osmosis Desalination

Clifford Dansoh

Abstract The use of renewable energy to power reverse osmosis desalination plants was investigated to provide potable water for around 50,000 people in Newhaven, in South East England, and in Massawa in Eritrea. The energy sources were specifically assessed in a variety of combination of wind power, wave power, solar power, tidal current power, hydrogen production, and storage and use in Fuel Cells. The following types of reverse osmosis plants were studied: (1) No Brine Stream Recovery (BSR) reverse osmosis plant; (2) Pelton Wheel BSR reverse osmosis plant; (3) Pressure Exchanger BSR reverse osmosis plant.

Modelling was conducted to derive the amount of water that each reverse osmosis plant would deliver from various combinations and amounts of renewable power input, at varying feedwater temperatures. Scenarios that were not able to deliver enough water to meet user need were scaled-up so that they could do so. The costs of the scaled-up scenarios that were able to meet user water demands were compared with the costs associated with the equivalent conventionally-powered scenario over a 25-year life. Specifically, a coal-fired plant with carbon capture and storage (CCS) at Newhaven and a diesel generator at Massawa were considered. This comparison was made with and without the external costs associated with conventional energy production and use.

Keywords Reverse osmosis • Renewable energy • Energy storage

10.1 Introduction

The scenario employed within this paper to investigate the technical and financial viability of renewable energy was its use to power reverse osmosis desalination plants to provide water for the personal use of 50,000 people.

C. Dansoh (✉)
CKD Consultants Limited, Worthing, West Sussex, United Kingdom
e-mail: ckd.consultants@ntlworld.com

M. Baawain et al. (eds.), *Recent Progress in Desalination, Environmental and Marine Outfall Systems*, DOI 10.1007/978-3-319-19123-2_10

10.1.1 *Locations for Desalination*

Eritrea was selected as a site at which to model renewable powered desalination due to its susceptibility to droughts and the consequential loss of life. Water supply using desalination at Newhaven in South East England was also investigated, as this is a particularly dry part of the United Kingdom.

10.2 The Modelling Exercise

The modelling exercise was conducted in four main stages using 270 scenarios to simulate varying amounts and types of renewable power being applied to various RO plants as shown below in Fig. 10.1.

The four stages of modelling development are explained in the following text.

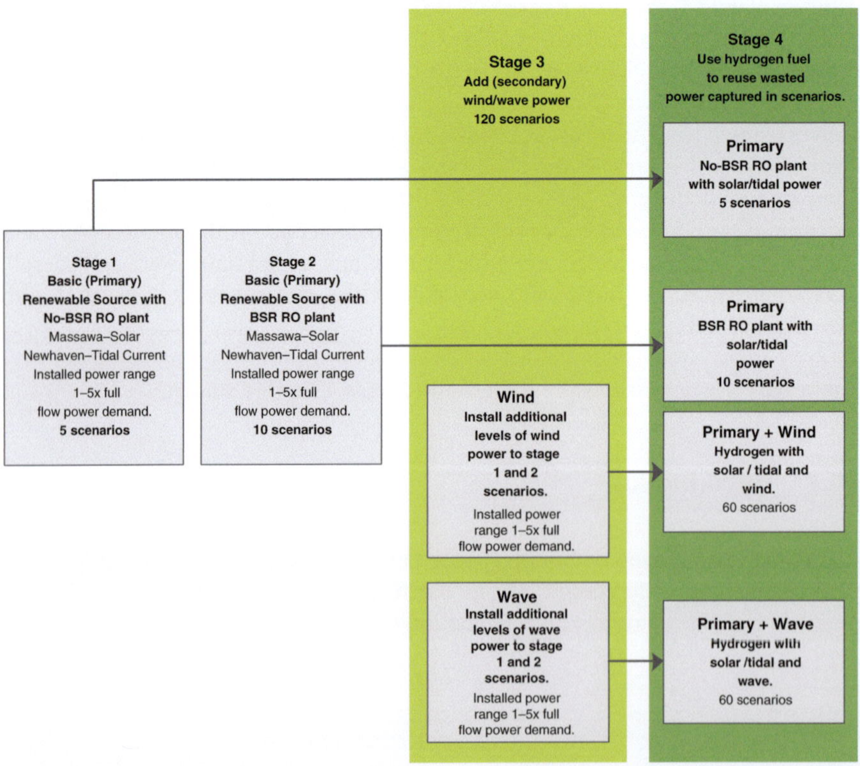

Fig. 10.1 Four stages of the modelling exercise

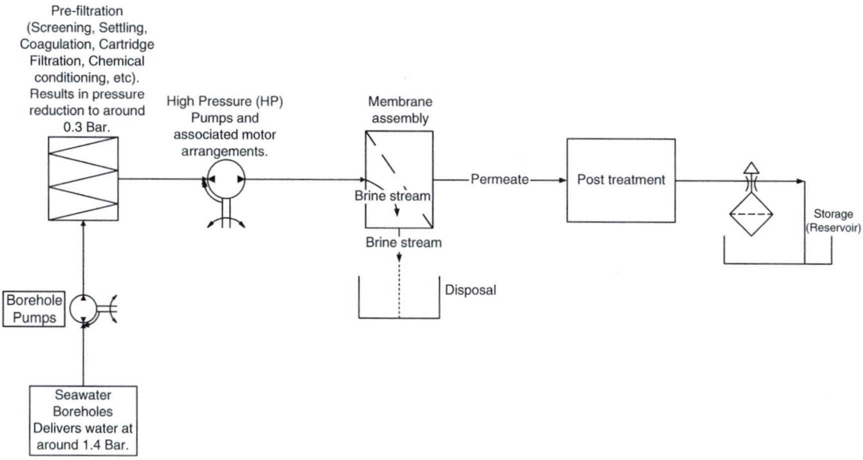

Fig. 10.2 No BSR Plant type used within modelling

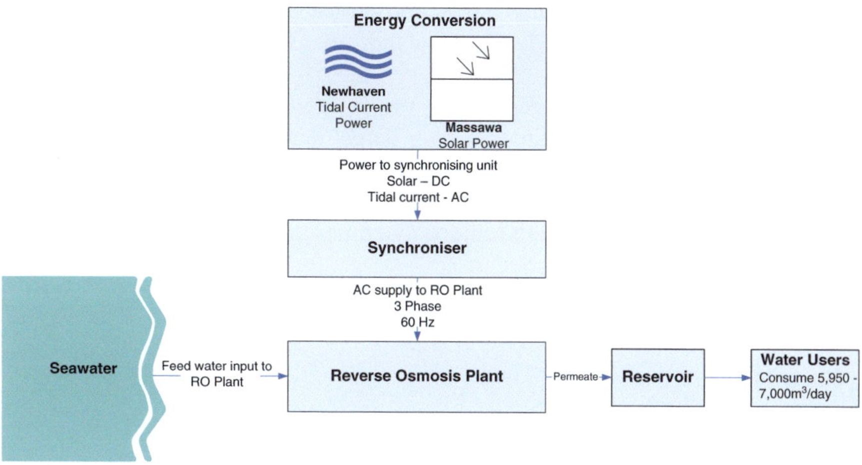

Fig. 10.3 Single source of renewable energy to power RO plant at both sites

10.2.1 Stage 1

Stage 1 employed the most reliable renewable resource at each of the sites in question (Solar at Massawa and Tidal Current at Newhaven) as shown below in Fig. 10.2 with the No Brine Stream Recovery (BSR) RO plant.

A schematic diagram of the No BSR plant employed for the modelling within this research is shown in Fig. 10.2.

An overview diagram of the water and energy processes modelled is shown in the following diagram, Fig. 10.3.

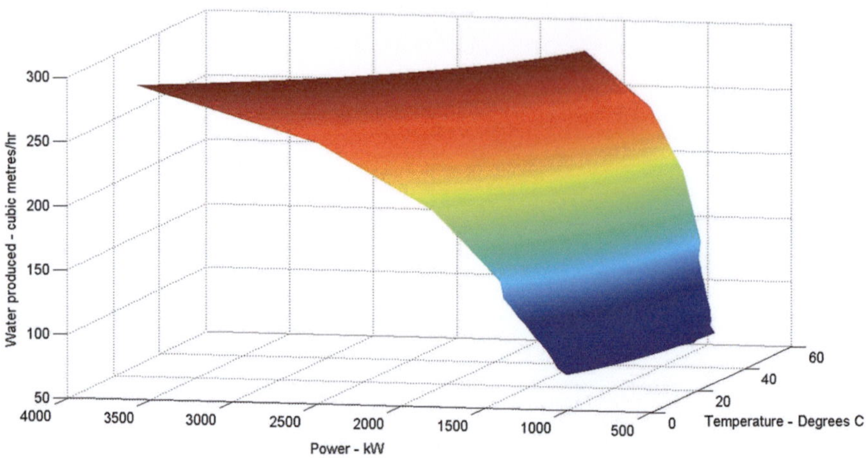

Fig. 10.4 No BSR RO Plant water production profile at varying power and feedwater temperature

The No BSR RO plant water production profile at varying input power and feedwater temperatures, derived using Dow Industries RO design software, ROSA, is shown below in Fig. 10.4.

Sufficient solar and tidal current power was installed at Massawa and Newhaven respectively, so that their maximum power output during the year would achieve the maximum flowrate of the RO plant, sized to deliver 7000 m^3/day when operated continuously. Additional power was then added in discrete levels, up to (and including), the power required to achieve five times the maximum flowrate of the RO plant.

10.2.2 Stage 2

Stage 2 employed the same methodology as Stage 1 (application of the most reliable power source at each site), but for the BSR RO plants (Pelton Wheel and Pressure Exchanger). The Pelton Wheel RO plant system modelled is shown below in Fig. 10.5.

As shown in Fig. 10.5, the Pelton Wheel BSR RO plant design utilises the brine/concentrate stream to power a Pelton Wheel turbine, which is mechanically linked to a high pressure pump (HP p/p) arrangement. The power produced from the Pelton Wheel is used to partially pressurise the incoming feedwater which reduces the external power required to raise the feedwater to an adequate pressure for desalination via the RO plant membranes. Due to the extraction of energy from the brine stream, the brine must be pumped away for disposal. The resulting Pelton Wheel BSR RO plant water production profile, for a plant capable of producing 7000 m^3/day at varying input power and feedwater temperatures, is shown below in Fig. 10.6.

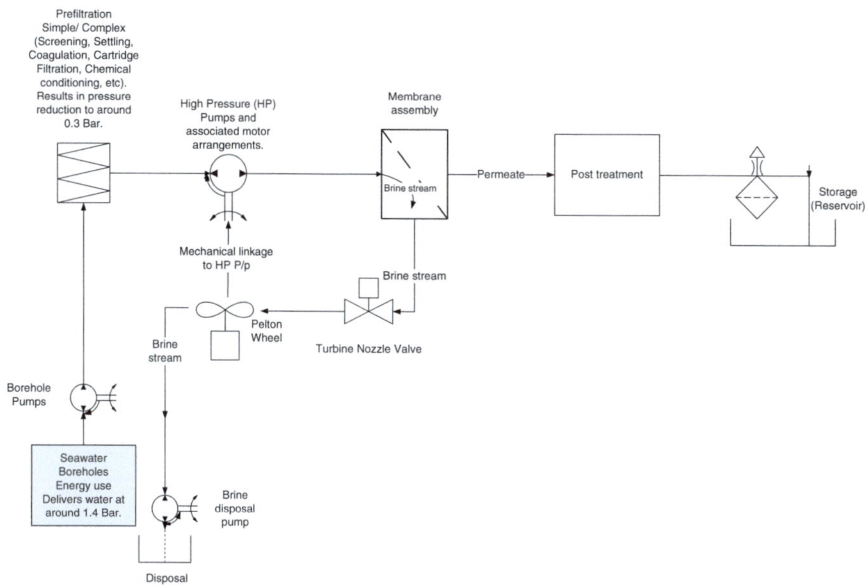

Fig. 10.5 Simple plant using Pelton wheel for BSR design

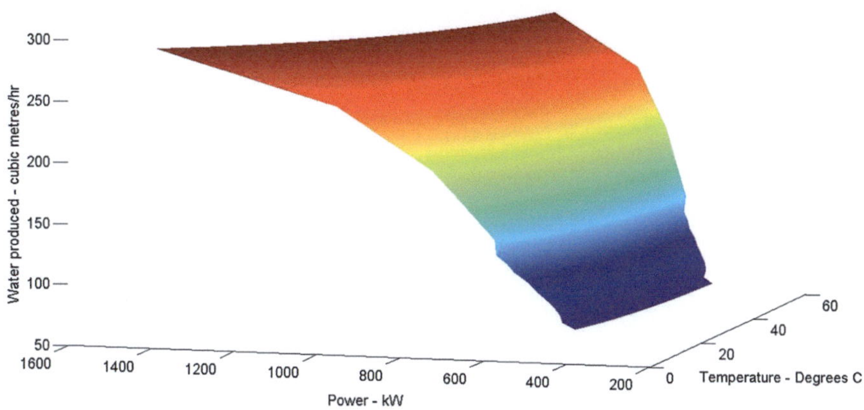

Fig. 10.6 Pelton Wheel RO Plant water production profile at varying power and feedwater temperature

The Pressure Exchanger RO plant system modelled is shown below in Fig. 10.7.

As shown in Fig. 10.7, the Pressure Exchanger BSR RO plant uses the brine/concentrate stream to pressurise a hydraulic chamber. This hydraulic chamber acts with a piston arrangement which in turn is used to partially pressurise the incoming feedwater. A booster pump then raises the now partially pressurised feedwater to the correct pressure to combine with the feedwater pressurised by the high pressure pump for desalination by the RO plant membranes. After pressurising the incoming

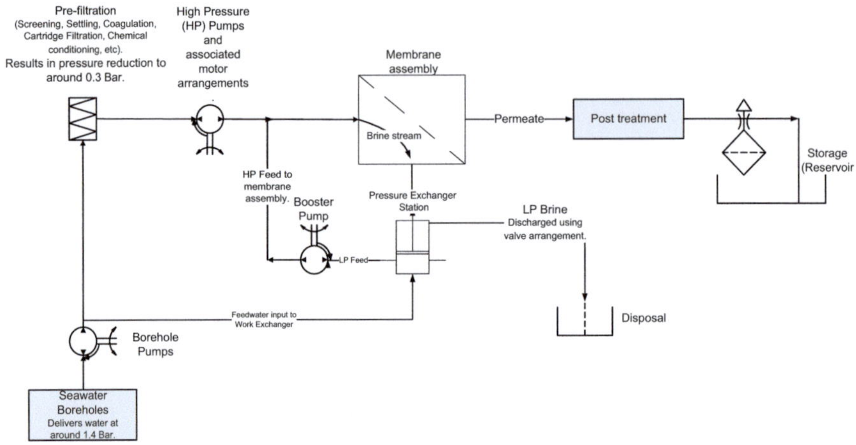

Fig. 10.7 RO plant using Pressure Exchanger for BSR design

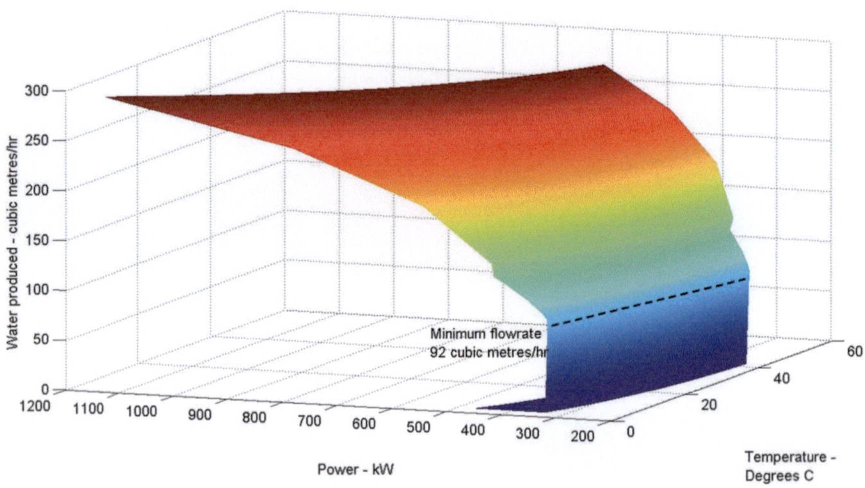

Fig. 10.8 Pressure Exchanger RO Plant water production profile at varying power and feedwater temperature

feedwater, the brine stream (which is still partially pressurised) is discharged using valve arrangements as a low pressure brine stream. The resulting Pressure Exchanger BSR RO plant water production profile, for a plant capable of producing 7000 m^3/day at varying input power and feedwater temperatures, is shown below in Fig. 10.8.

As was the case in Stage 1, additional solar and tidal power was added in discrete levels up to (and including) the power required to achieve five times the maximum flowrate of each of the RO plants.

10.2.3 Stage 3

There were two aspects to stage 3 as the model attempted to increase the scenario water production: addition of wind power and addition of wave power. Wind or wave power was added to the stage 1 and 2 scenarios in discrete levels up to (and including) the power required to achieve five times the maximum flowrate of each of the RO plants.

10.2.4 Stage 4

Stage 4 was the use of hydrogen storage of captured energy, and reuse with primary energy and hybridised (primary with wind or wave) power scenarios for No BSR and BSR RO Plants. This was to allow the normally wasted power captured during normal RO plant operation to be reapplied at times when insufficient power is available to maintain maximum RO plant water production. The efficiency of hydrogen production, storage and reuse via fuel cells was taken as 22 %. It was clear that if the scenarios were scaled-up (by extrapolation), competent scenarios could be identified where the full demand of the local water users could be met even without energy storage. Therefore, it was decided to increase the RO and Power plant by the ratio of water shortfall, i.e. if RO plant and Power scenario produced 50 % of the water required, both the RO plant and installed power would be doubled in size.

10.3 Costs

The first part of the section deals with CAPEX and OPEX costs and the later part of this section identifies the external costs associated with the conventionally powered scenarios.

These costs were taken from a variety of international sources, and the exchange rates employed to align with pounds sterling (£) are shown below in Table 10.1 taken from www.x-rates.com on Tuesday 14 December 2010.

Table 10.1 Exchange rates used for financial modelling

	$US	$GBP	€EUR
$US	1	1.57238	1.3669
£GBP	0.635976	1	0.843749
€EUR	0.75375	1.18518	1

Table 10.2 Capital and total costs (including O&M) over 25 years for RO plants

	No BSR	Pelton Wheel	Pressure Exchanger
Capital costs (£ × 10^6)	9.27	10.38	11.12
Total costs (£ × 10^6)	48.8	79.1	56.0

Table 10.3 Capital and O&M costs of renewable energy sources

	Solar PwC (2010)	Tidal Current (EPRI 2006)	Wind (PB1 2009)	Wave
Capital costs (£/kW installed)	3000	1288	1200	4000 (Wilson 2011)
O&M costs (£/kW/annum)	15	51	37	27 (PB2 2008)

10.3.1 CAPEX and OPEX Costs

10.3.1.1 RO Plant Costs

Table 10.2 above shows the CAPEX and OPEX costs associated with the unscaled RO plants employed at each site based on various sources (ADC 2011; Boswell 2011; Conlon 2011; Department of Environmental Protection 2010; Shackleton 2009).

10.3.1.2 Reservoir Cost

The reservoir was priced as holding 15 % of the annual production of a plant that produces 100 % of the annual water required by 50,000 people at each site (2,555,000 m^3). Fifteen percent is taken as 383,250 m^3 at the end of the year. This equates to a reservoir costing £82,115,200 based on extrapolation of various reservoir costs presented in 'Design, Construction and repair of potable water reservoirs' (Boyle 2008).

10.3.1.3 Renewables

Table 10.3 above shows the CAPEX and OPEX costs associated with the renewable energy sources employed at each site.

10.3.1.4 Hydrogen Fuel System

Table 10.4 above shows the CAPEX and OPEX costs associated with the modelling of hydrogen production, storage and reuse systems.

Table 10.4 Capital and O&M for hydrogen fuel system

	Electrolyser (Zoulias 2008)	Hydrogen storage (Hirscher 2010)	Fuel Cells
Capital costs (£/kW installed)	1320	30.15	3500 (Hordeski 2009)
O&M costs (£/kW/annum)	26[a]	2.43	45 (Shipley and Elliot 2004)

[a]A standing cost of 2 % of the CAPEX costs was taken as the annual O&M cost

10.3.1.5 Conventional Power Costs

The conventional power sources that were modelled as the options that the renewable energy sources needed to demonstrate viability against are as follows:

- Massawa – Local diesel generators
- Newhaven – Centralised coal fired plant with carbon capture and storage (CCS) facilities.

Table 10.5 below shows the CAPEX and OPEX costs associated with the conventional power plants modelled at each site, which were based on various sources.

10.3.1.6 Externalities of Energy Production and Use

The hidden costs, (the externalities), borne by society for the use of conventional fuels are not reflected in the CAPEX and OPEX figures above. The best available studies of externalities of power generation are the European Union's (EU) 'ExternE Project' (ExternE 2003), and its successor, 'New Energy Externalities Development for Sustainability (NEEDS)'.

10.3.1.7 Externalities Associated with Coal- Fired Plant with CCS Technology to Be Employed in Newhaven

ExternE presents external costs for the use of coal & lignite in the UK of 4–7€ cent (3.38–5.93p)/kWh, so the external cost associated with electricity generation via coal at Newhaven was taken as the lower end of the range at 3.4p/kWh. The conventional power plant used at Newhaven is modelled as having CCS, which is taken as being 90 % efficient at removing carbon dioxide. The final ExternE report provides a breakdown of the externalities of electricity production using coal in Germany, which is attributed with external costs for the use of coal & lignite of 3–6 € cent (2.54–5.08p)/kWh. This breakdown of externalities, presented for Germany, is taken to be reasonably applicable to the UK. The costs attributed to the avoidance of CO_2 produced are 63 % of the total external costs for producing electricity using coal. So, to take account of the coal-fired plant with CCS efficiency (taken as 90 %)

Table 10.5 Capital and O&M conventional power plants

	Massawa – diesel generator costs for 25 years (Gabriel et al. 2011)				Newhaven – coal fired plant with CCS for 25 years (Bauer et al. 2008 and Power Engineering 2010)			
	Installed power costs (£)	Fixed O&M costs (£)	Fuel costs (£ × 10^6)	Total scenario costs[a] (£ × 10^6)	Installed power costs (£ × 10^6)	Fixed O&M costs (£ × 10^6)	Fuel costs (£ × 10^6)	Total scenario costs[a] (£ × 10^6)
No BSR	380,000	40,000	100.2	232	9.5	4.6	66.7	212
Pelton Wheel	160,000	27,500	43.4	205	3.9	1.9	27.0	194
Pressure Exchanger	125,000	22,500	32.8	171	3.1	1.5	21.4	164

[a]Including RO plant and reservoir

at capturing CO_2, the cost due to externalities to power the RO plants at Newhaven is 1.47p/kWh.

This cost due to externalities is based upon 3.4p (total cost of coal externalities)–(benefit of CO_2 capture).

The benefit of CO_2 capture = 0.63 (portion of externalities associated with carbon dioxide production) × 0.9 (efficiency of CCS at removing carbon dioxide from exhaust) × 3.4 (total cost of coal externalities) = 1.928 p/kWh benefit due to CCS CO_2 capture.

Therefore, 3.4 – 1.928 = 1.47p/kWh.

10.3.1.8 Externalities Associated with Diesel Generation to Be Employed in Massawa

ExternE presents external costs for the use of oil in the UK at 3–5 € cent (2.5–4.2p)/kWh. So, for the purpose of this research, the external cost associated with diesel electricity generation to power the RO plants at Massawa will be taken as the lower end of the range at 2.5p/kWh.

Energy Security

The International Center for Technology Assessment (CTA) makes the point that an indeterminate portion of the defence budgets of the US and of other countries are concerned with protecting oil supplies by maintaining regional stability in the countries that produce oil. The estimated external costs associated with US military expenditure to protect the world's petroleum supplies range from $47.6 billion–$113.1 billion (£30.27 billion–£71.92 billion). This equates to $0.13–$0.20/ US Gallon (3.785 l).

There are other externalities associated with US petroleum use, including the Coast Guard and other municipal services, which bring the range of external costs associated with energy security paid by (predominantly American) taxpayers, up to $78.215 billion–$158.39 billion, or in specific terms between, $0.214 and $0.321/ US gallon. For the purposes of this paper, the cost of security for the fuel used to run the diesel generator in Massawa is the lower end of the range for military expenditure only, as it is assumed that municipal service costs are not incurred, and taken to be $0.13/ US Gallon which equates to 2.4p/l.

So, the total externalities due to use of diesel-powered generation at Massawa, is:

- 2.5p/kWh of energy produced, plus
- 2.4p/l of diesel fuel used.

10.3.1.9 Costs Associated with Conventional Power Scenarios

Based on the information above, the following sections provide detail of the costs used to model conventional power scenarios with externalities over the 25 year of the plant. It is noteworthy that the external costs used are the most conservative costs available from the information available (Tables 10.6 and 10.7).

Massawa

Table 10.6 External costs associated with power production using diesel generators at Massawa

	Total without externalities (£ × 10^6)	Cost of externalities (£ × 10^6)	Total cost with externalities (£ × 10^6)
No BSR	231.7	102.4	334
PW	204.8	44.4	249
PX	171.1	33.6	205

Newhaven

Table 10.7 External costs associated with power production using CCS at Newhaven

	Total without externalities (£ × 10^6)	Cost of externalities (£ × 10^6)	Total cost with externalities (£ × 10^6)
No BSR	211.7	8.98	221
PW	194.0	3.64	198
PX	164.1	2.89	167

10.4 Results

Shown below in Tables 10.8 and 10.9 are the results for the scaled-up scenarios capable of producing 100 % of the water required when externalities are applied at Massawa and Newhaven, respectively.

Scenarios that have become financially viable (are cheaper than the conventionally-powered equivalent) due to the application of externalities are highlighted in yellow in Table 10.8.

10.5 Conclusion

The objective of this research was to assess the viability of renewable energy to completely replace a conventional power source, and to provide a fundamental and significant human need. To make this assessment, this paper has modelled various

Table 10.8 Technically competent and most financially viable scenarios at Massawa when externalities are applied

Stage	Type of RO plant	Primary power (MW)	Secondary power	Secondary power (MW)	Hydrogen fuel	Ratio of renewable scenario cost against conventional	Ratio against conventional with externalities	Difference (%)
1	No BSR	37.2	None	0	No	1.46	1.01	30.82
2	Pelton Wheel	21.8	None	0	No	1.694	1.392	17.82
	Pressure Exchanger	17.4	None	0	No	1.643	1.373	16.43
3	No BSR	17.37	wind	9.93	No	1.227	0.85	30.73
	Pelton Wheel	3.69	wind	14.68	No	1.353	1.112	17.81
	Pressure Exchanger	2.98	wind	12.41	No	1.423	1.19	16.37
	No BSR	15.82	wave	22.14	No	1.289	0.894	30.64
	Pelton Wheel	4	wave	12.53	No	1.286	1.057	17.81
	Pressure Exchanger	2.98	wave	12.15	No	1.273	1.064	16.42
4	No BSR	33.16	None	0	Yes	1.475	1.023	30.64
	Pelton Wheel	21.82	None	0	Yes	1.44	1.183	17.85
	Pressure Exchanger	17.39	None	0	Yes	1.408	1.176	16.48
	No BSR	9.69	wave	19.73	Yes	1.316	0.913	30.62
	Pelton Wheel	3.41	wave	10.68	Yes	1.253	1.0303	17.77
	Pressure Exchanger	2.67	wave	8.14	Yes	1.237	1.033	16.49
	No BSR	13.45	wind	15.38	Yes	1.317	0.913	30.68
	Pelton Wheel	7.22	wind	7.43	Yes	1.257	1.033	17.82
	Pressure Exchanger	2.736	wind	4.69	Yes	1.284	1.073	16.43

Table 10.9 Technically competent and most financially attractive scenarios at Newhaven when externalities are applied

Stage	Type of RO plant	Primary power (MW)	Secondary power	Secondary power (MW)	Hydrogen storage	Ratio of renewable scenario cost against conventional	Ratio against conventional with externalities	Difference (%)
1	No BSR	135.3	None	0	No	2.969	2.848	4.08
2	Pelton Wheel	66.5	None	0	No	2.525	2.479	1.82
	Pressure Exchanger	54.38	None	0	No	2.469	2.426	1.74
3	No BSR	20.5	wind	29.48	No	1.527	1.465	4.06
	Pelton Wheel	5.59	wind	14.12	No	1.38	1.355	1.81
	Pressure Exchanger	4.28	wind	13.72	No	1.358	1.334	1.78
	No BSR	20.5	wave	19.62	No	1.443	1.384	4.09
	Pelton Wheel	2.479	wave	7.63	No	1.196	1.174	1.84
	Pressure Exchanger	1.82	wave	6.33	No	1.188	1.168	1.68
4	No BSR	122.5	None	0	Yes	3.118	2.991	4.07
	Pelton Wheel	54.38	None	0	Yes	2.474	2.429	1.82
	Pressure Exchanger	45.37	None	0	Yes	2.371	2.33	1.73
	No BSR	15.84	wave	15.41	Yes	1.607	1.542	4.043
	Pelton Wheel	4.88	wave	6.67	Yes	1.26	1.237	1.83
	Pressure Exchanger	4.28	wave	4.75	Yes	1.253	1.232	1.68
	No BSR	14.67	wind	21.09	Yes	1.651	1.577	4.48
	Pelton Wheel	6.11	wind	8.24	Yes	1.362	1.337	1.84
	Pressure Exchanger	4.06	wind	7.3	Yes	1.303	1.281	1.69

scenarios at Massawa in Eritrea and Newhaven in southeast England, with a view to address the water needs of 50,000 people using various RO plant types and combinations of renewable energy and energy storage. The results have demonstrated that a significant and fundamental human need can be addressed by using renewable energy.

All the combinations of renewable energy sources modelled were able to meet the water requirements of 50,000 people at each site. Although the scenarios that employed energy storage were more economic, with respect to installed energy capacity, in achieving the required water output, the most financially viable scenario was at Massawa, using solar and wind power and a simple No BSR RO plant. A relatively simple hybrid renewable energy plant such as this:

- Minimises the difficulties associated with the operation of very complex machinery required to implement many of the other modelled scenarios
- Improves the prospects for Massawa to have security over its water supply.

The financial attractiveness of the Massawa wind and solar hybrid scenario is due to:

- The relatively cheap costs associated with onshore wind power, even in the relatively poor wind climate modelled.
- The high current costs associated with hydrogen generation and re-use;
- The ability to scale up the RO plant allowed over production of water, using energy that would have otherwise been wasted. Storage in a reservoir would effectively store wasted energy as water relatively cheaply in comparison to hydrogen.
- The high cost of externalities associated with diesel fuel.

List of Figures

References

Affordable Desalination (ADC) (2011) See http://www.affordabledesal.com/home/test_data.html on 14 January 2011 for an RO plant using the Filmtec SW30HR-380 membrane

Bauer C, Heck T, Dones R, Mayer-Spohn O, Blesl M (2008) New Energy Externalities Developments for Sustainability (NEEDS) integrated project. Priority 6.1: sustainable energy systems and, more specifically, sub-priority 6.1.3.2.5: socio-economic tools and concepts for energy strategy. Deliverable No. 7.2 – RS 1a "Final report on technical data, costs, and life cycle inventories of advanced fossil power generation systems". Available at http://gabe.web.psi.ch/pdfs/Needs/NEEDS_RS1a_D7-2.pdf

Boyle JWC (2008) Design, construction and repair of potable water reservoirs. In: Proceedings of the British Columbia water and waste association annual conference, Whistler April 2008

Department of Environmental Protection (2010) Desalination in Florida: technology, implementation, and environmental issues. Division of Water Resource Management, Florida Department of Environmental Protection, April, 2010. Available at http://www.dep.state.fl.us/water/docs/desalination-in-florida-report.pdf

Discussion with Philip Boswell (International Technical Consultant) Accepta during February 2011. See website at http://www.accepta.com/ for details of Accepta

Discussions during January and February 2011 with William J. Conlon, P.E., BCEE, F.ASCE Technical Manager, Principal Professional Associate. Water Technical Excellence Center, Parsons Brinkerhoff Americas, Inc

Discussions with Daniel Shackleton – Director of Salt Separation Limited, on 9 Nov 2009. See website at http://www.saltsep.co.uk/ for greater details of Salt Separation Limited

Electric Power Research Institute Inc. (EPRI) (2006) System level design, performance, cost and economic assessment – San Francisco tidal in-stream power plant 10 June 2006. Available at http://oceanenergy.epri.com/attachments/streamenergy/reports/006_CA_06-10_-06.pdf

ExternE (2003) European Commission EUR 20198—external costs research results on socio-environmental damages due to electricity and transport. ISBN 92-894-3353-1 is available at http://www.externe.info/externpr.pdf

Gabriel et al. (2011) These sources include: (1) Cost of diesel generators – Discussion with Mike Gabriel – Sales Manager Power Electrics (Bristol) Ltd. See http://www.power-electrics.co.uk/ for greater detail of power electrics organisation; (2) Cost of diesel fuel $1.07 (£0.68) per litre taken from Data Preview – January 2011; (3) International Fuel Prices 2010/2011 by The Deutsche Gesellschaftfür Internationale Zusammenarbeit (GIZ) GmbH. Available at http://www.gtz.de/de/dokumente/giz2011-international-fuel-prices-2010-2011-data-preview.pdf. It is also noteworthy that this report indicated that Eritrea has the most expensive petrol in the world at £1.615 per litre

Hirscher M (2010) Handbook of hydrogen storage – new materials for future energy storage, WILEY-VCH. 2010. ISBN: 978-3-527-32273-2. Based on tanks capable of storing up to 700 bar as this system is assumed to operate around 300 bar

Hordeski MF (2009) Hydrogen and fuel cells: advances in transportation and power. Fairmont Press, Lilburn. ISBN 0-88173-562-0

PB1 (2009) Powering the future, full report – mapping our low-carbon path to 2050. December 2009 by Parsons Brinkerhoff. Available at http://www.pbpoweringthefuture.com/

PB2 (2008) Powering the nation, by Parsons Brinkerhoff, 2008

Power Engineering (2010) IGCC cost wrap. Power Engineering Webpage. Available at http://www.powergenworldwide.com/index/display/articledisplay.articles.powergenworldwide.coal-generation.new-projects.2010.03.igcc-cost-wrap.QP129867.dcmp=rss.page=1.html

PwC (2010) Approximation of costs based on the range presented in a study by consultants Price Waterhouse Coopers, in collaboration with researchers from the Potsdam Institute for Climate Impact Research (PIK), the International Institute for Applied Systems Analysis (IIASA) and the European Climate Forum (ECF). Available at http://www.pwc.co.uk/eng/publications/100_percent_renewable_electricity.html

Shipley AM, Elliot RN (2004) Stationary fuel cells – future promise, current hype. Report IE041, American Council for an Energy-Efficient Economy

Wilson P (2011) Based on e-mail correspondence on 14 January 2011 with Paul Wilson of Parsons Brinkerhoff author of 'Powering the Nation'. Available at http://www.pbpoweringthefuture.com/

Zoulias E (2008) 'Hydrogen-based autonomous power systems: techno-economic analysis of the integration of Hydrogen in Autonomous Power Systems'. By E. I. Zoulias with N. Lymberopoulos. Available to read online at http://www.springer.com/engineering/energy+technology/book/978-1-84800-246-3

Part II
Environmental Systems

Chapter 11
Desalination Integration with Renewable Energy for Climate Change Abatement in the MENA Region

Eman Hasan

Abstract The Middle East and North Africa (MENA) region is considered the most water-scarce region in the world. Renewable energy in this region has tremendous potential to provide energy security and reduce greenhouses gas emissions. Renewable energies for use in desalination processes include wind, solar thermal, photovoltaic and geothermal. The objective of this research is to review and select the best choice of a desalination technology combination with renewable solar energy in the MENA region from among different alternative systems, using the Analytic Hierarchy Process (AHP). These alternative systems are Solar still, Solar Humidification (Solar–MHE), Concentrating Solar Plant – Multi Effect Desalination (CSP-MED), Photovoltaic Reverse Osmosis (PV-RO) and Photovoltaic Electrodialysis Reverse (PV-EDR). The criteria that have been taken into consideration in the process of analysis are the energy demand, the capacity of the desalination plant (m^3/day), the cost of produced water (US $/$m^3$), the volume of saline water per cubic metre of fresh water, the volume of brine effluents per cubic metre of fresh water and the cost of brine disposal (US $/$m^3$). A specific weight has been given to each criterion according to its relative impact on the process of decision-making. The results from the presented methodology determine the relative suitability of renewable energy desalination systems in the MENA.

Keywords Climate change • Desalination • GHG • Renewable energy • Solar energy

---

E. Hasan (✉)
Misurata University, Bani Waled, Libya

National Water Research Center, Cairo, Egypt
e-mail: dr_eman30@hotmail.com

M. Baawain et al. (eds.), *Recent Progress in Desalination, Environmental and Marine Outfall Systems*, DOI 10.1007/978-3-319-19123-2_11

11.1 Introduction

The Middle East and North Africa (MENA) region is considered the most water-scarce region in the world. If current rates of growth continue and the global climate warms as expected, water demand in the MENA Region is expected to increase by 50 % by 2050 (World Bank 2012).

Desalination is the best solution for facing the increasing water demand in the MENA countries, as these countries have access to sea water as a source of water for desalination. Moreover, different sources of renewable energy are available in this region. Today's global desalinated water production amounts to about 65.2 million m^3 per day (about 24 billion m^3 per year), equivalent to 0.6 % of the global water supply (IEA-ETSAP and IRENA 2012). The MENA region accounts for about 38 % of the global desalination capacity, with Saudi Arabia being the country with the greatest desalinated water product. Major desalination technology options are based on thermal processes using both heat and electricity, and on membrane technologies using electricity only. The dominant technology is Reverse Osmosis (RO), which accounts for 60 % of the global capacity, followed by Multi Stage Flash (MSF), with a 26.8 % share. The larger desalination plants can reach a capacity of up to 800,000 m^3 per day or more (IEA-ETSAP and IRENA 2012).

The objective of this research is to review and select the best desalination technology combination with renewable energy (solar system) in the MENA region from among different alternatives based on a set of evaluation criteria using Analytic Hierarchy Process (AHP). This method of analysis assists decision-makers to simplify the problem by creating a hierarchy of decision criteria that best suits their goal and their understanding of the problem (Rinner 2007). The procedure for using AHP can be summarized by the following analysis. The problem is modeled as a hierarchy containing the decision goal, the options or alternatives for reaching it, and the criteria for evaluating the options. Furthermore, priorities are established among the elements of the hierarchy, and these judgments are synthesized to yield a set of overall priorities for the hierarchy. Finally, the consistency of the judgments is checked and a final decision arrived at based on the results of this process (Saaty 2008).

11.1.1 The MENA Region

The MENA region's virtually unlimited solar irradiance, which is several times larger than the total current world energy demand, coupled with proven solar energy power generation technologies will ensure an environmentally-sustainable desalinated water supply to the MENA and ensure energy security to the water sector. If Renewable Energies (RE) replace fossil fuels, except for peaking power, the MENA's annual Carbon Dioxide gas emissions could be reduced to 265 million tons by 2050, which is less than current emissions (World Bank 2012).

Desalination has proved to be a technically-feasible supply solution to the MENA's water gap and will continue to be so. Within the Gulf Cooperation Council (GCC) countries, dependence is high. However, dependence dwindles among the Maghreb countries in which even the biggest users, Algeria and Libya, rely on desalination for less than 5 % of their water supply.

11.2 Methodology

In this research the following procedures have been emphasized:

- Presentation and review of major desalination technologies in the world
- Presentation of existing renewable energy system (RES) in the MENA
- Proposals for the use of different alternative renewable solar energy desalination systems
- Assessment of suggested alternatives using AHP.

11.2.1 Major Desalination Technologies

Major desalination technologies consist of thermal processes using either thermal power or electricity as the energy input, or membrane-based processes using only electricity (Table 11.1). The dominant desalination processes in use today are based on Reverse Osmosis (RO) and Multi-Stage Flash (MSF) which constitute 60.0 % and 26.8 % of the worldwide capacity respectively. The feasibility of each technology depends on specific conditions such as energy price, water quality and the technical resources of the region.

A brief description of each desalination technology will be given below.

11.2.1.1 Multi-Stage Flash Distillation (MSF)

Water distillation in a vessel operating at a reduced pressure, thus providing a lower boiling point for water, has been used for over a century. In the 1950s, Weirs of Cathcart in Scotland developed this method to invent the MSF process which had significant developments and wide applications throughout the 1960s due to both to its economical scale and its ability to operate on low-grade steam (Halcrow Water Services 2007).

Table 11.1 Major desalination technologies

Thermal technologies	Membrane technologies
Multistage Flash Distillation (MSF)	Reverse Osmosis (RO)
Multi-Effect Distillation (MED)	Electrodialysis (EDR)
Vapour Compression (VC)	

11.2.1.2 Multi-Effect Distillation (MED)

The MED process is the oldest large-scale distillation method used for seawater desalination. At present, 3.5 % of the world's desalinated water is produced by MED plants (Wangnick 2000). High distilled water quality, high unit capacity and high heat efficiency are its most obvious characteristics (Al-Shammiri and Safar 1999; Wangnick 2000). The MED process, like MSF, takes place in a series of vessels or evaporators called effects, and it also uses the principle of evaporation and condensation by reducing the ambient pressure in the various effects.

11.2.1.3 Vapour-Compression Evaporation

The vapour compression distillation process is used in combination with other process like MED and single-effect vapour compression. In this process, the heat for evaporating the seawater comes from the compression of vapour. VC plants take advantage of the principle of reducing the boiling point temperature by reducing the pressure.

11.2.1.4 Reverse Osmosis (RO)

The RO process is relatively new in comparison to other technologies and was introduced as a successful commercialized technology in water desalination in the early 1970s. RO is a membrane separation process in which the water from a pressurized saline solution is separated from the solutes (the dissolved material) by flowing through a membrane without the need for heating or phase change. The major energy required is for pressurizing the feed water (Buros 2000). A typical large saline water RO plant consists of five major components, a saline water supply system, a feed water pretreatment system, high-pressure pumping, RO modules (membrane separation) and a post-treatment system (Khawaji et al. 2007). In recent years, the largest desalination plants using the RO process have been built in the Middle East, particularly in Saudi Arabia. A plant in Jeddah produces 15 million gallons per day (MGD), while the Al Jubail and Yanbu RO plants have capacities of 24 and 33.8 MGD, respectively (AL Mobayed and Balaji 2005; Khawaji et al. 2007).

11.2.1.5 Electrodialysis Reverse (EDR)

ED is an electrochemical separation process that employs electrically-charged ion-exchange membranes with an electrical potential difference as a driving force. The process depends on the fact that most salts dissolved in water are ionic, being either positively (cationic) or negatively (anionic) charged, migrating towards electrodes with an opposite electric charge. Membranes can be constructed to permit a selective passage of either cations or anions. Reversed Electrodialysis is based on the same principles as Electrodialysis (ED), except for the fact that the polarity of the electrodes is reversed several times per hour.

11.2.2 Renewable Energy System (RES) in MENA

The RES converts renewable sources of energy to useful energy vectors or carriers, such as heat, electricity, mechanical power or fuel (gaseous or liquid), which provide an alternative solution to the decreasing reserves of fossil fuels. The common renewable energy resources include hydro-electric, solar, wind, geothermal, biomass and ocean energy which are also termed as 'Green power technologies' (Fig. 11.1). The RES is able to provide reliable sources of electricity within a short and flexible time-frame. In contrast, most fossil fuel plants, large hydrodams, or nuclear plants take several years to develop from the planning to implementation stages.

Two technologies exist for converting direct solar energy to electricity: Concentrating solar power (CSP) and photovoltaic (PV) power. As reported by the World Bank in 2012, a CSP power plant generally consists of three parts: a solar

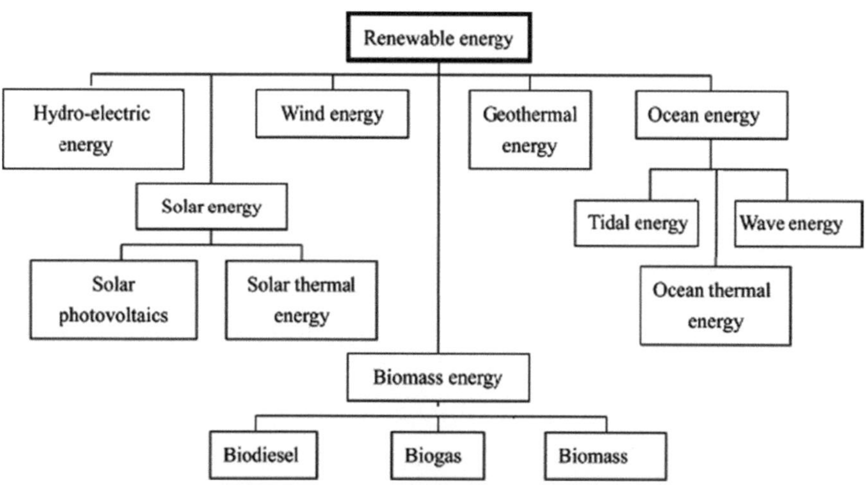

Fig. 11.1 Renewable energies systems (Sözen and Teksoy 2008)

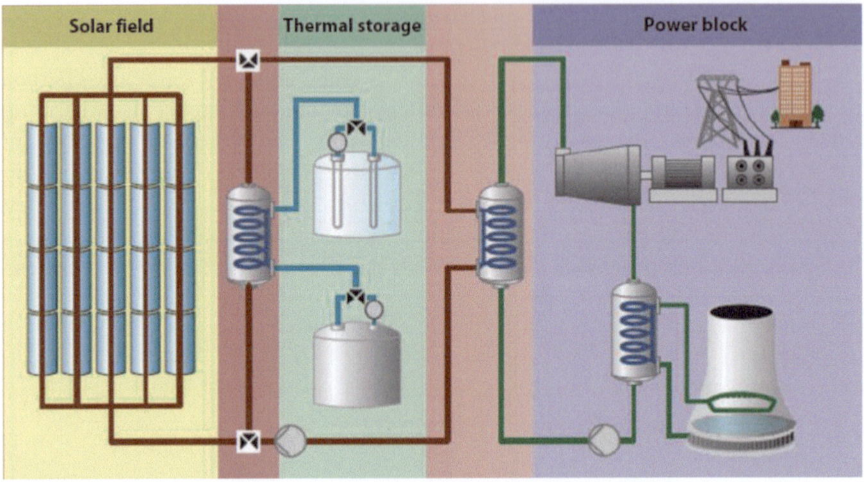

Fig. 11.2 Storage system in a trough solar plant (Solar Millennium 2011)

field, thermal energy storage, and a power system (block) that can produce electricity or heat or both (Fig. 11.2). Photovoltaic (PV) systems use semi-conductor materials, usually made of silicon to convert sunlight into electricity. Sunlight is converted to electricity using photovoltaic or solar cells (Sözen and Teksoy 2008).

11.2.3 Suggested Alternative to Renewable Solar Energy Desalination Systems

Renewable desalination is mostly based on the RO process (62 %), followed by thermal processes such as MSF and MED. The dominant energy source is solar photovoltaics (PV), which is used in some 43 % of the existing applications, followed by solar thermal and wind energy (Sözen and Teksoy 2008).

A wide variety of options are available to link Renewable Energy (RE) and desalination technologies. Each combination of technologies has its own merits in terms of the scope of water production, 24 h availability of RE sources to power desalination plants, and cost. Between 1974 and 2009, 131 RE desalination plants were installed worldwide (Solar Millennium 2011). When these 131 plants are categorized by energy source, solar heat is the most common, followed by PV. The primary reason that solar heat and PV are the preferred energy sources is that solar energy is more predictable.

In this study we will consider the following combinations of renewable solar energy systems and desalination technology: (1) Solar stills; (2) Solar Multi Effect Humidification (MEH); (3) Concentrating Solar Plant-Multi Effect Distillation

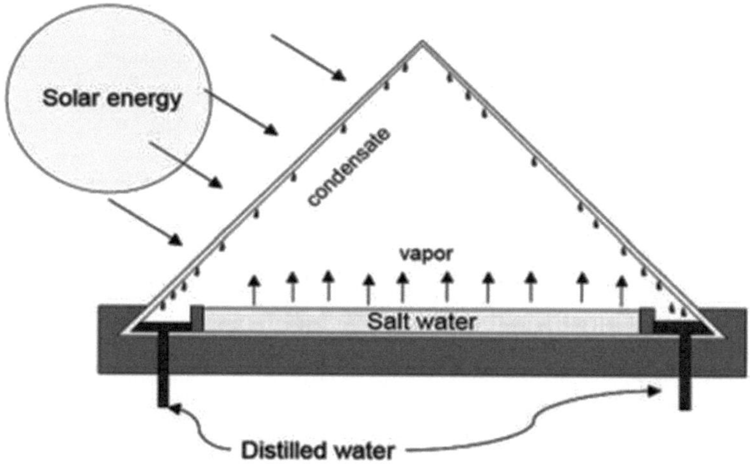

Fig. 11.3 Solar still unit (Miller 2003)

(CSP-MED); (4) Photovoltaic Electrodialysis Reverse (PV-EDR); and (5) Photovoltaic Reverse Osmosis (PV-RO).

11.2.3.1 Solar Stills

The basic design of a solar still, which is similar to a greenhouse, is shown in Fig. 11.3. Solar energy enters the device through a sloping transparent glass or plastic panel and heats a basin of salt water. The basin is generally black to absorb energy more efficiently. The heated water evaporates and then condenses on the cooler glass panels. The condensed droplets run down the panels and are collected for use as fresh water. Experience shows that 1 m^2 of ground will produce 3–4 l/day of freshwater. Because of this low production, it is important to minimize capital costs by using very inexpensive construction material. Efforts have been made to increase the efficiency of solar stills and the production per unit area (Buros 2000). Solar stills can be economically viable for small-scale production for households and small communities, especially where solar energy and low-cost labor are abundant (Buros 2000; Miller 2003).

11.2.3.2 Solar Multi-Effect Humidification (MEH)

The basic operation principle of the solar MEH process is the evaporation of seawater and condensation of water vapour from humid air taking place inside the unit at ambient pressure, while there is continuous humid air flow from the

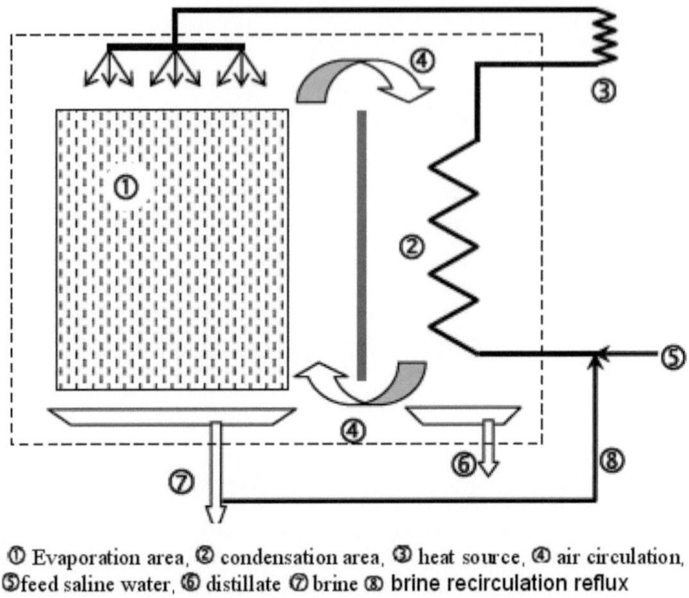

Fig. 11.4 General layout of a HD unit (MEDRC R&D Report 2002)

evaporator to the condenser, including the recovery of the latent heat of condensation for the preheating of the feed water (Fig. 11.4). More specifically, when circulating air comes into contact with hot saline water in the evaporator, a certain quantity of vapour is extracted by the air. A part of the vapour mixed with air may be recovered as condensate by bringing the humid air into contact with a cooling surface in another exchanger, in which the saline feed water is preheated by the latent heat of condensation (MEDRC R&D Report 2002).

11.2.3.3 Concentrating Solar Plant – Multi-Effect Distillation (CSP-MED)

This desalination plant consists of two parts (i.e. solar heat collector and distiller). The process is referred to as an indirect process if the heat comes from a separate solar collector or solar ponds, whereas it is referred to as direct if all the components are integrated into the desalination plant (Mahmoud and Ibrik 2006). Particularly attractive is the desalination associated with concentrating solar power (CSP) plants. CSP plants collect solar radiation and provide high-temperature heat for electricity generation.

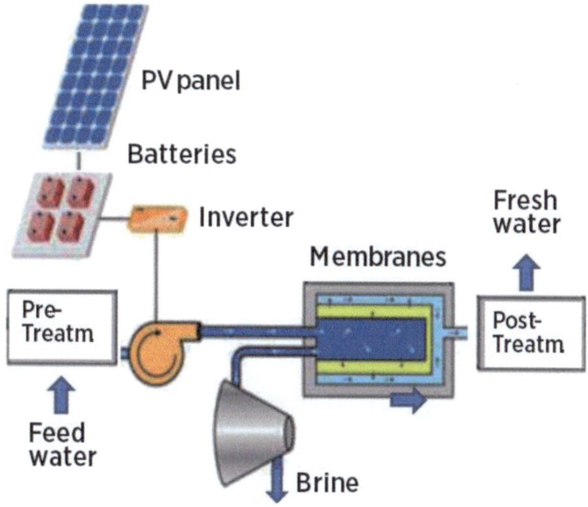

Fig. 11.5 Coupled PV and RO desalination plants (Al-Karaghouli and Kazmerski 2011)

11.2.3.4 Photovoltaic Reverse Osmosis (PV-RO) and/or (PV-EDR)

The PV generator can be connected either with RO or ED water desalination technology (Mahmoud and Ibrik 2006). Figure 11.5 shows the assembly of an RO desalination plant coupled with a PV-generator with multi-PV modules to produce electricity, a charge controller to protect the battery block from deep discharge and overcharge, a set of battery blocks to stabilize the energy input to the RO unit to compensate for solar radiation variations and an RO unit to desalinate the water. Many small PV-based desalination systems have been used throughout the world, especially in remote areas and on islands, including Gran Canaria, Canary Islands (PV-RO, seawater, 1–5 m^3/day), Riyadh, Saudi Arabia (PV-RO, brackish water, 5 m^3/day), and Ohshima Island, Japan (PV-ED, seawater, 10 m^3/day) (Kalogirou 2005). The main issue of PV desalination is the high cost of PV cells and batteries for electricity storage. The careful maintenance and operation of battery systems are also necessary.

11.2.4 Assessment of Suggested Alternatives System Using AHP

The criteria which are used to evaluate different alternatives are selected based on their importance to the process of desalination and energy required are as follows:

- Energy demand (Electricity and Thermal)
- Plant capacity (m^3/day)
- Water cost (USD/m^3)

Table 11.2 Relation between priorities and number in AHP rating procedure

Number	Priority
1	Equal importance
3	Moderate importance
5	Strong importance
7	Very strong
9	Extreme

- Volume of saline water per m^3 fresh water
- Volume of brine effluent per m^3 of fresh water and
- Cost of brine disposal (US\$/m^3)

The relative importance of one criterion over another can be expressed by using numbers in Table 11.2. The proposed alternatives to renewable energy desalination systems are reviewed and studied according to the criteria mentioned above and the findings are as written in Table 11.3. The computed eigenvector gives us the relative ranking of the criteria (Table 11.4). The preference of each alternative over another in terms of each criterion is determined (Table 11.5). Pairwise comparisons computing the eigenvector determine the relative ranking of alternatives under each criterion and gave the benefits of each alternative (Table 11.5).

11.3 Results

Values of the governing criteria for different alternatives are shown in Table 11.3 that contain the ranges of energy consumption, either electric or thermal, for different alternatives. In addition to plant capacity, which ranges from 0.1 to 5000 m^3/day, the cost of produced water and brine disposal ranges from US \$1–12/m^3 and US \$0.03–0.3/m^3 respectively. Based on pairwise comparison, the cost of the produced water criterion takes the first ranking among all the criteria, followed by energy demand and then desalination plant capacity (Table 11.4). With regards to the energy demand criterion and to the cost of produced water, the PV-RO alternative takes the first priority among different alternatives, while the Solar-MEH alternative occupies the first ranking with respect to plant capacity and the volume of saline water/m^3 of fresh water (Table 11.5).

Table 11.3 Values of governing criteria for different alternatives

Criteria	Units	RE source / Desalination technology/reference	Solar heat: CSP-MED	Solar heat: Stills	Solar heat: MEH	Photovoltaic (PV): EDR	Photovoltaic (PV): RO
Energy demand electricity	Kwh/m^3	Papapetrou et al. (2010)	1.5–2	Solar	1.5	1	4–5
		Sözen and Teksoy (2008)					
Thermal	kJt/kg	Papapetrou et al. (2010)	60–70	Passive	100	0	0
		Sözen and Teksoy (2008)					
Plant capacity	m^3/day	PRODES (2010)	5000	0.1	1–100	100	100
Water cost	US \$/m^3	PRODES (2010)	1.8–2.2	1–15	2–5	8–9	9–12
Volume of saline per m^3 fresh water		World Bank (2004)	3	NA[a]	NA	2–2.5	2–2.5
Volume of brine effluent per m^3 of fresh water		World Bank (2004)	2	NA	NA	1–1.5	1–1.5
Cost of brine disposal	(US \$/m^3)	Greenlee et al. (2009)	0.03–0.3				

[a] *NA* data Not available

Table 11.4 Pairwise comparison matrix of the main criteria with respect to the goal

Criteria	Energy demand	Plant capacity	Cost of produced water	Volume of saline/m^3	Volume of brine effluent/m^3	Cost of brine disposal	Priorities
Energy demand	1.0000	2.0000	1.0000	3.0000	5.0000	9.0000	0.2171
Plant capacity	1.0000	1.0000	1.0000	4.0000	6.0000	9.0000	0.1978
Cost of produced water	0.5000	1.0000	1.0000	4.0000	6.0000	9.0000	0.4149
Volume of saline/m^3	0.3300	0.2500	0.2500	1.0000	3.0000	2.0000	0.0525
Volume of brine effluent/m^3	0.2000	0.1600	1.6000	0.3300	1.0000	1.0000	0.0327
Cost of brine disposal	0.1000	0.1100	0.1100	0.5000	1.0000	1.0000	0.0851

Figure 11.6 presents the ranking of the suggested alternatives of the following renewable solar energy desalination systems: (1) Photovoltaic-reverse osmosis (PV-RO); (2) Photovoltaic-electrodialysis reverse (PV-EDR); (3) Concentrating solar plant – multi-effect distillation (CSP-MED); (4) Solar multi-effect humidification (MEH); (5) Solar stills.

11.4 Conclusion

Different alternatives system of renewable solar energy integrated with desalination techniques are suggested and evaluated based on a set of governing criteria using the Analytic Hierarchy Process (AHP). The analytic hierarchy process provides a logical framework to determine the benefits of each alternative. The governing criteria were the energy demand, the capacity of the desalination plant (m^3/day), the cost of the produced water (US \$/m^3), the volume of the saline water/m^3 of fresh water, the volume of the brine effluents/m^3 of fresh water and the cost of brine disposal (US \$/m^3). The results of this study determined the relative suitability of renewable energy desalination systems ranked in descending order, to help policy makers in the MENA region. Based on this study, the best alternative of solar energy integrated with the desalination system is Photovoltaic Reverse Osmosis (PV- RO).

Table 11.5 Synthesizing to obtain the final results

	Criteria	Energy demand	Plant capacity	Cost of produced water	Volume of saline/m^3	Volume of brine effluent/m^3	Cost of brine disposal	Overall alternative priority
	Criteria priorities	0.2171	0.1978	0.4149	0.0525	0.0327	0.0851	
Activities	CSP-MED	0.1798	0.3511	0.0357	0.2042	0.2923	0.2000	0.1606
	Solar stills	0.0597	0.0781	0.1934	0.1021	0.1461	0.2	0.1358
	Solar-MEH	0.1336	0.4292	0.2292	0.3063	0.1872	0.2	0.2482
	PV-EDR	0.2716	0.0660	0.2146	0.2042	0.1872	0.2	0.1949
	PV-RO	0.3552	0.0756	0.3271	0.1833	0.1872	0.2	0.2605

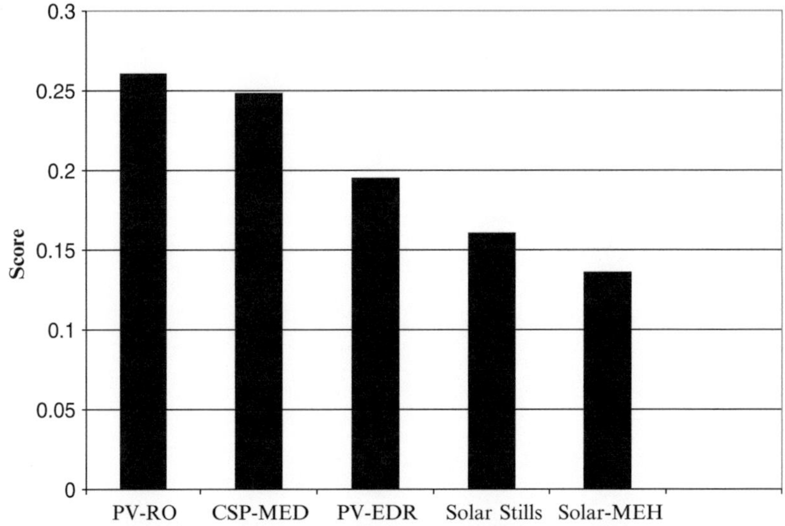

Fig. 11.6 Ranking of different renewable energy desalination systems

List of Figures

References

Al-Karaghouli AA, Kazmerski LL (2011) Desalination, trends and technologies. In: Michael Schorr (ed) Renewable energy opportunities in water desalination. ISBN: 978-953-307-311-8. pp 149–184. Available from: http://cdn.intechopen.com/pdfs/13758/InTech

AL Mobayed AA, Balaji S (2005) Successful operation of pretreatment in Al-Jubail SWRO Plant. In: IDA World congress on desalination & water reuse conference, USA

Al-Shammiri M, Safar M (1999) Multi-effect distillation plants: state of the art. Desalination 126:45–59

Buros OK (2000) The ABCs of desalting, 2nd edn. ASIN: B0006S2DHY, International Desalination Association, Massachusetts, USA, 30 pages

Greenlee LF, Lawler DF, Freeman BD, Marrot B, Moulin P (2009) Reverse osmosis desalination: water sources, technology, and today's challenges. Water Res 43(9):2317–2348

Halcrow Water Services (2007) Multi stage flash. Available from: http://www.hwsdesalination.com

IEA-ETSAP, IRENA (2012) Water desalination using renewable energy technology brief. Available from: http://www.irena.org

Kalogirou SA (2005) Seawater desalination using renewable energy sources. Energy Combust Sci 31:242–281

Khawaji AD, Kutubkhanah IK, Wie J-M (2007) A 13.3 MGD seawater RO desalination plant for Yanbu industrial city. Desalination 203:176–188

Mahmoud MM, Ibrik IH (2006) Techno-economic feasibility of energy supply of remote villages in Palestine by PV-systems, diesel generators and electric grid. Renew Sustain Energy Rev 10:128–138

MEDRC R&D Report (2002) A comprehensive study of solar desalination with a humidification-dehumidification cycle, ZAE Bayern, Munich

Miller J (2003) Review of water resources and desalination technologies. Materials Chemistry Department, Sandia National Laboratories. Available from: http://www.prod.sandia.gov/cgi-bin/techlib/access-control.pl

Papapetrou M, Wieghaus M, Biercampi Ch (2010) Roadmap for the development of desalination powered by renewable energy. Promotion of renewable energy for water production through desalination. PRODES Project, ISBN 978-3-8396-0147-1

PRODES (Promotion of Renewable Energy for Water Production through Desalination) (2010) Global renewable energy desalination by energy source. ProDes. Available from: http://www.prodes-project.org

Rinner C (2007) A geographic visualization approach to multi-criteria evaluation of urban quality of life. Geogr Inf Sci 21(8):907–919

Saaty TL (2008) Decision making for leaders: the analytic hierarchy process for decisions in a complex world. RWS Publications, Pittsburgh

Solar Millennium (2011) Parabolic through power plant with thermal storage. Solar Millennium AG. Available from: http://www.solarmillennium.de/includes/force_download.php

Sözen S, Teksoy S (2008) ADRIA handbook. Autonomous desalination system concepts for sweater and brackish water in rural areas with renewable energies – potential, technologies, field experience, socio–technical and socio-economic impacts. A guide to autonomous desalination system concept. Conceptual Framework of ADS and RES. MDEA WATER Co-funded by the European Union, p 1–7

Wangnick K (2000) IDA worldwide desalting plants inventory: Report No. 16. Wangnick Consulting for the International Desalination Association

World Bank (2004) Seawater and brackish water desalination in the Middle East, North Africa and Central Asia: a review of key issues and experiences in six countries. Final and main report prepared by a consortium of consultants, consisting of DHV Water BV, Amersfoort, the Netherlands, and BRL Ingénierie, Nimes, France. Available from: http://siteresources.worldbank.org/INTWSS/Resources/Desal_mainreport-Final2.pdf

World Bank (2012) MENA development report. Renewable energy desalination. An emerging solution to close the water gap in the Middle East and North Africa. Available from: http://water.worldbank.org

Chapter 12
Reducing Carbon Footprint of Desalination: The Australian Experience

Neil T. Palmer

Abstract One of the Funding Objectives of The National Centre of Excellence in Desalination Australia (NCEDA) is "*researching ways of efficiently and affordably reducing the carbon footprint of desalination facilities and technologies*". A number of NCEDA's 50 research projects have focussed on separating salt from water using renewable or waste heat energy sources. These include the Tjuntjuntjara project to demonstrate solar membrane distillation of hypersaline groundwater in a community in the Great Victoria Desert which is a novel boosted multi-effect distillation pilot plant using waste low grade industrial heat; and a solar-powered capacitive deionisation pilot plant desalinating brackish water in remote areas of the north of Australia. Another project is investigating the use of geothermal energy from deep brackish groundwater for desalination. NCEDA has also formed a relationship with Sundrop Farms, a new company in South Australia that has developed a commercially successful pilot greenhouse using solar powered seawater desalination in semi-desert near Port Augusta.

Massive desalination occurs every day in nature. Using renewable energy or waste heat from industrial processes to drive desalination makes sense, but is expensive. Ongoing research to reduce the unit cost of collecting and using these forms of energy will be the key to widespread adoption.

Keywords NCEDA • Desalination • Renewable energy • Solar • Wind • Wave • Innovative technologies

12.1 Introduction

It is self-evident that virtually all fresh water on Earth is derived from desalination of seawater by natural evaporation and precipitation. The solar energy (actually nuclear energy) used to drive this cycle is enormous and dwarfs the total of all human energy generation. Increasing expenditure on research to harness renewable

N.T. Palmer (✉)
National Centre of Excellence in Desalination Australia, Dixon Rd, Rockingham, WA 6168, Australia
e-mail: neil.palmer@murdoch.edu.au

M. Baawain et al. (eds.), *Recent Progress in Desalination, Environmental and Marine Outfall Systems*, DOI 10.1007/978-3-319-19123-2_12

Fig. 12.1 Location of Australian desalination and water recycling plants

energy from the sun, wind and waves is occurring worldwide. The National Centre of Excellence in Desalination Australia (NCEDA) was first mooted as an election promise of the-then incoming Rudd Labor Government in 2007. At the time, Australia was deep into the Millennium Drought, so-called because it was thought to be the worst drought in 1000 years. As a result of the real possibility that some Australian State capital cities might actually run out of water, State governments commenced building desalination plants in all mainland state capital cities, and an indirect potable water recycling system in Brisbane to ensure water security. The plant locations are shown in Fig. 12.1 and the details of the plants are shown in Table 12.1.

The new government promised funding for two research centres of excellence, for water recycling and for desalination. A competitive process was run and the Australian Water Recycling Centre of Excellence (AWRCE) was established in Brisbane and the NCEDA in Perth. The NCEDA comprises an unincorporated joint venture of 12 universities and the Commonwealth Scientific and Industrial Research Organisation. Murdoch University is the Administering Organisation and a pilot scale test facility was established on its campus at Rockingham, 45 km south of Perth.

The Australian Desalination Research Roadmap was compiled in 2010 following consultation with academia and industry and provides guidance to NCEDA's research program.

Table 12.1 Australia's mainland capital city major desalination and water recycling plants

State	Location	Owner	Process	Capacity m^3/d	Status	Completion date
WA	Kwinana	Water corporation	MMF[1]/RO[2]	145	Operating	2006
WA	Bunbury	Water corporation	UF/RO	300	Operating	2012
SA	Adelaide	SA water	UF/RO	300	Operating	2011
Vic	Wonthaggi	Aquasure	MMF/RO	450	Standby	2012
NSW	Kurnell	Sydney desalination plant	MMF/RO	250	Mothballed	2010
Qld	Gold Coast	SEQ[3] water	MMF/RO	125	Hot Standby	2009
QLD	Brisbane[4]	SEQ water	UF/RO	230	Mothballed	2010
			Total	1800		

Notes: (1) Multimedia filters (sand and coal); (2) Reverse osmosis; (3) South East Queensland Water; (4) Indirect potable water recycling – plants at Luggage Point, Gibson Island and Bundamba

The mandate given to NCEDA by the Australian government was to:

- Optimise and adapt desalination technology for use in Australia's unique circumstances.
- Develop suitable desalination technology for use in rural and regional areas.
- Efficiently and affordably reduce the carbon footprint of desalination facilities and technologies.

The NCEDA has also awarded 50 research projects, involving more than 300 researchers and more than 100 Australian and international project partners. It has awarded more than 60 student scholarships and established a Desal Discovery Centre to educate the community. It has also established a pilot scale test facility and held international technical workshops each year.

12.2 Reducing Carbon Footprint

12.2.1 Membrane and System Improvement

12.2.1.1 General

For membrane systems, progress since 1980 has reduced energy intensity for the desalination process from 16 kWh/m^3 to 1.8 kWh/m^3 in 2008 (Elimelech and Philip 2011). There is a theoretical minimum energy (dictated by thermodynamics) needed to separate dissolved salt from water, equivalent to 1.06 kWh/kL for seawater at 35 g/L salinity, with 25 °C and 50 % recovery. Practical conversion and the need for pre- and post-treatment mean that the total energy y intensity for modern desalination plants is generally greater than 3 kWh/kL. From this, approximately 1.4 kg CO_2 per kL is generated. Technology and membrane improvements can reduce energy consumption and the carbon footprint to a certain extent, but the separation of dissolved salt from water requires energy. Two recent improvements are NanoH_2O and Desalitech.

12.2.2 NanoH_2O

A recently formed US company (NanoH_2O) has successfully commercialised polyamide reverse osmosis membrane material containing aligned carbon nanotubes (Kurth et al. 2011). The company claims improved performance in terms of significantly higher normalised specific flux. Using standard 100 mm and 200 mm diameter membrane elements, the membranes can be retrofitted to existing plants with up to 10 % energy reduction.

The company has had commercial success. The largest plant to be equipped with NanoH_2O membranes is the 110 ML/day Palmachim seawater desalination plant in Israel. This represents successful implementation of nanotechnology on a commercial scale and is an example of a step change in membrane performance.

12.2.3 Desalitech

Batch desalination originated from an idea from Israel developed by Prof Avi Efraty and commercialised by Desalitech. It is a novel approach to a step reduction in desalination energy intensity by replacing continuous reverse osmosis desalination and energy recovery with a simple batch process. Conventional large scale RO processes use long pressure vessels (up to eight membrane elements in series) which have some intrinsic inefficiency as they require uniform feed pressure dictated by the last membrane which experiences the highest salinity.

The batch approach means that the membranes are subject to a uniform feed salinity and pressure, resulting not only in lower energy intensity but also obviating the need for energy recovery, as the system operates in "dead end" mode.

The Desalitech website describes the process (Desalitech 2014) with reference to Fig. 12.2. The high pressure pump (HP) feeds a closed loop of single stage membrane elements and a recirculation pump (CP) maintains cross flow where permeate flow equals feed flow. When the desired recovery is reached, brine is throttled out of the system and displaced by feedwater in a single plug flow sweep. The exchange of brine and feed water is achieved without stopping the HP pump or permeate production. The cycle then starts again.

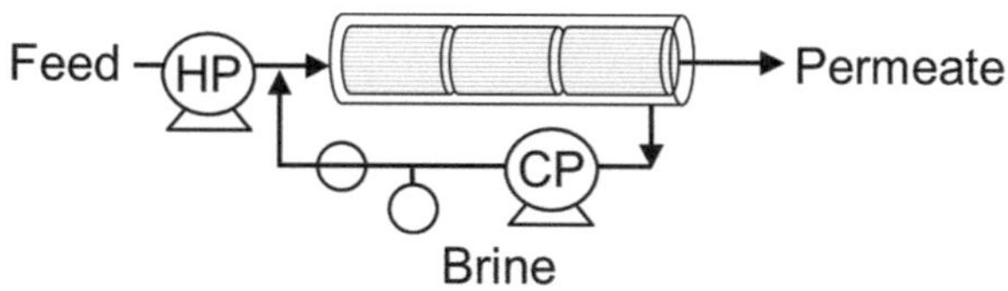

Fig. 12.2 Schematic representation of Desalitech Closed Circuit Desalination (CCD^{TM}) renewable energy sources

12.2.4 Membrane Distillation

Tjuntjuntjara is an aboriginal community in the Great Victoria Desert, 1300 km east of Perth. It is supplied with water from a brackish groundwater lens floating on a much larger ancient hypersaline groundwater resource. The lens is recharged by rainfall (average 150 mm per year) and is considered a fragile resource. The community has a population of around 150 and uses around 20 kL/day on average.

Murdoch University, in a project led by Dr Trevor Pryor and funded by NCEDA and the WA Department of Housing, is developing a solar-powered vacuum-assisted membrane distillation (VMED) pilot plant (Pryor 2013). A 0.5 kL/day VMED unit supplied by the German manufacturer memSYS has been tested at Rockingham and on site. Initially electrically heated, the system will be coupled to a CoGenra concentrated solar power collection system comprising photovoltaic panels and a cooling water loop. The cooling loop provides heat for membrane distillation and the electricity runs the heat loop and vacuum pumps (Fig. 12.3).

The hypersaline groundwater feed is saturated with calcium and magnesium carbonate and also contains soluble iron. Concentration through the distillation step tends to cause bulk precipitation and aeration causes ferric hydroxide precipitation. The key has been to supply the groundwater directly to the VMED unit, preventing aeration. It has been found that regular flushing with distillate and periodic acid flushing is needed to maintain production.

The intention of the trials (due for completion in June 2013) is to obtain design data for a full scale facility of around 20–30 kL/day to double the fresh water supply to the community.

Development of an affordable system for replication more widely across the Australian outback (and internationally) is the ultimate goal of the NCEDA (Fig. 12.4).

12.2.5 Boosted Multi Effect Distillation

Prof Hui Tong Chua of the University of Western Australia leads a project with industrial partner BHP Billiton at an alumina plant in WA to use waste process heat in a modified multi-effect distillation plant (Chua and Christ 2013). Using Alfa-Laval small scale MED units, the modification inserts a second stage MED to use heat that would otherwise be wasted. The result is a 30–40 % increase in efficiency.

Fig. 12.3 The memSYS VMED unit on site at Tjuntjuntjara

Fig. 12.4 The CoGenra concentrating solar panel under test at Rockingham

A pilot scale unit has been constructed and tested in the Rockingham facility and will be relocated to the processing plant site in the south of WA later in 2014.

The pilot scale plant produces up to 1.8 kL/day. The ultimate aim of the refinery is to have a full-scale plant produce fresh water from tailings, reducing the water balance and costs associated with managing spill risk as well as reducing fresh

Fig. 12.5 Boosted multi effect distillation pilot plant

water intake from external sources (as it is a scarce resource), and to do all this with readily available low grade waste heat (Fig. 12.5 and 12.6).

12.2.6 Capacitive De-ionization

Prof Linda Zou from UniSA has been researching capacitive deionisation and looking specifically at graphene as an electrode material. CDI is particularly suited to remote operations as the unit is tolerant to variable feed water, is not subject to silica scaling, and as it requires direct current electricity, is ideal to be driven by photovoltaics. A trailer-mounted demonstration plant (using standard mesoporous carbon electrodes - Fig. 12.7) has been constructed and tested in outback Australia (Zhang et al. 2013).

12.2.7 Geothermal Power

Prof Klaus Regenaur-Lieb of the University of Western Australia is undertaking a project to examine the potential for using large reserves of geothermally-heated water to drive energy and desalination projects.

There are plentiful reserves under the Perth metropolitan area which are deep (1 km or more) but there are also widespread reserves in regional and remote areas. It is these areas which are likely to see developments as the higher costs associated

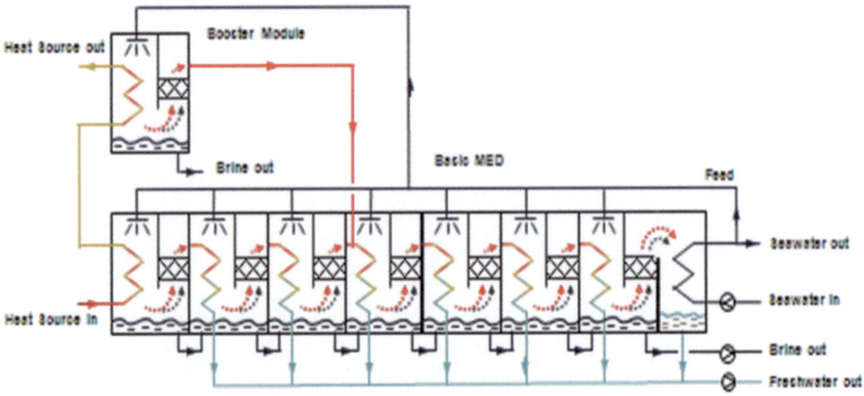

Fig. 12.6 Boosted MED schematic (Chua and Christ 2013)

Fig. 12.7 Solar powered capacitive deionisation unit

with using geothermal water (principally the high cost of drilling) would most likely not be competitive with current conventional desalination technology. However, the NCEDA research efforts going into small-scale distillation techniques would be appropriate to utilise the heat from such sources. The project's first stage is to identify promising areas and examine the feasibility of utilizing the geothermally-heated source water.

Fig. 12.8 Greenough river 10 MW solar electricity farm near Geraldton. All of its production is purchased by WA water corporation to offset part of the energy used at Southern seawater desalination plant

12.2.8 Wind and Solar Power Offsets for Major Australian Desalination Plants

In order to counter the perception that desalination plants are extremely energy-intensive, the mainland state governments decided to purchase renewable energy equivalent to the total annual energy consumption of the desalination plants. Primarily sourced from wind farms, the exception is the purchase of all the output from the 10 MW Greenough River solar farm by the WA Water Corporation which offsets part of the energy consumption at Southern Seawater Desalination Plant at Binningup near Bunbury (Fig. 12.8).

12.2.9 Growing Food with Solar Powered Desalination

Sundrop Farms Pty Ltd established a pilot-scale greenhouse (0.2 ha) in 2011. Located in arid country near Port Augusta in South Australia, the greenhouse derives heat energy from a concentrating solar collector to generate steam which drives a 10 kL/day multi-effect distillation plant, an electrical generating set and also provides heat (which is rarely needed in sunny Port Augusta) (Fig. 12.9).

Hypersaline groundwater which originates from nearby Spencer Gulf is used as feed for the desalination plant. Evaporative cooling and greenhouse humidification is achieved using the salt water. The unit cost of the desalinated water is high, but the relative cost of water as an input is offset by the use of a very efficient precision hydroponic watering system.

The promoters have big goals as there is an increasing gap worldwide between supply and demand for high quality foods. Their sustainable solution (with very low carbon footprint) is attractive to modern investors with an eye for sustainable

Fig. 12.9 Capsicums growing on desalinated seawater at Sundrop farm near Port Augusta, South Australia

projects, and more such greenhouses are likely to be built in a number of suitable arid geographical areas. CSIRO, with partner Valoriza Agua and led by Dr Olga Barron has completed an NCEDA- funded project to assess inland Australia for areas which might be suitable for agricultural development using desalinated water as a direct supply or by augmenting existing but fragile freshwater supplies. The study concludes that there are many suitable locations, but the cost of desalination suggests that high value, intensive food crops would be more economical to grow (Barron 2013).

12.2.10 Wave Energy

A wave energy system is being constructed off Garden Island, south of Perth in Western Australia, by Carnegie Wave Energy Ltd. The full-scale trial plant comprises a number of buoyant actuators which float below the surface and actuate high pressure piston pumps from the swell. The high-pressure liquid is piped to shore to run an electrical generator and a desalination plant. Power and water will be supplied to the nearby naval base. The system is expected to be commissioned by mid-2014.

Fig. 12.10 Australian made Carocell wick distillation units at Kendenup near Albany in Western Australia

12.2.11 Wick Distillation

Desalination using solar-wick distillation has been known for more than 30 years (Sodha 1981). FCubed manufactures Carocell wick-distillation units in Melbourne, Australia. They have been examined by the NCEDA, and the largest Australian installation is at a tourist farm in Kendenup near Albany, Western Australia. Each 3 m^2 unit produces up to 17 L/day at about 50 % recovery (Fig. 12.10).

The 30 units at Kendenup are arranged in two stages, increasing recovery to 75 %. In summer, the brackish groundwater feed (1400 mg/L) produces 500 L/day of fresh water to supplement the farms' rainwater system. The plant is shut down in winter when rain falls and sunshine is limited.

12.2.12 District Cooling and Multi-effect Distillation

Hughes et al. (2013) describes the use of multi-effect distillation driven by waste heat from large regional chillers which are common in the United Arab Emirates. Although marginal now, the paper predicts that with rising energy costs, a combined regional chiller and MED will significantly reduce the cost of operation of both cooling systems and desalination compared to the independent operation of each method.

12.2.13 Australian Backyard Innovator Prize

In 2013, a national competition for innovation was run by the national newspaper "The Australian". Trevor Powell of Impact Building Systems in Brisbane had developed a novel solar panel and entered the Backyard Innovator section by combining his solar energy system to generate heat with the boosted multi-effect distillation technology described earlier in this paper. There were seven finalists in this section and Trevor won first prize, illustrating the high level of public interest in desalination powered by renewable energy (Parker 2013).

12.3 Conclusion

Separation of salt from water requires energy. Membrane separation processes have steadily progressed towards the minimum energy required (around 1 kWh/kL). For energy which comes from burning fossil fuel, future improvement is limited, although there may be some opportunity to reduce the unit energy consumption of pre- and post-treatment.

To keep things in perspective, it should be remembered that desalination is not a significant user of energy compared to other activities. In Australia, if all the water for a typical house came from a desalination plant, the energy used to produce this water is only equivalent to the energy used by a typical two-door household refrigerator.

Further reductions in carbon footprint must rely on the use of renewable energy. This form of energy costs more and some renewable sources are intermittent. Considerable research effort is being directed at simple and robust systems to extract fresh water from salt water.

In a bold move to reduce carbon footprint, all major Australian seawater desalination plants purchase renewable energy (wind and photovoltaic) equivalent to total energy consumed and can be considered to have almost zero operating carbon footprint.

List of Figures

References

Barron O (2013) Opportunities for desalination in Australian Agriculture. In: NCEDA international desalination workshop 6, Melbourne, 28–29 November 2013

Chua HT, Christ A (2013) Development of an advanced low grade heat technology. In: NCEDA international desalination workshop 6, Melbourne, 28–29 November 2013

Desalitech (2014) http://desalitech.com/technology/

Elimelech M, Philip WA (2011) The future of seawater desalination: energy, technology and the environment. Science 333:712–717

Hughes BR, Rezazadeh F, Chaudry HN (2013) Economic viability of incorporating multi-effect distillation with district cooling systems in the United Arab Emirates. Sustain Cities Soc 7:37–43

Kurth CJ, Burk R, Green J (2011) Utilizing nanotechnology to enhance RO membrane performance for seawater desalination. In: IDA World Congress, Perth, September 2011

Parker D (2013) Trevor Powell – the solar/desalination power plant promises a low cost solution to several challenges. The Weekend Australian Innovation Challenge. Supplement 30 November, p 6

Pryor T (2013) Tjuntjuntjara remote inland indigenous community solar/waste energy groundwater desalination project. In: NCEDA international desalination workshop 6, Melbourne, 28-–29 November 2013

Sodha M (1981) Simple multiple wick solar still: analysis and performance. Sol Energy 26 (2):127–131

Zhang W, Mossad M, Zou L (2013) A study of the long-term operation of capacitive deionisation in inland brackish water desalination. Desalination 320:80–85

Chapter 13
Feasibility of Using Desalination for Irrigation in the Souss Massa Region in the South of Morocco

Abdelaziz Hirich, Redouane Choukr-Allah, Abdellatif Rami, and Mohamed El-Otmani

Abstract The region of Souss Massa in the south of Morocco is considered the most productive in terms of horticultural products especially destined for export. The part of the region in exportation of fruit and vegetable is about 90 % at national level. A greenhouse cropping system is the most common with more than 15,000 ha of crops grown in greenhouses. However, this region suffers from a serious problem of water scarcity, where the annual rainfall does not exceed 200 mm, and the water deficit is more than 260 Mm^3. In addition to this, agriculture in this region consumes about 90 % of the water resources. Over-pumping of groundwater is among the practices aggravating the situation by lowering the water table and consequently increasing pumping costs and groundwater salinisation due to sea water intrusion, especially in the coastal areas. Using desalination of sea water for irrigation of rentable crops as tomato and berries and other vegetables crops could be a judicious solution to continue producing horticultural products and saving water.

Pumping cost in Souss Massa region is about US$0.3, and the average desalination cost is equal to US$0.5 with a little change depending on desalination

A. Hirich (✉)
International Center for Biosaline Agriculture, P.O. Box 14660, Dubai, United Arab Emirates

Complex of Horticulture, Hassan II Institute of Agronomy and Veterinary Medicine, Agadir, Morocco
e-mail: hirich_aziz@yahoo.fr

R. Choukr-Allah
Complex of Horticulture, Hassan II Institute of Agronomy and Veterinary Medicine, Agadir, Morocco
e-mail: redouane53@yahoo.fr

A. Rami
Faculty of Science, University IbnZohr, Agadir, Morocco
e-mail: rami.abdel.horti@gmail.com

M. El-Otmani
International Center for Biosaline Agriculture, P.O. Box 14660, Dubai, United Arab Emirates
e-mail: elotmani.mohamed@gmail.com

M. Baawain et al. (eds.), *Recent Progress in Desalination, Environmental and Marine Outfall Systems*, DOI 10.1007/978-3-319-19123-2_13

technology. A study has been conducted surveying about 847 farmers representing farms of a cropping area of 12,770 ha, in order to find out their views regarding a desalination project for irrigation. 92 % of surveyed farmers accepted this project with a water demand equal to 32 Mm^3, 61 % of them accepted to participate in the project investment. 42 % of the farmers agreed to pay US$0.59–0.83 for 1 m^3 of desalinated water. 15 % and 4 % accepted to pay US$0.95 and US$1.18 respectively for 1 m^3 of desalinated water.

This survey indicates that there is a great potential of using desalination for irrigation of rentable crops in the Souss Massa region. Farmers are aware of the water scarcity problem and they accept the use of desalinated water for irrigation. Furthermore, the progress in desalination technology will make the price of desalinated less costly in the future.

Keywords Water price • Greenhouse • Tomato • Pumping

13.1 Introduction

With worldwide concern about water scarcity, agriculture is under pressure to improve water management and explore available options to match supply and demand (FAO 2004). Morocco is a nation heavily dependent on agriculture, which is responsible for 19 % of the country's gross domestic product (GDP) (Addison et al. 2012). The agricultural production in Morocco, as in most Mediterranean countries, suffers from water scarcity, enhanced by increasing demand and climate change impacts (Schilling et al. 2012).

Morocco has currently implemented several strategies to cope with water scarcity, such as the National Green Plan which is a big project launched by the government to improve agricultural production and the livelihood of famers. The development of the agricultural sector has allowed the Souss Massa region to play an important role in the socio-economic development of the kingdom. However, this situation is accompanied by a significant depletion of groundwater reserves. Indeed, the increased demand for irrigation water and the reduction in renewable water resources due to a succession of drought years are currently the major constraints in the management of groundwater resources which are considered to be the main source to satisfy water requirements (ABHSM 2008). This situation calls for speedy intervention by both the public and private sectors to find a permanent solution as well as an alternative water resource. Among those resources which have a big potential in the Souss Massa region are wastewater and desalinated sea water. The latter could be a very judicious choice as the region is close to sea, but desalination costs include investment, operation and maintenance all of which vary according to the type of desalination technology used.

In this paper, the water resources situation and the horticultural production in the Souss region are highlighted. A survey about the acceptability of farmers to use desalinated water is presented with a general view about the desalination plant project which will be implemented in the coming years.

13.2 Water Resources Situation in the Souss Massa Region

13.2.1 Groundwater

The Souss Massa region is characterized by an arid climate with low precipitations (<200 mm/year). The region has two important aquifer systems, the Souss and the Chtouka aquifer which have a surface equal to 4150 and 940 km^2, respectively (Fig. 13.1).

A survey was carried out by the Agency of the Hydraulic Basin of Souss Mass (ABHSM), the Office of Agricultural Development of Souss Massa (ORMVASM), National Office of Potable Water (ONEP) and Associations of Water Users (AUEA). The results obtained indicated that the water withdrawal in the region was equal to 738 Mm^3 of which 5 % is used for drinking and industrial water and 95 % for irrigation as shown in Table 13.1 (ABHSM 2005).

13.2.2 Surface Water

The large hydraulic structure built in the Souss Massa basin allowed for a water storage of 330 Mm^3/year to irrigate more than 50,000 ha, 12 Mm^3 for drinking water and 162 Mm^3 for the artificial recharge of aquifers. Table 13.2 presents the existing dams, their capacity and use objectives in the Souss Massa region (MEMEE 2014).

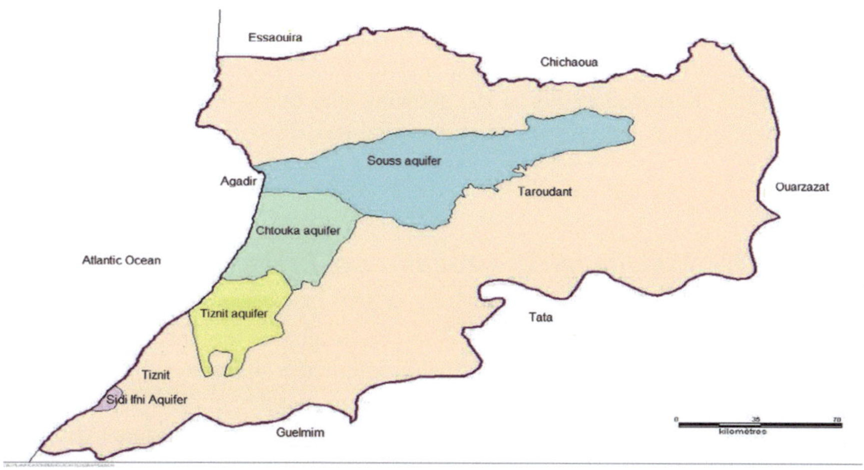

Fig. 13.1 Aquifers map of Souss Massa Region

Table 13.1 Water withdrawal in the Souss and Chtouka aquifers

Aquifer	Irrigation			Drinking water		
	Number of farms	Number of water sources	Water withdrawal (Mm^3/year)	Number of entities	Number of water sources	Water withdrawal (Mm^3/year)
Souss	7034	8597	608	513	545	28.7
Chtouka	2482	2758	85	221	244	7
Total	9516	11,355	693	734	789	35.7

Table 13.2 Existing hydraulic structures

Dam	Setup year	Capacity (Mm^3)	Water use
Youssed Ben Tachfine	1972	304	Irrigation, drinking water
Abdelmoumen	1981	214	Irrigation, drinking water, protection
Dkhila	1986	0.7	Compensation
Aoulouz	1991	108	Irrigation, protection
Imin El Kheng	1993	11	Irrigation
MokhtarSoussi	2001	50	Irrigation
MoulayAbdellah	2002	110	Drinking water
AhlSouss	2004	5	Irrigation, drinking water
Total		800	

13.3 Constraints Related to Water Resources Management

13.3.1 Drought

The annual rainfall is very variable and precipitation in a humid year sometimes can reach 3 times that of an average year and up to 15 times of a dry year. This irregularity is illustrated in Fig. 13.2, showing that the region was subjected to four drought periods and only one humid period (ABHSM 2005).

13.3.2 Overexploitation of Groundwater Resources

The reduction in the piezometric level of the water table of the Souss and Chtouka aquifers due to over-pumping and drought exhibited a decrease of water deficit varying from 100 to 370 Mm^3/year for the Souss aquifer and 60 Mm^3 for the Chtouka aquifer. Figure 13.3 shows the water balance for both aquifers (MEMEE 2014).

According to ABHSM 2005, the piezometric level analysis in the Souss aquifer between 1968 and 2003 shows a reduction in water table level of about 15 m in the Souss upstream, more than 30 m in the middle Souss and 20 m in the Souss

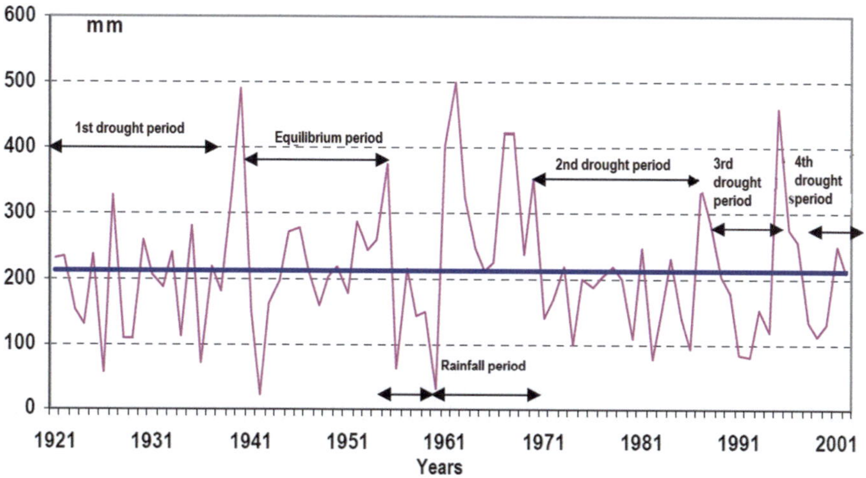

Fig. 13.2 Variation of precipitation in the Souss Massa region

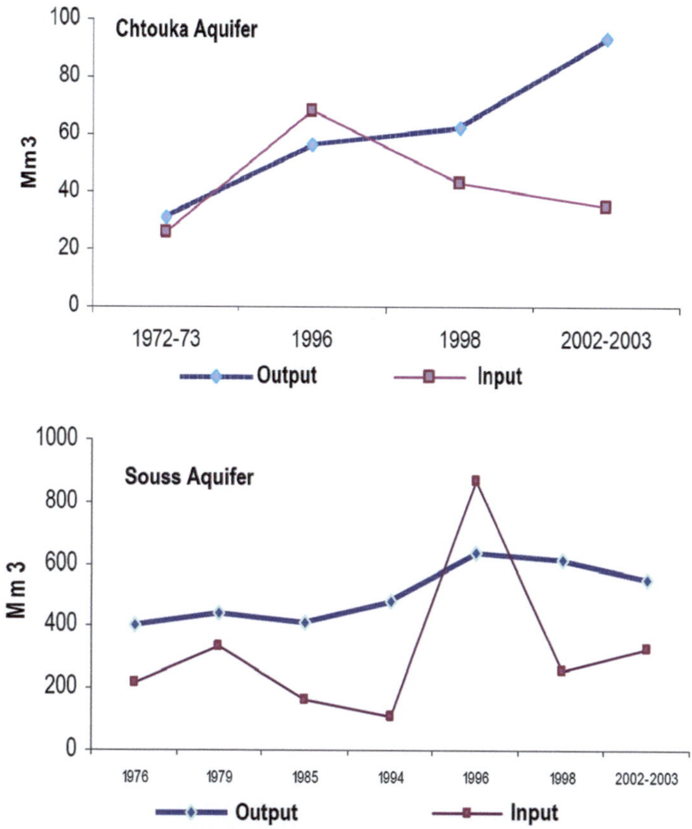

Fig. 13.3 Water balance of Souss and Chtouka aquifers

downstream, while in the Chtouka aquifer, the piezometric level was reduced by more than 20 m (Fig. 13.4).

13.4 Horticultural Production in the Souss Massa Region

According to the Ministry of Agriculture and Fisheries, the production of the agricultural sector should experience an increase from 11,838 Million DH in 2010–17,669 Million DH in 2020. The created working days increased from 30,804 working days in 2010 to a predicted 36,845 working days in 2020. Table 13.3 shows the agricultural production of the Souss Massa region as a proportion of national agricultural production, indicating that the region produces 77 % of the country's vegetables and 40 % of its citrus fruit (MAPM 2014). Those crops require a great amount of water. However, the available water resources cannot sustain increased development in the agricultural sector. Thus, the need for new water resources has become a priority for all agricultural producers and authorities, as well as for the water authorities.

13.5 Cost of Water Pumping

The direct annual cost of water pumping varies from 11,000 to 27,000 DH/ha (US $1326–US$3256/ha) depending on the energy used (electric or thermic). These costs include:

- Energy which varies from 8000 to 22,000 DH/ha (US$964–US$2653/ha) depending on thermic or electric energy as well as aquifer depth.
- Well-digging represents on average of 21,000 DH (US$2533) per well, that is 1500–3000 DH/ha (US$181–US$362/ha) at an average of one well per 7 to 14 ha.
- Pump maintenance, which is important in the case of thermic pumps, varies from 20,000 to 30,000 DH/well/year (US$2412–US$3618/well/year) and 1500 to 2000 DH/ha (US$181–US$.241/ha).

For an average water consumption of 12,000 m^3/year. The total cost of 1 m^3 varies between 0.9 DH (0,11 US$) and 2.25 DH (0.27 US$).

In some farms in the Chtouka zone, the pumping cost exceeds 3 DH/m^3 (0.36 US $/$m^3$), especially when the depth of the well becomes higher. In addition to that, there is also the problem of sea water intrusion due to an increase in the aquifer depth which in some cases exceed 300 m (Faquir et al. 2007). The rise in the cost of pumping water and the increase of sea water intrusion that causes ground water salinisation makes it necessary to look for alternative water resources such as desalinated water.

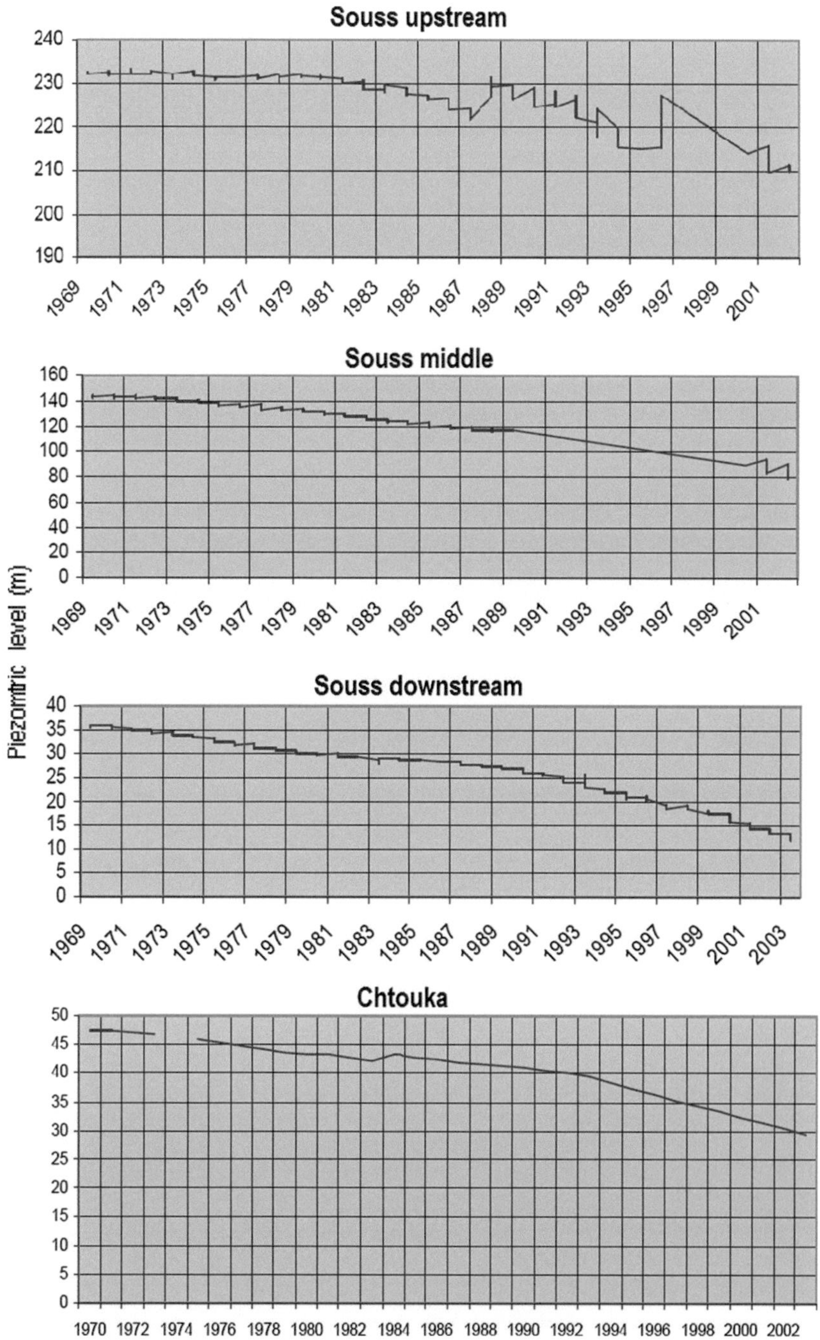

Fig. 13.4 Variation of piezometric level in the Souss and Chtouka aquifers between 1968 and 2005

13.6 Cost of Desalination

The cost of desalination depends mainly on the type of desalination process used, the quality of the intake and product waters, the output capacity of the plant, and the available options for waste disposal. They include:

- Investment costs (cost of land, equipment, civil works, etc.)
- Operation and maintenance (O&M) costs (energy, chemicals, labour, etc.)
- Environmental costs (water intake and environmental externalities, safe brine disposal, etc.)
- Other indirect costs (insurance, etc.).

As an example, Table 13.4 details the installation (investment costs without land costs) and the O&M costs of some types of desalination plants (FAO 2004). The primary operating cost of desalination plants is power, which typically accounts for 44 % of the O&M costs of a seawater RO plant (considered less expensive than thermal distillation). Thermal distillation processes for desalinating very highly saline waters and seawater are relatively expensive because of high operating temperatures and high construction costs. In contrast, RO processes for desalinating brackish water are less expensive because they are modular in setup and simpler to operate. However, a reduction in the costs of high-capacity seawater desalination plants has been observed over time. The costs of desalinated water are relatively high so that its major use is urban rather than irrigated agriculture.

Table 13.3 Cereals and horticultural production in the Souss Massa region (MAPM 2014)

Product	2010	National part (%)	2013	2020
Cereals	385	5	283	287
Vegetables	1480	77	1778	2140
Olive	27	2	28	43
Citrus fruit	646	39	893	1070
Other fruit	562	18	571	677

Table 13.4 Installation and operational and maintenance costs of various desalination plants

Desalination plant	Installation costs	Water production cost
	US $/m^3	
Multistage flash distillation	1200–1500	1.10–1.25
Multistage flash distillation (Singapore)	2300	1.50
Multiple-effect distillation	900–100	0.75–0.85
Multiple-effect distillation (Metropolitan Water District, California, USA)	660	0.46
Vapour compression distillation	950–1000	0.87–0.95
Reverse osmosis	700–900	0.68–0.92

13.7 Desalination Project

Before the installation of this project, it is necessary to study the following key points:

- Identify the user groups in order to determine the optimal level of desalinated water production as well as the plant capacity, which is an important indicator for its design
- Identify the potential sites of desalinated water production. For this purpose, it is important to take deposits from wells close to the sea if the hydrogeological conditions are convenient. The available studies at the Agency of the Hydraulic Basin of Souss Mass and the National Office of Potable Water will be the subject of all work to be carried out within this project. This work must be coupled with identification of brine disposal sites.
- Determine the optimal approaches related to funding, organization and management
- Ensure the availability of an adequate electricity network for the desalination site

13.7.1 Greenhouses Farmers Survey

The Agrotech[1] association has conducted a survey, in collaboration with the ORMVASM, among greenhouse producers in the Chtouka zone. The objectives of this survey are:

- To identify the area equipped with greenhouses in order to determine their spatial distribution
- To determine the total water volume used by farmers and the potential desalinated water volume required
- To canvass the opinion of farmers about the use of desalinated water for irrigation and their interest in participating as stakeholders in the project

The survey covered 847 greenhouses farms with a total area of 12,770 Ha. The obtained results are presented in Tables 13.5, 13.6 and 13.7 (AGROTECH 2010).

During the running of the survey it was observed that the surveyed farms can be divided to three categories:

[1] The Agrotech association of the Souss Massa Drâa (Agrotech SMD) is an association of institutions created in 2006 in order to create an activity advanced center in the agro-technology field in the Souss Massa Drâa region, to help companies and institutions in the field of research and development of food biotechnology. The association is public and based in the Agronomic and Veterinary Medicine Hassan II Institute, Complex of Horticulture in Ait Melloul City.

Table 13.5 Main results element of the survey

Number of surveyed farms	847
Number of farms who want to join the project	782
Percentage (%)	92 %
Total surveyed greenhouses area (ha)	12,770
Actual greenhouse covered area (ha)	7986
Consumed water volume (m^3)	54,729,087
Total area adhered to the project (ha)	12,186
Actual greenhouse covered area adhered to the project (ha)	7555
Required water volume (m^3)	31,722,349

Table 13.6 Survey results in terms of area and volume

	%	Required volume (m^3)	Greenhouse area (ha)
Percentage of project membership	95	31,672,549	7555.10
Participation in the investment	67	23,724,146	5350.39
Pay 5–7 DH/m^3 (0.6–0.85 \$/m^3)	50	17,459,891	3996.30
Pay ≥ 8 DH/m^3 (≥0.97 \$/m^3)	20	6,183,842	1581.71
Pay ≥ 10 DH/m^3 (≥1.21 \$/m^3)	6	1,419,055	497.10

Table 13.7 Survey results in terms of farm number and volume

	%	Required volume (m^3)	Number of farms
Percentage of project membership	92	31,672,549	782
Participation in the investment	61	23,724,146	515
Pay 5–7 DH/m^3 (0.6–0.85 \$/m^3)	42	17,459,891	354
Pay ≥ 8 DH/m^3 (≥0.97 \$/m^3)	15	6,183,842	127
Pay ≥ 10 DH/m^3 (≥1.21 \$/m^3)	4	1,419,055	36

- Group 1: farms located in the irrigated zone of Massa
- Group 2: farms located in the zone of SidiBibi and Khmiss Ait Amira
- Group 3: farms located in the zone of Biougra and Sebt Ait Milk (Fig. 13.5)

13.7.2 *Importance of the Desalination Project*

The Ministry of Agriculture was engaged in an irrigation project using desalinated sea water in the Chtouka zone in the Souss Massa.

This project has a particular importance in Morocco as it is implemented within the framework of the National Green Plan. The project will support agricultural development in the Souss Massa region as it is the premier region in terms of production and export of agricultural products with high added value. This project will respond to the needs of vegetable and fruit producers in the Chtouka zone where the aquifer is being used for irrigation and consequently being subjected to

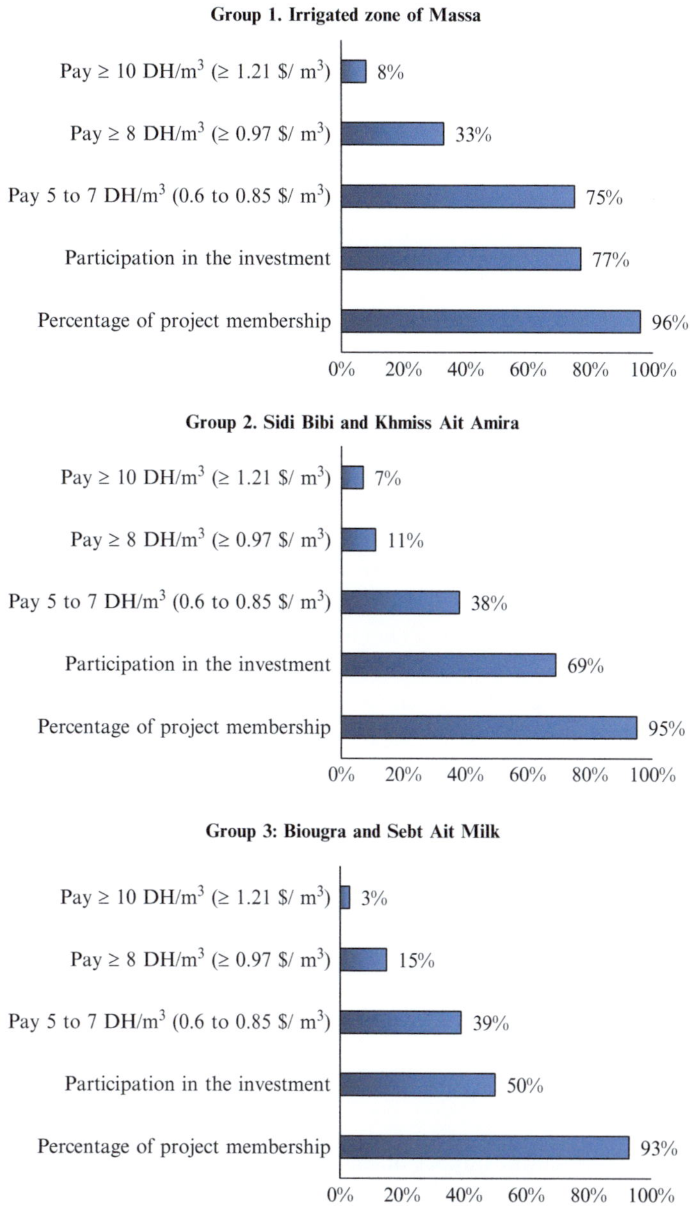

Fig. 13.5 Greenhouses farmers survey results

overexploitation due to over-pumping. This has resulted in the reduction of the piezometric level and sea water intrusion. Furthermore, the surface water resources in this area are very scarce. The surveys carried out in this area by the Ministry departments showed that farmers were willing to join this project.

The worst-case scenario in the project zone predicts the disappearance of the aquifer and a loss of about 9 billion DH (US $1.1 billion) of added value and 3 billion DH (US $363 million) of capital and more than 2830 permanent jobs by 2035 (Arrifi 2013).

13.7.3 Technical Components of the Desalination Project

The concerned area: 9000 ha (at the beginning of the project run) and 12,500 ha (by 2035).

Required desalinated water: 55 Mm^3/year (the aquifer water pumping should not exceed 25 Mm^3/year so that it is within the limits of renewable groundwater).

13.7.3.1 Desalination Plant

- *Possible location*: coastal site in the National Park of Souss Massa (between Tifnit and Douira)
- *Capacity*: 111,000 m^3/day in the beginning of the project run and 166,500 m^3/day by 2035
- *Desalination technique*: reverse osmosis with double pass Modular and advanced plant to follow as the project expands in scope

13.7.3.2 Infrastructures of Pumping and Distribution of Desalinated Water for Irrigation

- Pumping station to pump the desalinated water into a regulation and control basin
- Pumping from the regulation basin to the mean distribution network
- Mean distribution network (Fig. 13.6)
- Pressuring station

13.7.3.3 Environmental Impact of the Desalination Project

A major environmental problem of water desalination is the production of a flow of brine that contains the salts removed from the intake water that needs to be disposed of. In addition, this brine may be polluted. This brine represents a significant fraction of the intake water flow. Seawater desalination typically yields a brine flow of 50–65 % of the intake water flow, with about twice the initial concentration (FAO 2003).

The project will have a positive impact on the environment as it will contribute significantly in the reduction of the overexploitation of the aquifer and

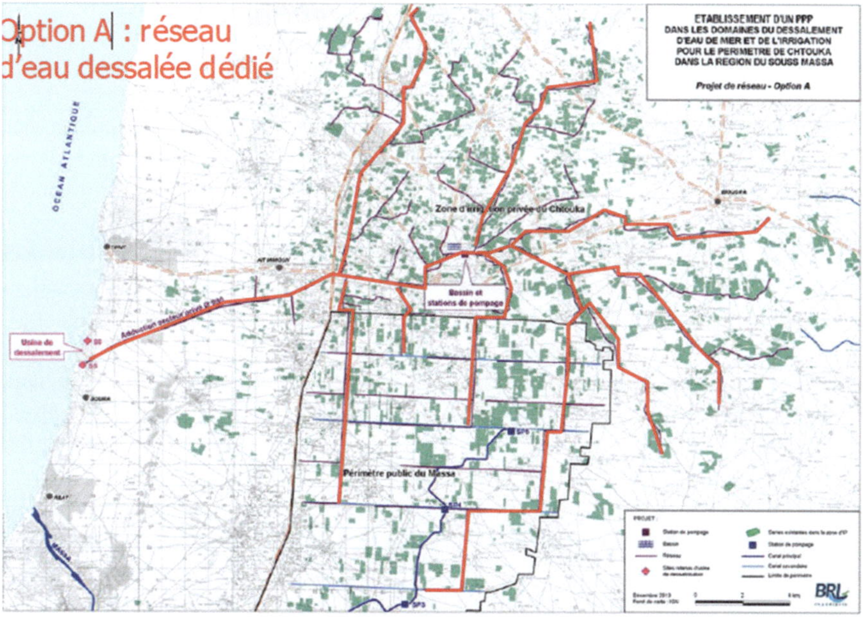

Fig. 13.6 Network distribution of the desalinated water

consequently the project will achieve equilibrium in terms of groundwater balance as well as the sustainable use of the aquifer in the Chtouka zone.

The implementation site has been chosen to allow a zero impact especially on the ibis protected in the National Park of Souss Massa where the project will be installed. The infrastructure has been designed to control the brine disposal through the use of modern technology. Marine cartography and mapping will be applied in order to minimize the impacts and risks as much as possible.

13.7.3.4 Institutional Issues of the Project

The project will be built within the framework of a public service in partnership with a private operator (Public-Private Partnership: PPP) who will be charged to build, design and manage all the infrastructure related to sea water desalination and irrigation.

- All the concerned members of the management (state, regional consul, users) will provide a financial contribution to the initial investment. The contribution of the members will be related to membership fees.
- Users will pay a membership fee to the operator in charge to use the desalinated water for irrigation
- These financial contributions to the initial investment, their terms and conditions, and also the price of a cubic meter of desalinated water will all be determined by a study of the structure and devolution of the PPP project.

13.7.4 Structuring and Implementation of PPP Project of Chtouka

- Due to the success of the El-Guerdane project, the Ministry of Agriculture has selected the International Finance Corporation (IFC), a member of the World Bank Group, as the main consultant to assist in the structuring and implementation of the transactions related to the project.
- The structuring study will allow the determination of the feasibility of the project and will designate, by international tender, the operator who will proceed to build up and install all the desalination and irrigation infrastructure and who will further the project management for a period of operation of 30 years.
- The structuring study is necessary to analyse the risks related to the project (especially financial risks, profitability and availability of potential operators, etc.). It is a complex project, innovative, with a high technology level and the first in Morocco.
- Currently, the first phase of the project structuring study, which consisted of carrying out the feasibility study and identifying the strategic options available to realize the project, has been achieved. The Moroccan government is in the process of preparing the funding of the project and making some decisions regarding some technical aspects.
- Once the decisions on the strategic options are made, the government will proceed to launch the selection process by an international tender in order to choose the operator who will be in charge of bringing the project to completion.

List of Figures

References

ABHSM (2005) Stratégie de préservation des ressources en eau souterraine dans le bassin du Souss Massa- Plan d'action 2005–2020. Agence du Bassin Hydraulique du Souss Massa, Agadir

ABHSM (2008) Situalyion hydrologique du bassin hydraulique de Souss Massa. Agence du Bassin Hydraulique de Souss Massa, Agadir

Addison B, El Korchi T, Rosenstock J, Badran K, Baker J, Collins B (2012) Water management and conservation in Rural Morocco. Worcester Polytechnic Institute, Worcester

AGROTECH (2010) Projet de dessalement de l'eau de mer pour l'irrigation. http://www.agrotech.ma/index.php?option=com_content&view=article&id=39&Itemid=5. Accessed 20 Dec 2013

Arrifi E (2013) Projet d'irrigation a partir de dessalement de l'eau de mer dans la plaine de chtouka – maroc. Direction de l'irrigation de l'amenagement de l'espace agricole, Rabat

FAO (2003) Desalination of brackish waters and seawater, status in California and the USA. Draft report. Food and Agriculture Organization, Rome

FAO (2004) Water desalination for agricultural applications. In: Beltrán JM, Koo-Oshima S (eds) FAO expert consultation on water desalination for agricultural applications, 26–27 April 2004, Food and Agriculture Organization, Rome

Faquir M, Abainou D, Moumadi H., Hachimy L (2007) Sauvegarde de la zone agrimucole El Guerdane: une nouvelle génération de projet d'irrigation au Maroc. HTC Anafide 109:35–45

MAPM (2014) Production végétale de la région de Souss Mass, Minitère de l'Agriculture et de la Pêche Maritime. http://www.agriculture.gov.ma/pages/regions/r%C3%A9gion-de-souss-massa-daraa. Accessed 8 Jan 2014

MEMEE (2014) Les bassins hydraulique du Maroc. Ministère de l'Energie, des mines, de l'Eau et de l'Environnement, Rabat

Schilling J, Freier KP, Hertig E, Scheffran J (2012) Climate change, vulnerability and adaptation in North Africa with focus on Morocco. Agric Ecosys Environ 156:12–26

Chapter 14
Desalination from an Integrated Water Resources Management Perspective

Farhad Yazdandoost

Abstract With more than half of the world's population living close to the oceans and a high percentage of the world's megacities located near their shores, seawater desalination is an attractive option for coastal water supply systems. Water scarcity in arid and semi-arid regions has lately driven decision makers and planners to investigate the viability of transfers of desalinated sea water to water scarce locations. Arid and semi-arid zones are defined as areas where the rainfall patterns are inherently erratic and where, generally, precipitation is below potential evapotranspiration. The hydrological regime in these areas is extreme and highly variable, and globally, these areas face the greatest challenges to deliver and manage freshwater resources. Problems are further exacerbated by population growth, increasing domestic water use, expansion of agriculture, pollution and the threat of climate change. These areas have often embraced the notion of water transfers from other catchments as an added water security element. However, sustainable development considerations remain as a source of concern under such circumstances. Limited access to water resources associated with climatic conditions of arid and semi-arid regions have often been the source of competition and conflict amongst stakeholders. Integrated Water Resources Management (IWRM) is seen worldwide as an appropriate means of conflict resolution. IWRM is the response to the growing pressure on water resources systems as a result of growing population and socio-economic developments. Major considerations associated with desalination are those related to the environment and the economy with respect to energy consumption. The increasing competitiveness of desalination may be related to considerable improvements in technical advances, thereby alleviating the problem to a certain extent as far as the above considerations are concerned. However, challenges remain in the face of greater water resources management systems where the output from desalination is one of a number of inputs affecting the system in an interactive way. Many criteria and objectives may need to be taken into consideration at the decision-making level which would in turn necessitate the use of Decision Support Systems (DSS). The DSS approach, in the framework of the utilisation of Multi Criteria Analysis (MCA) tools to assess the ranking of

F. Yazdandoost (✉)
Department of Civil Engineering, K N Toosi University of Technology, Tehran, Iran
e-mail: yazdandoost@kntu.ac.ir

M. Baawain et al. (eds.), *Recent Progress in Desalination, Environmental and Marine Outfall Systems*, DOI 10.1007/978-3-319-19123-2_14

potential development scenarios, would become essential as part of a comprehensive decision-making process.

Keywords Desalination • IWRM • Arid and semi arid • MCDM • DSS

14.1 Desalination

Many countries in the world suffer from a shortage of natural fresh water. Increasing amounts of fresh water will be required in the future as a result of the rise in population rates and enhanced living standards, together with the expansion of industrial and agricultural activities. Available fresh-water resources from rivers and groundwater are presently limited and are being increasingly depleted at an alarming rate in many places. The oceans represent the earth's major water reservoir. About 97 % of the earth's water is seawater while another 2 % is locked in icecaps and glaciers. Available fresh water accounts for less than 0.5 % of the earth's total water supply. Vast reserves of fresh water underlie the earth's surface, but much of it is too deep to access in an economically efficient manner. Additionally, seawater is unsuitable for human consumption and for industrial and agricultural uses. By removing salt from the virtually unlimited supply of seawater, desalination has emerged as an important source of fresh water. Today, some countries depend on desalination technologies for the purpose of meeting their fresh water requirements. In the Middle East in particular, seawater desalination is a vital and dependable fresh water resource in countries such as Saudi Arabia, the United Arab Emirates and Kuwait. Furthermore, it is likely that desalination will continue to grow in popularity in the Middle East. Figure 14.1 depicts the sea water desalination capacity on a global scale. Overall, it is estimated that over 75 million people worldwide obtain fresh water by desalinating seawater or brackish water. The IDA Desalting Inventory 2004 Report shows that at the end of 2002, installed and contracted brackish and seawater desalination plants worldwide totaled 17,348 units in 10,350 desalination plants with a total capacity of 37.75 million m^3/day of fresh water. The five leading countries in the world by desalination capacity are Saudi Arabia (17.4 %), USA (16.2 %), the United Arab Emirates (14.7 %), Spain (6.4 %), and Kuwait (5.8 %). In 2001, seawater and brackish water accounted for about 60 % and 40 %, respectively, of all desalinated water sources in the world (Khawaji et al. 2008). At the end of 2002, MSF and RO accounted for 36.5 % and 47.2 %, respectively, of the installed brackish and seawater desalination capacity. For seawater desalination MSF accounted for 61.6 % whereas RO accounted for 26.7 %. It should be noted that MSF holds the lead in all plants producing over 5000 m^3/day units. The current world desalination plant capacity is 40 million m^3/day and the annual average growth rate for the last 5 years has been 12 %.

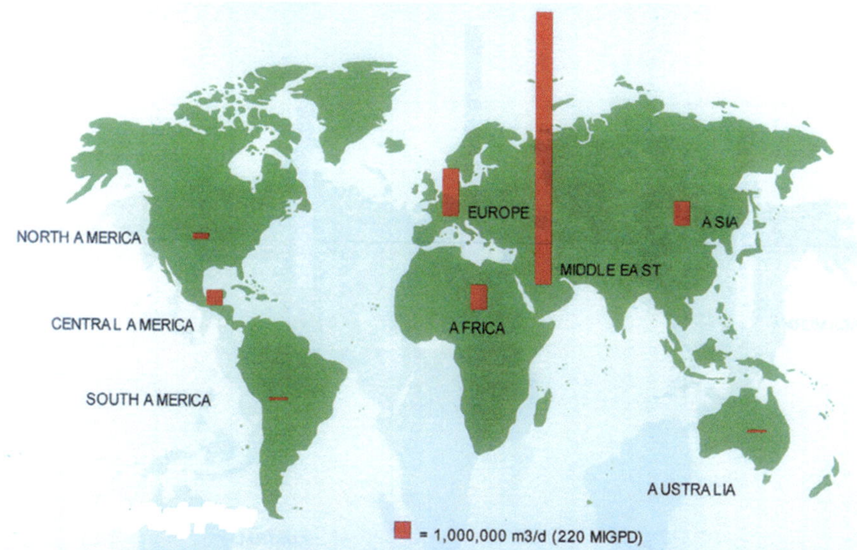

Fig. 14.1 Sea water desalination capacity in the world

14.1.1 Integrated Water Resources Management

Arid and semi-arid zones are defined as areas where rainfall patterns are inherently erratic and where, generally, precipitation is below potential evapotranspiration. Due to limited access to shallow water, lack of freshwater resources and presence of often polluted ground water, water is at its most scarce in such areas. The hydrological regime in these areas is extreme and highly variable and globally, these areas face the greatest pressures to deliver and manage freshwater resources (Barrow 2001, Rubin et al. 2006). Problems are further exacerbated by population growth, increasing domestic water use, expansion of agriculture, pollution, and the threat of climate change (Ospina-Noreña et al. 2009). However, there is little guidance on the decision support tools that are needed to underpin water resources management.

Water shortage has often been the source of competition and conflict amongst stakeholders. Integrated Water Resources Management (IWRM) is seen worldwide as an appropriate means of conflict resolution (Loucks 2005). IWRM is the response to the growing pressure on water resources systems as a result of growing population and socio-economic developments. Water resources management has undergone a drastic change world-wide, moving from a mainly supply-oriented, engineering-based approach towards a demand-oriented, multi-sector approach, often called Integrated Water Resources Management. The concept of IWRM lets us move away from 'water master planning', which focuses on water availability and development, towards 'comprehensive water policy planning' which addresses the interaction between different sub-sectors, seeks to establish priorities, considers institutional requirements, and deals with the building of capacity. IWRM should not be seen as a 'model' that has to be enforced upon the country (Biswas 2001) but

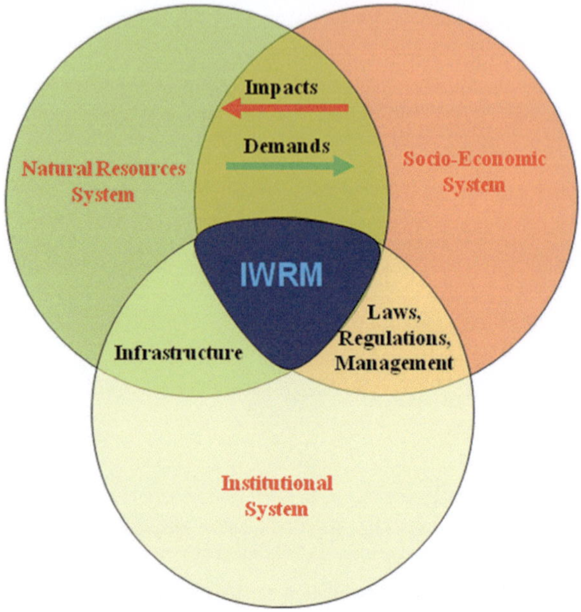

Fig. 14.2 Concepts of IWRM

more as a process. According to GWP 2000, IWRM is a process which promotes the co-ordinated development and management of water, land and related resources, in order to maximize the resultant economic and social welfare in an equitable manner without compromising the sustainability of vital ecosystems (Diep et al. 2008, Loucks et al. 2005]. Figure 14.2 shows the concepts of an IWRM approach.

14.1.2 The Proposed Approach

Given the scarcity associated with arid and semi-arid regions, all sources of water supply, both in terms of quality and quantity, have to be carefully identified. Water system planning is often associated with too many uncertain parameters due to ambiguities in defining objectives, forecasts and assumptions. To achieve coherence, a broad perspective is therefore required to effectively establish the relationship between certain and uncertain parameters. The identification of parameters may then be followed by setting up an allocation simulation model (Labadie 1995, Labadie 2007, Mike Basin 2003, Ribasim Version 6.33 2005, Sieber 2005, Wil and van der Krogt 2005]. Making decisions based on IWRM may be assisted by decision support systems encompassing technical, economic, environmental and social needs in an integrated framework [Janssen et al. 2000, Srdjevic et al. 2004]. The base/reference framework for quantity and quality of water supply and demand may be deduced based on available data. A sequence of events envisaged logically and coherently would form the basis of scenarios anticipated for future development of the area under study. Scenario generation may then be performed for socio-economic, political and environmental parameters and verified against criteria

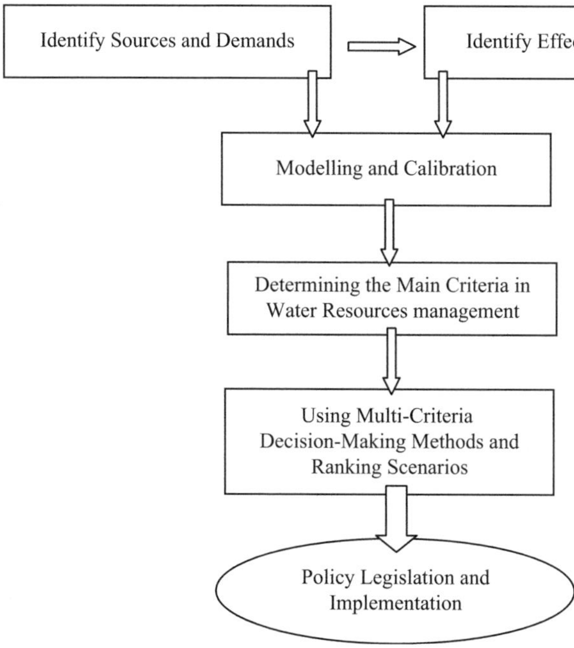

Fig. 14.3 The proposed IWRM toolbox

based on sustainable development. Generally speaking, scenario analysis is an appropriate method for dealing with uncertainty parameters in decision-making (Mannina and Viviani 2009). While these are prioritised based on their impacts on physical, social, economic and environmental characteristics, the best criteria-based scenario (ranked highest) may be selected as a solution. In order to use different and a larger number of criteria in the decision-making process and to simplify complex situations in making decisions, multi-criteria decision-making (MCDM) methods may be adopted (Janssen et al. 2000). Finally, practical policies for the prevailing natural conditions of the study area may be proposed for approval and implementation. Figure 14.3 depicts an algorithmic procedure of the developed IWRM toolbox for this research.

14.1.3 Case Study

To evaluate the performance of the proposed toolbox, it has been used practically for one of the dry zones of Iran, Kashan (Fig. 14.4). Kashan has a rich cultural and historical background, as well as being a globally-renowned centre for the production of the finest rosewater and hence a hotspot for the tourism industry. Situated in the central Iranian plateau and at the boundary of the central desert, this region has continually faced water resources limitations resulting in excessive pressures on groundwater resources whilst heavily depending on trans-boundary water transfers.

Fig. 14.4 Kashan and Aran Bidgol area

14.2 Application of the Toolbox

All necessary data for the preparation of an initial integrated model of the region were collected and classified. Given the suitable span of the existing data, 2006 was chosen as the base year. The main demands in the study area were identified as drinking water, industry and agriculture with the last reaching a high of about 387.29 MCM in the year 2006, while industry had the lowest water demand of around 5.89 MCM (Table 14.1).

In terms of resources, there are three types of water resources in the study area comprising groundwater, surface water and inter-basin water transfer (proposed water transfer through desalination). Overall, the study area has no noticeable permanent rivers, and only two ephemeral rivers, Ghohrood and Bonrood, are present in the entire basin and their flows are mostly used for agriculture when available. Due to surface resources limitation, groundwater supply plays an important role in this area and there is extreme pressure on groundwater, resulting in an average loss of approximately 0.5 m in the Kashan aquifer annually. Besides the natural sources, around 6 MCM/year is transferred to Kashan through a pipe system from the neighboring basin solely for drinking purposes. Table 14.2 gives a breakdown of the annual volume of different water resources in Kashan.

Table 14.1 Water demand in 2006 (MCM/Year)

Demand type	Water demand
Drinking water	33.98
Industry	5.89
Agriculture	387.29

Table 14.2 Water supply volumes in 2006 (MCM/Year)

Type of sources	Volume
Groundwater	314.13
Surface water	29.55
Golab tunnel	6

It is evident that with about 90 % of water demand in agriculture, 8 % for drinking water demand and 2 % for water demand in industry compared to the total supply available, there is no balance between supply and demand in the region.

Effective parameters for the study area were next identified. Kashan has specific hydrological conditions, and the limitations of surface water imply that groundwater resources play an important role in its integrated water resources management and as such, aquifer storage and water balance are the important factors in the development of decision support systems.

Wastewater and unaccountable water are the other effective parameters associated with the allocation simulation in this area. Unaccountable water includes actual loss (leakage of water supply and distribution), visual loss (unauthorized usage, measurement errors and data records) and allowable expenses without income. The amount of this water for the urban area was estimated at around 30.93 % of water production and it increased significantly to 55 % for the rural area.

Wastewater from sewage treatment plants or sanitary wastewater may be consumed for other purposes such as agriculture, green space irrigation, etc. This type of water can provide some parts of non-potable demand. In Kashan, municipal wastewater stands at about 14.8 %, whereas this figure may be 88–91 % for villages. Investigation of wastewater is therefore important both in terms of quality and quantity.

Pursuant to the previous steps of the toolbox, the WEAP allocation model (Sieber et al. 2005) was chosen to simulate a system of supply and demand. The model was prepared for the year 2006 and calibrated for the amount of water needed in 2010. In order to generate appropriate scenarios for deliberations and ranking, a number of criteria were considered, such as allocation priority, agriculture conditions, rate of population growth, demand management (reducing consumption per capita), wastewater reuse and loss management, increasing the amount of transferred water through desalination and changes in the industry sector.

The resulting scenarios were considered individually as well as in overlapping formats to evaluate their rankings using the MCDM approach based on both quantitative and non-quantitative criteria upon which implementation measures may be deduced and adopted.

14.3 Results

Individual scenarios were obtained for a projection to 2041, using the criteria mentioned above. Using Definite 2 Software (Janssen et al. 2000) these were ranked based on weights (priorities) allocated to unmet demands. Table 14.3 shows the weights given to different conditions for unmet demand within the MCDM procedure.

Results obtained indicate a displacement of rankings for each individual scenario, depending on the order of priority for unmet demand. However, it is anticipated that MCA2 would be the logical approach in light of prevailing water scarcity.

Table 14.4 shows the ranking obtained for practical scenarios for a given weight preference (MCA3) where sustaining the current conditions for 2041 (the reference scenario) is ranked 8th.

A combination of individual scenarios into groups of feasible practical actions provides a clear planning picture for decision makers and planners (Table 14.5).

Table 14.3 Weight effects based on unmet demand, Definite 2

	Label			
Order	MCA1	MCA2	MCA3	MCA4
1	Drinking water and industry unmet demand	Drinking water unmet demand	Drinking water unmet demand	Agriculture unmet demand
2	Agriculture unmet demand	Industry unmet demand	Agriculture unmet demand	Drinking water and industry unmet demand
3	–	Agriculture unmet demand	Industry unmet demand	–

Table 14.4 Scenario ranking for unmet demand in 2041 (MCM)

Rank	Drinking water	Agriculture	Industry	Scenario
1	0.38	226.6	6.49	Wastewater reuse and loss management
2	0.93	221.52	6.49	Wastewater reuse
3	26.32	311.87	21.49	Demand management
4	39.59	310.4	0.00	Reduce industrial activity
5	39.59	35.53	21.49	Reduce cultivated regions
6	33.27	306.92	21.49	Increasing the amount of transferred water through desalination to 12 MCM
7	36.27	308.94	21.49	Increasing the amount of transferred water through desalination to 9 MCM
8	39.59	311.4	21.49	Initial condition (Reference scenario)
9	39.59	311.4	32.92	Develop industry sector
10	39.59	1357.5	21.49	Increase cultivated regions
11	72.91	311.17	21.49	Higher population growth

Table 14.5 Scenarios combinations ranking in 2041 (MCM), MCA3

Rank	Drinking water	Agriculture	Industry	Combined Scenarios
1	0.35	29.20	11.52	Higher population growth + Demand Management + Increasing the amount of transferred water through Desalination to 12 MCM+ Develop industry sector + Reduce cultivated regions + Wastewater reuse and loss management
2	0.88	220.95	6.49	Higher population growth + Demand Management + Increasing the amount of transferred water through Desalination to 12 MCM+ Wastewater reuse
3	0.35	1054.04	6.49	Higher population growth + Increase cultivated regions + Demand and loss Management + Increasing the amount of transferred water through Desalination to 12 MCM
4	0.37	1054.6	6.49	Increase cultivated regions + loss management
5	0.93	1049.5	6.49	Increase cultivated regions + Wastewater reuse
6	39.59	311.4	21.49	Reference scenario (Initial condition)
7	26.32	1357.62	21.49	Increase cultivated regions + Demand Management
8	46.04	306.92	21.49	Higher population growth + Demand Management + Increasing the amount of transferred water through Desalination to 12 MCM
9	52.04	311.17	21.49	Higher population growth + Demand Management

14.4 Conclusions

In recent decades, water scarcity, particularly in arid and semi arid regions of the word, has imposed serious challenges on communities. This in turn has necessitated an integrated water resources management approach to ascertain sustainable development while allocating resources both for increasing current demands as well as safeguarding precious resources for future generations. Development of an IWRM toolbox has hence been the aim of this study to facilitate appropriate decision-making. The toolbox has various modules covering data preparation, use of allocation models and multi-criteria decision-making and has been utilized for the city of Kashan in central Iran. Investigations based on the toolbox have indicated that the projection of current conditions to the year 2041 will result in severe scarcity. Three main sectors, namely drinking water needs, agricultural water needs and industrial water needs have been considered and different criteria have been introduced for the regions sustainable development. Viable water resource scenarios generated using the toolbox have indicated that although reduction in agricultural activities will alleviate water scarcity, it will while introduce socio-economical impediments. Other feasible measures, such as demand management, wastewater reuse, loss management, industrial water demand management and inter-basin water transfer through desalination have been investigated individually and in combinations to generate a multi-scenario situation based on varying priorities of water use. A multi-criteria decision-making tool has been employed as part of the IWRM toolbox to further investigate and rank different combinations of measures in an ascending order of most viable measures based on sustainable development of the region. This exercise has been very worthwhile and conclusive in the presentation of an integrated approach in a community faced with many challenging and contradicting needs in the face of limited resources. The toolbox approach has provided the generality for application of the approach in other regions with individual specific characteristics.

List of Figures

References

Barrow CJ (2001) A review of the global experience of integrated river basin development planning and management. Integrated river basin management. Oxford University Press, Oxford

Biswas A (2001) World water forum: in retrospect. Water Policy 3:351–356

Diep VN, Khan HN, Son MN, Hanh VN, Huntjens P (2008) Integrated water resources management in the Red River Basin – problems and cooperation opportunity. In: Hydrological modeling in arid and semi-arid Area. Cambridge University Press, Cambridge

Janssen R, Herwijnen VM, Beinat E (2000) Decision support system for a finite set of alternatives (DEFINITE) manual. Institute for Environmental Studies Vrije Universiteit, Amsterdam

Khawaji AD, Kutubkhanah IK, Wie JM (2008) Advances in seawater desalination technologies. Desalination 221:47–69

Labadie JW (1995) MODSIM: river basin network flow model for conjunctive stream-aquifer management program user manual and documentation. Department of Civil Engineering, Colorado State University, Fort Collins

Labadie JW (2007) MODSIM 8.1: river basin management decision support system. User manual and documentation. Department of Civil and Environmental Engineering, Colorado State University, Fort Collins

Loucks DP, Stedinger JR, Dijkman PM (2005) Water resources systems planning and management. An introduction to methods, models and applications. United Nations Educational, Scientific and Cultural Organization (UNESCO), Paris

Mannina G, Viviani G (2009) Parameter uncertainty analysis of water quality model for small rivers. Dipartimento di Ingegneria Idraulica ed Applicazioni Ambientali, Università di Palermo, Palermo

Mike Basin (2003) Guide to getting started tutorial; DHI water and environment, Denmark

Ospina-Noreña JE, Gay-García C, Conde AC (2009) Analysis of the water supply-demand relationship in the Sinú-Caribe basin, Colombia, under different climate change scenarios, México

Ribasim Version 6.33 (2005) River basin simulation model; Quick start guide; Delft Hydraulics

Rubin H, Rubin A, Reuter C, Köngeter J (2006) Sustainable Integrated Water Resources Management (IWRM) in a semi-arid area. Int J Environ Cult Econ Soc Sustain 2(3):165–180

Sieber J, Swartz C, Huber-Lee A (2005) User guide for WEAP21. Stockholm Environment Institute/Tellus Institute, Boston

Srdjevic B, Srdjevic Z, Zoranovic T, Potkonjak S (2004) Advanced decision support tools in agricultural and water management, Faculty of Agriculture, University of Novi Sad, Serbia and Montenegro. In: International conference on sustainable agriculture and European integration processes Novi Sad, Serbia and Montenegro

Wil NM, van der Krogt (2005) User manual; RIBASIM Version 6.32; Delft Hydraulics

Chapter 15
Governance Strategies for Addressing Sustainability Risks of Seawater Desalination in the Arabian Gulf

Aliyu Salisu Barau

Abstract Presently, seawater desalination seems to be the most reliable source of clean water in the Arabian Gulf. However, multiple brine discharges from desalination plants endanger marine organisms. Similarly, high energy consumption and greenhouse gas emissions raise sustainability concerns. By and large, lack of an integrated regional governance framework hampers realising sustainable desalination industry. This paper seeks to unravel complexities in sustainability dimensions of the Gulf's desalination projects. It also identifies pathways for overcoming risks of fragmented governance of the desalination industry. The study applied principles of complexity theory, the DPSIR model, and the Earth System Governance paradigm to prop its theoretical and methodological underpinnings. The paper identified 29 factors (natural, institutional, technological, policy and demographic) that underscore dynamics of the Gulf desalination industry. The recommendations made for sustainable pathways were based on principles of the Earth System Governance and interdisciplinary team building. This study stressed that sustainable future for the desalination industry in the Gulf region would depend on building a unified, and comprehensive desalinated water governance framework.

Keywords Seawater desalination • Earth system governance • Sustainability • Interdisciplinarity

15.1 Introduction

According to Reynolds (1993), the Arabian Gulf covers an area of 239,000 km^2 encompassing the Gulf of Oman and the Straits of Hormuz. Its length extends to some 1000 km; while its maximum and minimum widths are approximately 338 and 56 km respectively. The six countries that form its political geography through the Gulf Cooperation Council (GCC) are: Bahrain, Kuwait, Oman, Qatar,

A.S. Barau (✉)
Research Fellow, Earth System Governance Project, IPO, Lund University, Lund, Sweden

Faculty of Built Environment, Universiti Teknologi Malaysia, Johor Bahru, Johor, Malaysia
e-mail: aliyubarau1@yahoo.co.uk

M. Baawain et al. (eds.), *Recent Progress in Desalination, Environmental and Marine Outfall Systems*, DOI 10.1007/978-3-319-19123-2_15

Saudi Arabia and the United Arab Emirates. Apart from Saudi Arabia, about 90 % of the Gulf states population live in urban areas whose urbanisation rate is 3.5–6 % per annum (World Bank 2005; Malecki and Ewers 2007). At the same time, some 15 million people live along the Gulf coastal zones and this trend keeps rising (Burt et al. 2013; Nadim et al. 2008). The Gulf is famous for its huge oil resource deposits; it is equally infamous for its acute scarcity of water resources.

Ackerman (1969)'s typology of global population-resource-technology regions marked the Gulf region as *Arctic-Desert type*. According to this classification, the Gulf is a technology-deficient region characterised by a very low population and technology. More recently, scientists vouched that most countries in the Gulf have high water debts (amount of groundwater withdrawal higher than natural inflow); while their absolute water scarcity value is estimated at <500 m^3/capita/year (Eales and Clifford 2013). The Gulf also has the world's poorest water per capita, and the highest per capita cost of water supply (Alsharhan et al. 2001). Ironically, water consumption in the region is one of the highest in the world. According to statistics of the World Bank (2005), the Gulf States have average daily water consumption per capita that ranges between 300 and 750 litres.

As Ackerman (1969) postulated, technology transfer could aid and improve life in the technology-deficient regions. Thus, seawater desalination technology was introduced to the Gulf in 1950s and became a more prominent alternative source of water for the region in 1970s (Alsharhan et al. 2001). For instance, desalination plants provide about 55 % of urban water needs in eastern Saudi Arabia. The GCC member states desalination plants have a total capacity of producing more than 11 million cubic metres per day or about 45 % of the global total desalinated water production (Latteman and Hopner 2008). Addressing water scarcity scenarios and sustainability risks of desalination plants would need alternative approaches for future stability and sustainability in the Gulf. For instance, it is a common knowledge that brine discharges cause serious ecological damages (Alsharhan et al. 2001). Naturally, the Gulf region has a very high temperature and salinity (Sheppard et al. 2010), and thus, its desalination plants exacerbate this through continuous discharge of high saline brine (Raouf 2009). This means the region needs to meet water and sustainability through sustainable technologies and environmental governance approach.

The Dublin Principles underscore the need for integrated water management approach (United Nations 1992). Recently, researchers espoused integrated approach in managing water resources (World Bank 2005; Khan 2008; UNEP 2012). Integrated approach in water resources management may not suffice without linking it to the concept of environmental governance. Parkes et al. (2010) proposed an integrated governance of water resources in a way that fosters sustainability within socio-ecological prism. The term governance according to Biermann [with Turney] (2013) may refer to "decentralised policies, non-hierarchical decision-making, and the involvement of both public and private actors" for steering society towards sustainability of the Earth system. In addition, he observed that social scientists are not comfortable with notions like environmental policy because it does not adequately address human interactions with planetary system, just as the term environmental management connotes technocratic, top-down and centralised management. As alternative, the new paradigm of the Earth system governance underscores need for engagement, holistic approach, conflict resolution, decision making, and collaboration with all stakeholders in order to achieve efficient, legitimate and equitable transitions to sustainability (Biermann 2012; Carvalho and Fidélis 2013).

The seawater desalination plants of the Gulf region create opportunities for water security and pose sustainability risks. At present, holistic and effective governance approach in tackling desalination risks are lacking at national let alone regional level. Previous efforts made through the GCC include *System for Conservation of Water Resources* policy draft signed in 1998. However, this policy draft was neither holistic, effective nor implemented (Alsharhan et al. 2001). Prior to this, the Regional Organisation for the Protection of Marine Environment (ROPME) was established in 1979 to address marine pollution in the Gulf littoral areas (El-Habr and Hutchinson 2008). In principle, ROPME has an integrated scope that covers coastal management, marine pollution by oil, urban and industrial activities. Unfortunately, ROPME is constrained by financial and technical challenges, absence of sustainability standards to check upstream and downstream impacts of industrial discharges, as well as lack implementation of many rules and regulations that establish the ROPME (Nadim et al. 2008). The present situation makes the risk of desalination plants more visible and daunting particularly by considering the future in terms of population growth, urbanisation, and increasing sustainability risks.

The Gulf region attracts attention of global desalination research community among others through the Middle East Desalination Research Center (MEDRC) which has made giant strides in building global partnerships and collaborations for state of the art scientific research in water desalination (Delyannis and Belessiotis 2010). However, this hardly makes any meaningful long-term sustainability impact considering scores of challenges that Sheppard et al. (2010) highlighted. These include resistance to holistic approach and very limited culture of sharing information among agencies across the Gulf States. This situation sharply contrasts with what obtains in other major regional groupings such as the European Union where environmental governance approach is strongly established. For instance, the European Water Framework Directive is a very comprehensive framework set up to support sustainable water resources governance in Europe (Borja et al. 2006).

It is obvious that, the Gulf countries will continue to depend on desalination plants as principal sources of water security particularly for urban areas. Many researchers from different disciplinary backgrounds have conducted numerous studies on prospects of innovative technologies in mitigating ecological hazards as well as costs of desalination technologies (Altaee et al. 2014; Ghaffour et al. 2013; Shatat et al. 2013 etc.). However, sustainability of seawater desalination is a complex issue that goes beyond alternative technologies or cost reduction. Regional integrated regulatory issues, population change and urbanisation dynamics are important determinants of water security for the future. Unfortunately, many physical scientists and policymakers often relegate such issues to the background.

It is bothersome that desalination plants in most of the Gulf countries are located within close proximity to each other on the shallow shores of the Gulf. Thus, the risk of convergence and redistribution of discharges into the waters from all countries is very high. In addition to that, water footprints of many important towns and capital cities in the Gulf are likely to continue rising. In view of this, some researchers argued that governance or policy issues are crucial to addressing complex issues surrounding water desalination in the Gulf (Dawoud and Al Mulla 2012; Merwe et al. 2013). The present policy response for addressing desalination plants discharge in the Gulf can be described as rather patchy and lacking in adequate governance strategies for sustainable future water demand and supply. It is obvious that there is no common policy among the Gulf countries for tackling sustainability dimensions of the desalination plants. Perhaps, this is a common problem of the Asia-Pacific region, where, in the opinion of King et al. (2012), regional environmental institutions are weak and unconsolidated.

Based on the insights drawn from the above paragraphs, it is apparent that there is interplay of different issues surrounding desalination industry in the Gulf. Thus, complex system theory is needed to disentangle issues and might be helpful for achieving innovative solutions as well. Ay et al. (2011) defined complex systems as collective behaviour of systems whose parts entail emergence of properties that can hardly, if not at all, be inferred from properties of the parts. Another view on complexity suggests that socio-economic, infrastructural, cultural, energy and material stock are interdependent facets that create complex systems in space and time (Bettencourt 2013; Zellner et al. 2008). Because complex systems are embedded into each other they are also referred to as complex adaptive systems. Keshavarz et al. (2010) listed the five characteristics of complex adaptive systems as: nested systems, continuous, networks, emerging, and unpredictable.

This study seeks to identify complexities in sustainable governance of seawater desalination in the Gulf region. The dual objectives of this study are: (1) to map out complexities that characterise desalination industry in the Gulf; (2) to identify environmental governance strategies for long-term sustainable solutions. The Earth system governance approach is a promising strategy that can complement existing sustainability research on managing the intake-and outfall locality, concentrated brine discharges and their effects at local, national and the Gulf region as a whole.

15.2 Methodology

The methodology of this research reflected the objectives of the study. From the experiences of developed regions, understanding complexity of water resources dynamics is crucial to planning for sustainable water governance. Sustainability researchers applied some frameworks to unravel complexities in water resource management initiatives. For instance, Borja et al. (2006) applied the Driving force-Pressure-State-Impact-Response (DPSIR) model to assess ecological and policy issues in the context of the European Water Framework Directive (WFD). Similarly, Gregory et al. (2013) also coupled the DPSIR model with complex adaptive systems theory to structure socio-ecological problems associated with marine ecosystems management challenges. According Tscherning et al. (2012), researchers applied the DPSIR to among others develop indicators to identify complexities. In meeting the first objective of this study, the researcher applied the DPSIR model in order to map out complexities in water desalination industry and sustainability. Thus, for this study, *Driving forces* in question are socio-economic or geographical conditions that create increasing demand for water; how did that exert *Pressures* on the ecosystems; which cause measurable change in the *State* of the environment; what *Impacts* did that create; and what measures were taken as *Reponses* from society, scientists or policymakers.

The second leg of the methodology identified solution pathways which Ignaciuk et al. (2012) stressed was imperative for science and policy for sustainability. Hence, the researcher employed the five research tools of the Earth system governance namely, *Architecture*, *Agency*, *Accountability*, *Access and Allocation*, *Adaptiveness* and four crosscutting issues (*Knowledge*, *Power*, *Norms* and *Scale*) in order to identify key questions in the Gulf water desalination industry and the needed solutions. These Earth system research paradigm tools and the relevant questions raised for this research are explained in Table 15.1 based on Biermann et al. (2009). As common with most DPSIR studies, this study drew its views through peer-reviewed desktop sources.

15.3 Results and Discussion

Complexity associated with the Gulf's desalination industry was a function of numerous factors. Based on the DPSIR model applied (Table 15.2), this study identified a total of 29 different factors that constituted complex picture of the operation of the desalination industry in the Gulf region. Each of these factors directly or indirectly associates with others to affect the said complexity. In other words, there is interdependency or inter-linkages among the factors that surround the desalination industry (Table 15.2).

This is consistent with the underpinnings of the complexity theory and complex adaptive systems (Ay et al. 2011; Keshavarz et al. 2010). Thus, it is important to point that, the geography of the Gulf determines the region's hyper-aridity which Ackerman (1969) identified as *Desert-Arctic* typology. Nevertheless, technological innovations and water desalination in particular has helped the Gulf to attain a certain level of water supply for the present and future (Alsharhan et al. 2001; Latteman and Hopner 2008). However, considering the fact that the region has one of the highest water debt figure (Eales and Clifford 2013) and one of highest average daily water consumption (World Bank 2005), then it could be said that the Gulf water crises is from being over. These explanations are part of complexities of water desalination, population water needs, policy issues and governance of

Table 15.1 Earth system governance research analytical tools in the context of sustainability in the Gulf desalination industry

ESG analytic tools	Issues in sustainability of Gulf desalination industry
Architecture	What arrangements and governance hierarchies available for tackling desalination challenge?
Agency	Which public/private/international organisation is responsible for addressing desalination challenges and how does it discharge the responsibility?
Accountability	What is the quality and effectiveness of agencies and policies for managing existing challenges?
Access and allocation	How do public/private organisations plan to achieve balance between people water needs and ecosystem security
Adaptiveness	How does institutions adapt to new innovations for sustainability and how does that affect or helps environmental sustainability of regional water
Crosscutting themes	**Crosscutting themes questions**
Knowledge	How effective and knowledge based are decisions about development and management of desalination industry?
Norms	What role norms and values play in designing sustainability change issues and Gulf people and governments?
Scale	What is scale of institutions local, regional, scale of influence of stakeholders, environmentalists, bureaucrats etc.
Power	Where does power to change ecological threats of desalination industry lie? People? Agencies? Governments? Businesses? Individuals?

water resources as a whole. Already the GCC states have established a number of desalination plants along their shallow shores where many important urban areas are also closely located. Hence, it is essential to develop a regional framework for governance of water resources. Such framework should fuse social, economic, environmental, economic, demographic, institutional and ecological variables. This will go a long way in addressing the numerous challenges and disconnections between the Gulf desalination industry and sustainability imperative as many researchers pointed (Nadim et al. 2008; Raouf 2009; Dawoud and Al Mulla 2012). Based on the findings of previous studies, it is agreeable that integrated and interdisciplinary approaches help researchers and institutions to unravel complexities in governance of natural resources (Borja et al. 2006; Tscherning et al. 2012; Gregory et al. 2013). This study identified priority joint actions needed for the Gulf region desalination plants (Table 15.3).

The recommendations presented above (Table 15.3) merely identified solution pathways that might guide and facilitate achieving sustainable governance. Seawater desalination seems to be one of the most viable options for portable water supply in the Gulf. Presently, most of the Arabian Gulf countries face what Burt et al. (2013) call “rapid expansion of coastal urbanization” which increases water unprecedented fresh water demand. Based on the current state of scientific knowledge, seawater desalination remains the most reliable source of fresh water for the

Table 15.2 DPSIR framework derived factors of complexity in the Gulf desalination industry

Components	Indicators	References
Driving forces	(a) Natural water scarcity due to hyper aridity	(Ackerman 1969; Alsharhan et al. 2001; Nadim et al. 2008; Mohammad and Sidaway 2012; Burt et al. 2013)
	(b) Increasing coastal urbanisation	
	(c) Unprecedented urbanisation + population growth	
	(d) Government policy to attain water security	
Pressures	(a) High energy consumption	(Alsharhan et al. 2001; Arconada et al. 2013; Younos 2005; Nadim et al. 2008)
	(b) Increasing Brine concentrators	
	(c) Marine ecosystems degradation	
	(d) Energy consumption	
	(e) Land, noise, air and sub-surface pollution	
State	(a) The Gulf harbours 50 % of the world desalination plants	(Dawoud and Al Mulla 2012; Latteman and Hopner 2008; Merwe et al. 2013; Karagiannis and Soldatos 2008; Raouf 2009)
	(b) Process: Multi-stage flash distillation (MSF: 88 %) Reverse Osmosis (RO <10 %)	
	(c) Fragmented governance of water resource development	
	(d) Poor information sharing	
	(e) Lack of holistic approach	
	(f) Plants located on shallow shoreline of the Gulf	
	(g) Cities depend on desalination of plants for clean water	
	(h) Cost of desalination is very expensive	
	(i) There is lack of accurate and reliable scientific data on intake- and outfall locality, brine concentrate discharge and cumulative effects of desalination plants in the Gulf	
	(j) Desalination-specific criteria and monitoring are lacking in the Gulf region	
Impact	(a) Increased salinity, temperature and dissolved oxygen fluxes affect marine species distribution	(Raouf 2009; Burt 2013; Dawoud and Al Mulla 2012; Merwe et al. 2013)
	(b) Corrosion of plants discharge heavy metals	
	(c) Endangering of marine ecosystem	
	(d) Sharp decline in coral reefs	

(continued)

Table 15.2 (continued)

Components	Indicators	References
Response	(a) Environmental Impact Assessment	(El-Habr and Hutchinson 2008; Delyannis and Belessiotis 2010; Lattemann and Amy 2013, Ghaffour et al. 2013)
	(b) Best Available Technology	
	(c) Monitoring programmes	
	(d) ROPME	
	(e) Use of hybrid systems	
	(f) Utilisation of state of the art technologies and scientific research	

Table 15.3 Recommendations for sustainability pathways for desalination industry in the Gulf

ESG Analytic Tools	Recommendations
Architecture	Establish broad based sustainable desalination policy for the GCC states to ensure uniform compliance to agreed-upon sustainable rules and best practices (local, regional, global).
Agency	Establish inter-agency and interdisciplinary agency and working groups to ensure compliance and support research documentation and dissemination of information among all stakeholders.
Accountability	Compile comprehensive annual reports on the state of Gulf waters and impact of desalination plants and other human related impacts and share such reports to all regional and global stakeholders.
Access and allocation	All the GCC countries should ensure that public and industrial water resources needs are met with full compliance to principles of wise use and minimal wastage.
Adaptiveness	Research and development should promoted holistic and participatory approach to respond to future sustainability uncertainties.
Crosscutting Themes	
Knowledge	It is essential to undertake science communication projects to bridge gaps between scientists, policymakers, civil societies and the public on sustainability dimensions of desalination plants.
Norms	Norms and values of the Gulf people who are Arab Muslims under monarchy system of government must be taken into consideration in building sustainable solutions.
Scale	It is imperative to balance water demand and supply in the context of costs and limited availability of water.
Power	The sphere of influence should be opened in a way decision making concerning desalination plants and the water they produce involves the public, businesses, researchers, civil society etc.

population. It is interesting that, this technology in spite of its sustainability risks is also being considered as a good alternative for some of advanced cities (Perth-Melbourne) in a bid to achieve water security (Werbeloff and Brown 2011). Nevertheless, seawater desalination in these Australian cities is designed based on integrated approach to water resources governance. The recommendations given in Table 15.3 are based on the Earth system governance paradigm and complex adaptive system theory. The prospect of this approach for the problem of fragmented seawater desalination governance might foster future sustainability in the Gulf if it involves people, private sector, scientists, civil society, governments, and international charters. The problem of fragmentation of institutions and policies on water resources development in most of the Gulf States is detrimental to sustainability of the Gulf water resources including the desalinated water (Raouf 2009; Dawoud and Al Mulla 2012). Considering the shallowness of the Gulf waters and other pressures exerted by the oil industry and increasing urbanisation and the natural salinity of the Gulf, it is clear that the future of desalination industry might not last to meet the increasing demands, or withstand the sustainability challenges.

Admittedly, technological advancement has made the cost of desalination to go lower than it has been in previous years (Karagiannis and Soldatos 2008; Ghaffour et al. 2013). However, this does not mean that unlimited supply of clean water in the Gulf is guaranteed. Innovative adaptation strategies are needed in order to identify priority action and strategies that could help in realising sustainable water resources governance in the region. For instance, as an oil rich region, the Gulf does need to waste so much water resources on agriculture. The Gulf's agricultural sector consumes 75–85 % of its water supply and that up to 30 % of total water supply is loss through pipe network distribution lapses (Alsharhan et al. 2001). Seemingly, this figure ignores pressures exerted by rising urbanization in the Gulf. Nevertheless, this is where the role of crosscutting issues play out. For instance, traditional food systems of the rural people of the Gulf are part of their age-long traditions and value systems. Hence, these food systems cannot easily be replaced by imported foods due to their cultural values. Importantly, it is important to observe that the traditional food production systems may not be the real sources of excessive water consumption when compared with urban household footprints. Based on this argument, all stakeholders need to understand that sustainability depends on synergistic approach (Mezher et al. 2011; Dawoud and Al Mulla 2012).

In order to implement the suggestions offered (Table 15.3), this study recommends the need to pool experts from various relevant disciplines in order to develop a hackathon for sustainable desalination industry in the Gulf. According to Gaffney (2013), the concept of hackathon entails an intense burst of real-time collaborative and interdisciplinary sharing of ideas to promote resilience and sustainability. The sustainability hackathon concept is a new practical initiative pioneered by the Stockholm Resilience Centre through its project captioned Stockholm Innovation for a Resilient Future (SHIFT). The pioneer hackathon captioned *Switched on Nature* was held at the Stockholm Resilience Centre and it involved input of computer, social, and natural scientists, media experts and entrepreneurs who jointly modelled urban sustainability pathways within 48 h of collaborative work in December 2013. In view of the success of the Stockholm hackathon, this study recommends adaptation of the same approach in order to build a hackathon for sustainable desalination industry for the Gulf region.

15.4 Conclusion

In the opinion of this study, the search for sustainability of desalination industry should not be limited to mitigation or reducing destructive impacts of brine pollution, emissions, energy consumption, or impacts on marine biodiversity. Though, sustainable desalination industry should target all of these, but it also need to go beyond these issues. This is because sustainability is complex and entails urban and population growth issues, policies, ecosystem vulnerability, cleaner technologies, information-sharing, collaboration, public and private involvement, knowledge-based decision-making, and conflicts management. Though the Gulf region is constituted by sovereign countries, its shores and waters are insensitive to political boundaries. As the future of water supply in the Gulf remains in the desalination industry, it becomes imperative to design effective and unified strategies that encompass technology and environmental governance in broad way.

References

Ackerman EA (1969) Population, natural resources, and technology. Ann Am Acad Polit Soc Sci 369:84–97

Alsharhan R, Nairn AEM, Bakhit DW, Alhajari SA (2001) Hydrogeology of an arid region: the Arabian Gulf and adjoining areas. Elsevier, Amsterdam

Altaee A, Mabrouk A, Bourouni K, Palenzuela P (2014) Forward osmosis pretreatment of seawater to thermal desalination: high temperature FO-MSF/MED hybrid system. Desalination 339:18–25

Arconada B, Delgado P, García A (2013) Minimizing environmental risks on constructing marine pipelines: Aguilas desalination plant. Desalin Water Treat 51(1–3):246–261

Ay N, Olbrich E, Bertschinger N, Jost J (2011) A geometric approach to complexity. Chaos 21 (3):037103

Bettencourt LMA (2013) The origins of scaling in cities. Science 340(6139):1438–1441

Biermann F (2012) Planetary boundaries and earth system governance: exploring the links. Ecol Econ 81:4–9

Biermann F (2013) What is earth system governance? Interview with Jon Turney. http://www.futureearth.info/2013-jul-15/what-earth-system-governance. Viewed on 17 July 2013

Biermann, F, Betsill MM, Gupta J, Kanie N, Lebel L, Liverman D, Schroeder H (2009) Earth system governance: people, places and the planet. Earth System Governance Report 1, Lund

Borja A, Galparsoro I, Solaun O, Muxika I, Tello EM, Uriarte A, Valencia V (2006) The European water framework directive and the DPSIR, a methodological approach to assess the risk of failing to achieve good ecological status. Estuar Coast Shelf Sci 66(1–2):84–96

Burt JA (2013) The growth of coral reef science in the Gulf: a historical perspective. Mar Pollut Bull 72(2):289–301

Burt JA, Feary DA, Cavalcante G, Bauman AG, Usseglio P (2013) Urban breakwaters as reef fish habitat in the Persian Gulf. Mar Pollut Bull 2(2):342–350

Carvalho TM, Fidélis T (2013) The relevance of governance models for estuary management plans. Land Use Policy 34:134–145

Dawoud MA, Al Mulla MA (2012) Environmental impacts of seawater desalination: Arabian Gulf case study. Int J Environ Sustain 1(3):22–37

Delyannis E, Belessiotis V (2010) Desalination: the recent development path. Desalination 264 (3):206–213

Eales A, Clifford M (2013) Sustainability and engineering. University of Nottingham, Smashwords

El-Habr HN, Hutchinson M (2008) Efforts of regional and international organisations in reducing levels of pollution in the Gulf. In: Abuzinada AH, Barth HJ, Krupp F, Böer B, Al Abdessalaam TZ (eds) Protecting the Gulf's marine ecosystems from pollution. Birkhäuser Verlag AG, Basel

Gaffney O (2013) Hacking for urban sustainability. http://www.futureearth.info/2014-jan-9/hacking-urban-sustainability. Accessed 1/2/2014

Ghaffour N, Missimer TM, Amy GM (2013) Technical review and evaluation of the economics of water desalination: current and future challenges for better water supply sustainability. Desalination 309:197–207

Gregory AJ, Atkins JP, Burdon D, Elliott M (2013) A problem structuring method for ecosystem-based management: the DPSIR modelling process. Eur J Oper Res 227(3):558–569

Ignaciuk A, Rice M, Bogardi J, Canadell JG, Dhakal S, Ingram J, Leemans R, Rosenberg M (2012) Responding to complex societal challenges: a decade of Earth System Science Partnership (ESSP) interdisciplinary research. Curr Opin Environ Sustain 4(1):147–158

Karagiannis IC, Soldatos PG (2008) Water desalination cost literature: review and assessment. Desalination 223(1–3):448–456

Keshavarz N, Nutbeam D, Rowling L, Khavarpour F (2010) Schools as social complex adaptive systems: a new way to understand the challenges of introducing the health promoting schools concept. Soc Sci Med 70(10):1467–1474

Khan NY (2008) Integrated management of pollution stress in the Gulf. In: Abuzinada AH, Barth HJ, Krupp F, Böer B, Al Abdessalaam TZ (eds) Protecting the Gulf's marine ecosystems from pollution. Birkhäuser Verlag AG, Basel

King P, Mori H, Kipp R (2012) Asia-pacific, green economy, and institutions for sustainable development. In: Greening governance in Asia-Pacific. Institute for Global Environmental Strategies – IGES White Paper IV, Hayama

Latteman S, Hopner T (2008) Impacts of seawater desalination plants on the marine environment of the Gulf. In: Abuzinada AH, Barth HJ, Krupp F, Böer B, Al Abdessalaam TZ (eds) Protecting the Gulf's marine ecosystems from pollution. Birkhäuser Verlag AG, Basel

Lattemann S, Amy G (2013) Marine monitoring surveys for desalination plants – a critical review. Desalin Water Treat 51(1–3):233–245

Malecki EJ, Ewers MC (2007) Labor migration to world cities: with a research agenda for the Arab Gulf. Prog Hum Geogr 31(4):467–484

Merwe R, Lattemann S, Amy G (2013) A review of environmental governance and its effects on concentrate discharge from desalination plants in the Kingdom of Saudi Arabia. Desalin Water Treat 51(1–3):262–272

Mezher T, Fath H, Abbas Z, Khaled A (2011) Techno-economic assessment and environmental impacts of desalination technologies. Desalination 266(1–3):263–273

Mohammad R, Sidaway JD (2012) Spectacular urbanization amidst variegated geographies of globalization: learning from Abu Dhabi's trajectory through the lives of South Asian men. Int J Urban Reg Res 36(3):606–627

Nadim F, Bagtzoglou AC, Iranmahboob J (2008) Coastal management in the Persian Gulf region within the framework of the ROPME programme of action. Ocean Coast Manag 51(7):556–565

Parkes MW, Morrison KE, Bunch MJ, Hallström LK, Neudoerffer RC, Venema HD, Waltner-Toews D (2010) Towards integrated governance for water, health and social–ecological systems: the watershed governance prism. Glob Environ Chang 20(4):693–704

Raouf MA (2009) Water issues in the Gulf: time for action. The Middle East Institute policy brief no. 22. The Middle East Institute, Washington, DC

Reynolds RM (1993) Physical oceanography of the Gulf, Strait of Hormuz, and the Gulf of Oman -results from the Mt Mitchell expedition. Mar Pollut Bull 27:35–59

Shatat M, Worall M, Riffat S (2013) Opportunities for solar water desalination worldwide: review. Sustain Cities Soc 9:67–80
Sheppard C et al (2010) The Gulf: a young sea in decline. Mar Pollut Bull 60(1):13–38
Tscherning K, Helming K, Krippner B, Sieber S, Paloma SG (2012) Does research applying the DPSIR framework support decision making? Land Use Policy 29(1):102–110
UNEP (2012) The UN-water status report on the application of integrated approaches to water resources management. UNEP, Nairobi
United Nations (1992) Declaration of the international conference on water and the environment, Dublin, Ireland, 26–31 January 1992. http://www.un-documents.net/h2o-dub.htm. Accessed 2 Feb 2014
Werbeloff L, Brown R (2011) Security through diversity: moving from rhetoric to practice. Water Sci Technol 64(4):781–788
World Bank (2005) A water sector assessment report on the countries of the Cooperation Council of the Arab States of the Gulf. World Bank, Washington, DC
Younos T (2005) Environmental issues of desalination. J Contemp Water Res Educ 132:11–18
Zellner ML, Theis TL, Karunanithi AT, Garmestani AS, Cabezas H (2008) A new framework for urban sustainability assessments: linking complexity, information and policy. Comput Environ Urban Syst 32(6):474–488

Chapter 16
Impact of the Coastal Intake Environment on the Operating Conditions of Thermal Desalination Plants: A Case Study in the United Arab Emirates

W.E. Elshorbagy and A.H. Basioni

Abstract The energy required to operate coastal thermal desalination plants and thus the relevant costs are greatly affected by coastal hydrodynamics. A multi-layered hydrodynamic model was developed in this study for a coastal area in the United Arab Emirates to investigate such an effect. The simulation was carried out using a curvilinear grid model with sigma layers in the vertical direction which incorporates the transport of salt and temperature interactively with water dynamics. Several simulations were carried out to investigate the impact of the brine and warm cooling water discharges released from the desalination plant as well as other nearby industrial facilities using a three-dimensional advection-dispersion surface formulation. The model output was used to determine suitable locations and configurations for water intakes as well as outlets to maintain the temperature and salinity of the water introduced to the plant at optimum acceptable levels, so that maximum efficiency and minimum operation cost are achieved. A number of alternative scenarios were also considered to fully assess the problem. This includes extreme desalination operation scenarios in summer and winter, possible maximum release of warm water by other industrial facilities, and scenarios of the future expansion of plant production. Three alternatives were investigated including shifting the intake to new offshore locations, moving the outfall away from the intake area, and having the outfall discharge its effluent into further and deeper zones. Cost analysis was carried out for two scenarios to evaluate the operation cost in terms of chemical and energy cost. The first alternative that involved shifting the intake location about 1 km offshore was found to be the best option as it achieved the maximum reduction of chemical and energy costs for all tested scenarios when compared with the existing configuration. 2.5 % of the total annual cost, equivalent

W.E. Elshorbagy (✉)
Department of Civil and Environmental Engineering, United Arab Emirates University, AlMaqam, AlAin, United Arab Emirates
e-mail: Walid.shorbagy@uaeu.ac.ae

A.H. Basioni
ADCO, AADNOC, Abu Dhabi, United Arab Emirates
e-mail: abasioni1974@yahoo.com

M. Baawain et al. (eds.), *Recent Progress in Desalination, Environmental and Marine Outfall Systems*, DOI 10.1007/978-3-319-19123-2_16

to US $1.2 million is saved, taking into consideration a major expansion to the existing industrial facilities which will produce 10 times the existing effluent levels.

Keywords Coastal hydrodynamics • Desalination • Intakes • Outlets salinity • Temperature

16.1 Introduction

Sufficient precautionary measures are usually taken in desalination plants to maintain maximum operational efficiency when transforming seawater to safe drinking water. However, depending on the local marine environment and layout of the inflow outflow water lines, the performance of the plant can be affected considerably. In the coastal water of the UAE, the natural level of salinity is very high (more than 40 parts per thousand (ppt)), while the water temperature in the shallow areas rises up to more than 30 °C during the summer (Elshorbagy et al. 2008, 2013). The discharges of brine wastewater and warm cooling water from the plants increases the water salinity and temperature in the vicinity of the plant intakes. Coastal flow dynamics and natural processes like evaporation, diffusion and dispersion, and mixing, all affect the transport of brine and warm water into the seawater that is subjected to tides, winds, and locally-generated eddies. Cooling the water used by coastal industrial facilities near many desalination plants in the UAE can cause a localized increase in water temperature. A careful selection of the intakes and outlets of these plants should therefore take into account the above considerations. The objective of the study is to determine the effect of the brine and the warm cooling water discharges released from the desalination plants as well as other nearby facilities by using a three-dimensional advection-dispersion surface model. The model outputs were used to determine suitable locations and configurations for the water intakes and outlets.

16.2 Methodology

A multi-layered hydrodynamic model has been developed for the coastal area, an industrial petroleum complex located about 230 km west of Abu Dhabi City that has a medium-size desalination plant. The study has been conducted using the Delft3D model, which considers a curvilinear grid model with sigma layers in the vertical direction, incorporating the transport of salt and temperature interactively with water dynamics. The boundary conditions of the model have been obtained from the simulation results of a regional model that simulates the dynamics of the entire Arabian Gulf. The bathymetry information has been collected from available universal charts known as "Admiralty Charts" (ATT 2001). Meteorological and other oceanographic information has been collected from various relevant authorities and literature. The average rate of water intake and discharge in the sea has been collected from the Petroleum Port Authority (2003), and any missing data has been estimated. The simulation was carried out for the summer and winter seasons by considering the change in solar radiation, variation of wind patterns, and warm water discharges from other nearby industries.

Field measurements for salinity and temperature, provided by recent studies, were collected at a number of locations in the coastal area. The water level and currents measured at selected locations for a reasonable period of time have been employed in the model. The hydrodynamic and transport models have been calibrated against the measured data. Comparison is made with water level data, current data, time-dependent salinity and temperature data. A number of optional scenarios are considered to fully assess the problem. These include extreme desalination operation scenarios in the summer and winter, possible maximum release of warm water by other industrial facilities, and scenarios of future expansion of desalination plant production. Other scenarios assess various intake and outlet configurations including their locations, depths, coastal angles and cross-sections. The salinity and temperature at the intake of the desalination plant calculated from each scenario have been utilized in estimating the operation costs of the multi-stage flash (MSF) desalination plant using an EXCEL sheet program. The calculated costs for the scenario under consideration allow the identification of the best configurations for intakes and outlets that satisfy maximum operation at a minimum operational cost.

16.3 Regional Model (Gulf Model)

The hydrodynamics of the southern Arabian Gulf plays has a significant effect on the coastal environment of United Arab Emirates where large cities and industries are located. The coastal waters are also busy with movements of oil tankers and their loading-unloading operations. Understanding coastal flow dynamics is essential for an investigation of the environmental processes of this region. The high rate of evaporation, water exchange with the Arabian Sea at the east and the seasonally varying wind field, all significantly contribute to the flow dynamic of the sea which is subject to tidal forcing through the straits of Hormuz. However, present knowledge of hydrodynamic behaviour of the southern gulf and its response to meteorological and oceanographic forces is relatively poor. The present study employs a three-dimensional numerical model to examine the sensitivity of the sea water flow in the UAE coast to tides, wind fields, salinity and to temperature.

The strategy adopted by this study is to employ a simulation of the entire Gulf area, calibrating it at a reasonable level, with the local model nested from the regional one. Since there are no available time-dependant data at the local boundary, the boundary conditions for the coastal model are extracted from the Gulf Model. The Arabian Gulf (approximately 1000 km by 200–300 km) slopes from the shallow United Arab Emirates coast to Iran at a depth of 80–100 m, narrowing at the

Qatar peninsula. Oceanographic features in the northern and southern regions of the Gulf are remarkably different. Complex physical dynamic phenomena are evidenced from a number of modelling and survey studies carried out in the Arabian Gulf. High solar radiation, the exchange of fresh water and its circulation driven by the wind and astronomical forces are the main contributory factors for such phenomena. Excessive evaporation occurs in the shallow coastal region of UAE leading to the creation of highly saline water (up to 46 ppt).

16.4 Local Model

The coastal flow pattern around the coastal industrial complex has been studied using the same 3-D hydrodynamic model. The study area is about 264 km^2 and partially sheltered from the open sea by islands and salt marshes. Such a configuration increases the risk of marine pollution near the industrial site. The obvious implications of such a development are the increased potential threats of pollution to the coastal ecosystem, as well as the effect of effluents, produced by the MSF desalination plant, on salinity and temperature which would affect the plant's performance. Discharge of industrial effluents, spillage in the port and the discharges of brine and of warm water may have a considerable impact on the marine environment. Hence, understanding the hydrodynamics of the area is crucial in order to assess the impact of ongoing activities. A three-dimensional model study has been conducted to understand the baseline hydrodynamic conditions of the coastal study area.

In order to establish the local model, a nesting process was carried out from the original overall Gulf model. A nesting process is used where the boundary conditions of a model are generated by a larger (overall) model onto a nested model. In principle, the nested boundary conditions are generated by bi-linear interpolation of computational results at monitoring stations of the overall model. The study area for the model is selected from the south of Sir Baniyas Island as illustrated in Fig. 16.1. The seaward extent of the model is about 15 km. The utilized model has an overall grid size of 87×60 and employs a curvilinear sigma system. Curvilinear grids should be smooth in order to minimize errors in the finite difference approximations. The vertical dimension is modelled in sigma co-ordinates with three layers. The bathymetry of the model is digitized from the navigational Admiralty Chart #3780. The central part of the area has a depth of up to 22 m while the areas to the east and the west are very shallow. The deep sections provide suitable entrance for large tankers. The model is calibrated against the water level and current data.

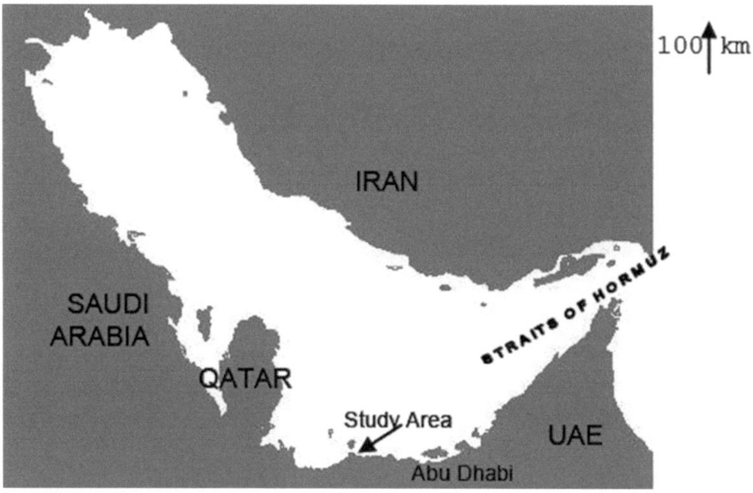

Fig. 16.1 The Arabian Gulf and the study area

16.5 Calibration

The model is calibrated by adjusting parameters within practical ranges to attain agreement with measured hydrodynamic data. All measurements are obtained from recent studies (Elshorbagy et al. 2006). Comparison of the measured and simulated water level data at coastal location near the study area is shown in Fig. 16.2, where a satisfactory agreement can be noticed. The match of the estimated water level with observation appears to be satisfactory. The current is generated under the influence of tidal forcing on the bathymetric variation, wind force and density gradient. A stream enters the local study area from the west towards the centre, south and east of the study area at the near shoreline. An outflow occurs in the east and northeast. As the tidal force and the gradient of temperature-salinity remain almost unchanged in both seasons (summer and winter), seasonal variation of wind force is apparently the prime source of such a flow pattern (Azam et al. 2006). It was observed that the current flow pattern is dominated by the wind direction.

Temperature-salinity dynamics is a three-dimensional process and as the atmospheric heat exchanges with the water-mass, evaporation occurs at the surface and the evolved density variation moves the water under the influence of gravity. It was observed that the higher salinity and temperature near the eastern shoreline are intensified due to the brine discharge and effluents from industries, especially from the eastern industrial facilities effluents.

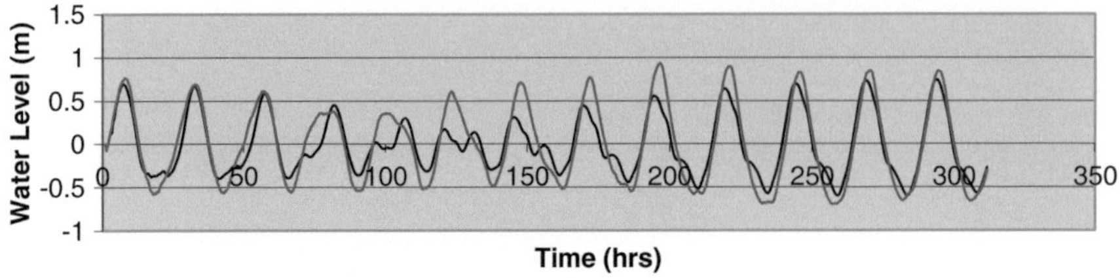

Fig. 16.2 Comparison of measured and simulated water level near the study area

16.6 Simulation Results

16.6.1 Considered Scenarios

The study evaluated the salinity and temperature at the desalination intake generated from several scenarios, the effect of wind magnitude and direction on the simulation results, and the impact on the MSF plant performance in terms of technical and commercial parameters. There are three main scenarios to be considered for both the summer and winter seasons. The first scenario uses the existing facilities, including all the basic data/information created to formulate the basic design model that considers the wind effect as the one caused by the most commonly-observed prevailing wind direction (north and north-west).

The second scenario considered the assumption of moderate expansion to the existing facilities, which leads to increase in the effluent discharge of the existing facilities by about five times the original base model. Third scenario considered the assumption of major expansion to the existing facilities, which leads to increase in the effluent discharge of the existing facilities by about 10 times the original base model. The salinity and temperature were investigated at the General Utilities Plant (GUP) desalination intake area. A summary of the above mentioned scenarios is illustrated in Table 16.1. The capacity of existing desalination plant at coastal area is about 64,000 m^3/day in the basic scenario considered herein. The discharge increases relatively with the increase in plant distillate capacity. It is assumed that the relation between plant capacity represented as distillate flow rate required and the effluent discharge representing the salinity and temperature is almost leaner. On the other word, the increase in distillate (water need) in addition to the expansion of adjacent industrial facilities increases the effluent discharge from all these industrial facilities into the sea. Consequently, this will lead to an increase in temperature and salinity at the shoreline area.

The capacity of the existing desalination plant in the coastal area is about 64,000 m^3/day in the basic scenario that has been mentioned above. The discharge increases relative to the increase in plant distillate capacity. It is assumed that the relationship between plant capacity, represented as Md (Distillate flow rate required), and the effluent discharge, representing the salinity and temperature, is almost leaner. In other words, the increase in distillate (water need) in addition to

Table 16.1 List of scenarios considered for the evaluation of the MSF desalination plant performance

S/N	Scenario	Description	Season
A	Basic	Current discharges from existing facilities (Q)	Summer
			Winter
B	Moderate expansion	Discharge increased five times (5Q)	Summer
			Winter
C	Major expansion	Discharge increased ten times (10Q)	Summer
			Winter

the expansion of adjacent industrial facilities increases the effluent discharge from all these industrial facilities into the sea. Consequently, this will lead to an increase in temperature and salinity in the shoreline area.

16.6.1.1 Hydrodynamic Results for Dominant Wind

Summer Results

The summer temperature and salinity that cover the majority of the year (about 7 months) range from 34.5 to 35.2 °C and from 45.5 to 46 ppt, respectively. The contour profile shows a higher temperature in the south close to the shoreline. The temperature in the west is the lowest. Similarly, the salinity is slightly higher in the south and becomes lower in the west. Such findings indicate that the basin receives less saline and cooler water from the north and the west near the open boundaries, while the more saline and warmer water forms at the shallower eastern side where industrial facility discharge brine water with a high temperature and greater salinity. The highest temperature and salinity are observed close to the shore in the east where the GUP and industrial facilities outfalls are located. Density force, convection and dispersion generate such movement of warm and highly saline water. While tidal circulation brings fresher water from the west and northwest which mixes with the central and lower basin water before leaving through the north, the tidal flow close to the east helps in mixing the water with industrial facility effluent where this mixtures moves towards the north of the basin. The impact of the industrial facilities is clearly noticed at the close area that surrounds the outfalls of these facilities, where the temperature and salinity concentration increases in that area.

On the other hand, a moderate expansion of the existing facilities is considered and its effect has been clearly observed. The temperature and salinity in the southeast are increased and even spread towards the north, where the possibility of adversely affecting the desalination intake (the area of concern to be studied) is more pronounced. In summer, an extreme scenario of major expansion of the exiting industrial facilities is considered, where the discharge was increased by ten times the discharge of all existing facilities. This scenario shows a very high temperature and salinity, not only at the shoreline, but the effect of this scenario covers a major part of the coastal shoreline and even spreads towards the north at the centre of the basin, yielding minimum success to the opportunity of proposing alternative intake and outfall configurations to overcome such a situation. It is noticed that a moderate expansion of the facilities increases the salinity and temperature at the intake from an average of 45.24–45.36 ppt and from 34.8 to 35.3 °C, respectively. The major expansion increases the salinity and temperature to an average of 45.48 ppt and 35.9 °C, respectively.

Winter Results

Winter salinities and temperatures at the desalination plant intake for the three scenarios are presented in Figs. 16.3 and 16.4.

The winter temperature and salinity, which spans over about 30 % of the year, ranges from 21.8 to 23.2 °C and from 45 to 46 ppt respectively. Again, the western side shows a relatively lower temperature and salinity as in the summer. As illustrated previously with the case of summer conditions, the existing industrial facilities had an influence on the salinity/temperature distribution, especially at the surface level of the shoreline. The wind plays a significant role in the mixing process, and the prevailing north-western wind greatly enhanced the overall circulation. The change in salinity for all three scenarios is found to be insignificant. However, the temperature at the intake desalination area increased from an average of 22.1 °C (existing facilities) to an average of 23 °C. The salinity increased from 45.10 to 45.65 ppt considering the basic and major expansion models.

16.6.1.2 Hydrodynamic Results for Reversed Wind Scenario

Eastward wind direction is considered in this scenario as east wind direction (90° Deg.). The objective of this scenario is to evaluate the effect of the reverse direction of the wind on the desalination plant operation cost. It is noticed that the temperature and salinity at the intake increased slightly in the reverse wind direction. This is attributed to the wind direction that plays a major role in transporting the effluent of the industrial and GUP facilities towards the intake location. However, the increased flow coming from the eastern boundary helps in diluting that effluent and alleviates the effect of brine discharge of industrial facilities on the intake. The reverse wind direction also reduces the impact of industrial facilities effluent on the intake condition. Since the effect of a reverse wind direction is limited, and it occurs for only about 23 % over the whole year, the alternative proposed scenarios discussed later will consider only the most dominant wind direction that represents the most prevailing conditions during the year.

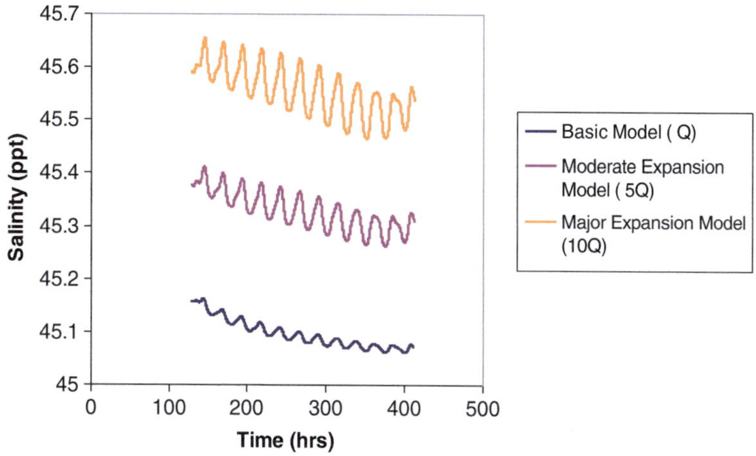

Fig. 16.3 Comparison of the salinity results considering the three different scenarios at the desalination intake in winter

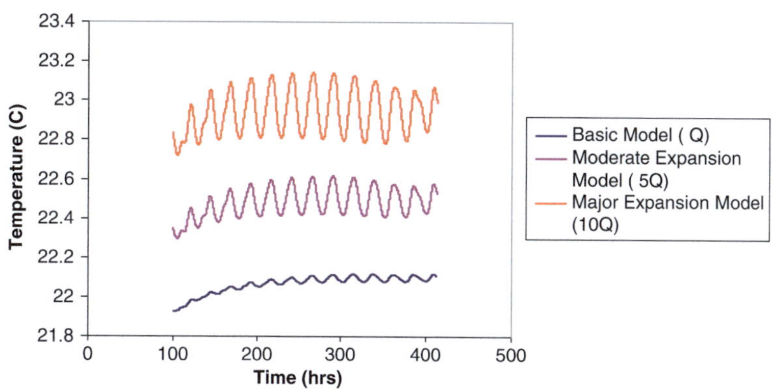

Fig. 16.4 Comparison of the temperature results considering the three different scenarios at the desalination intake in winter

16.7 Proposed Alternative Configuration

In order to minimize the negative impact of brine circulation towards the intake for the proposed scenarios, several proposals are investigated for the intake and outfall configurations. These proposals can be summarized as follows:

- Alternative 1 (Alt.1): Change the intake configuration by extending the intake location offshore to an area receiving cooler and less saline water.
- Alternative 2 (Alt.2): Change the outfall configuration by extending the desalination outfall in a location that has a lower impact on the intake location.

- Alternative 3 (Alt.3): Change the outfall configuration to discharge the effluent in a deeper zone.

In addition to the above alternatives, the salinity and temperature at different depths close to the intake location are evaluated. Results of the hydrodynamic simulations conducted for the three alternatives are summarized below.

16.7.1 Results of Alternative 1

Table 16.2 lists the average salinity and temperature as a result of adopting Alternative 1 and existing conditions. Table 16.3 shows the operation cost calculation for chemical and electrical power expenses (assuming constant pumping cost) for summer and winter. Table 16.4 shows the reduction in cost when Alternative 1 is employed for the three scenarios. It is noticed that the reduction in cost is insignificant. Salinity and temperature in the summer and winter are decreased. However, it is observed that the reduction in temperature in winter is quite noticeable compared to summer temperature.

16.7.2 Alternative 2

Extending the outfall location by about 1000 m away from the nearby zone in summer and winter improves the intake water properties slightly in terms of receiving less saline water which is cooler (See Table 16.5). Although the reduction difference in salinity and temperature is insignificant, the Alternative 2 emphasizes the importance of identifying the most suitable location and configuration for the outfall system, especially when an expansion of the existing industrial facilities is being planned. The results show that moving effluents away from the shoreline when considering the circulation process, wind direction and current movement, the new location helped in reducing the intake water salinity and temperature. It is worth mentioning that the bloom that flows from the easterly direction helped in lowering the effects of the brine effluent from the industrial facilities due to an effective mixing process.

Table 16.6 shows the comparison between existing conditions with the old outfall location and Alternative 2. Selecting the outfall location is considered to be an important factor in setting out the MSF location. By applying the two alternatives (Alternatives 1 and 2), the effect of the brine discharge will be minimized.

Table 16.2 Salinity and temperature results for Alternative 1 compared to existing conditions

	Summer Alternative 1		Winter Alternative 1	
Scenario	Salinity (ppt)	Temp. (°C)	Salinity (ppt)	Temp. (°C)
Basic	45.23	34.6	45.06	21.06
Moderate expansion	45.32	35.1	45.25	22.3
Major expansion	45.42	35.5	45.50	22.6
	Summer existing conditions		Winter existing conditions	
	Salinity (ppt)	Temp. (°C)	Salinity (ppt)	Temp. (°C)
Basic	45.24	34.8	45.07	22.1
Moderate expansion	45.36	35.3	45.30	22.5
Major expansion	45.48	35.9	45.60	23.0

Table 16.3 MSF desalination plant annual chemical and electric cost

	Basic model		Moderate expansion		Major expansion	
Scenario	Summer	Winter	Summer	Winter	Summer	Winter
Intake salinity (ppt)	45.24	45.07	45.36	45.30	45.48	45.60
Temperature (°C)	34.8	22.1	35.3	22.5	35.9	23.1
M_f (kg/s)	2092	2077	10,511	10,485	21,125	21,229
M_{cw} (kg/s)	8397	970	48,778	5098	111,914	11,046
M_b (kg/s)	1352	1337	6811	6785	13,725	13,829
M_{total} (kg/s)	11,841	4384	66,100	22,368	146,764	46,104
Power (KWh)	1871	692.89	10,447	3535	23,196	7287
Pumping cost (M$/year)	1.23	0.46	6.86	2.32	15.24	4.79
Chemical cost (M$/year)	3.59	3.56	18.02	17.98	36.22	36.40
Subtotal (M$/year)	4.82	4.02	24.89	20.30	51.46	41.19
Total (M$/year)	4.58		23.51		48.38	

Table 16.4 Total annual costs in US $ for three scenarios considering Alternative 1

Scenario	Existing conditions	Alternative 1	Difference in cost	Saving (%)
Basic	4,576,840	4,553,330	23,500	0.51
Moderate expansion	23,511,996	23,221,017	291,000	1.24
Major expansion	48,382,096	47,189,741	1,193,000	2.46

16.7.3 Alternative 3

Changing the outfall depth to discharge at a lower sea water level had no significant impact at the intake location since the effluent discharge had transported to saturate the full vertical alignment of the water profile especially in the coastline area.

Table 16.5 Salinity and temperature results in existing conditions for Alternative 2

Scenario	Summer		Winter	
	Salinity (ppt)	Temp. (°C)	Salinity (ppt)	Temp. (°C)
Basic	45.23	34.8	45.06	21.8
Moderate expansion	45.36	35.3	45.25	22.4
Major expansion	45.48	35.9	45.50	22.6
	Summer existing conditions		Winter existing conditions	
	Salinity (ppt)	Temp. (°C)	Salinity (ppt)	Temp. (°C)
Basic	45.24	34.8	45.07	22.1
Moderate expansion	45.36	35.3	45.30	22.5
Major expansion	45.48	35.9	45.60	23.0

Table 16.6 MSF plant total annual cost in US $ (chemical and electric cost) in winter for Alternative 2

Scenario	Basic conditions	Alternative 2	Difference in cost	Saving (%)
Basic	4,576,840	4,573,609	3,231	0.07
Moderate expansion	23,511,996	23,496,427	15,569	0.07
Major expansion	48,382,096	48,305,722	76,374	0.16

Changing the outfall location in terms of vertical alignment should be studied carefully and the non-hydrostatic aspects considered, especially in areas where many industrial facilities exist and where brine discharge is very high.

16.8 Effect of Salinity and Temperature on MSF Performance

Salinity and temperature at the desalination intake are considered one of the determining factors that affect the operation process of a desalination plant. Change in salinity may affect feed water requirements and change in intake seawater temperature may affect cooling water requirements. These changes in the feed and cooling water (M_f & M_{cw}) have an impact on pumping and chemical treatment requirements. The increase in salinity concentration at the intake, and the effect of this change on the MSF parameters, is illustrated in the previous model. The following relationship was obtained showing that the increase in the intake salinity concentration (X_f) will lead to an increase in feed water (M_f).

16.9 Cost Evaluation

Costs can be divided into capital and operational costs known to be largely site-specific. The performance ratio selected affects not only the investment cost, but also the subsequent operating cost. Calculations of unit product cost depend on the process capacity, site characteristics and design features. The system capacity specifies the size for various processing equipment, pumping units and for the required membrane surface area. Site characteristics have a strong influence on the type of pre-treatment and post-treatment equipment and on the consumption rate of chemicals. Process design features affect the consumption of electric power, of heating steam and of chemicals. Production cost is divided into direct and indirect capital costs and annual operating costs.

16.10 Conclusion and Recommendation

The current study highlighted the importance of selecting suitable locations for the intakes and outfalls of the MSF desalination plants. The selection was made so that the effects of environmental conditions of the intake feed water, mainly the salinity and temperature, upon the overall MSF performance is optimized and the operational cost is eventually minimized. Water dynamics and circulation prevailing in the coastal area dictate the status of temperature and salinity of the intake feed water. To achieve these goals, two hydrodynamic analyses were conducted in the study; the first for the entire Arabian Gulf and the second for the local study area of the coastal zone. That modelling was carried out using the well-known Delft3D model. The model results were favourably calibrated against the water levels of the Abu Dhabi coast. The local hydrodynamic model of the selected coastal area was also calibrated against water observations as well as against salinity and temperature field measurements. There was significant agreement between the simulated and observed variables.

The study evaluated the direct effect of intake salinity and temperature upon the operational cost of a MSF desalination plant. The operational costs were subdivided into two major groups; chemical costs and energy costs. Chemical costs were associated with the chemicals added at the feed point, such as anti-scalants, antifoams, etc. The energy costs were associated with lifting and pumping the feed water and the cooling waters throughout various desalination processes. The conducted cost analysis showed that an increase in salinity of about 33 % led to an increase in the chemical cost by about 33.5 % while an increase in the intake water temperature of about 20 % led to an increase in the energy cost by about 78 %. Such results indicated that the intake water temperature has a more pronounced effect on the annual operational cost of MSF desalination plants, thus showing energy costs are much higher than chemical costs.

In order to evaluate the impact of high salinity and temperature on the MSF overall performance and associated operational cost, three scenarios were considered. The first scenario considered existing conditions, while the second and third scenarios assumed moderate and major expansions, of the existing facilities as well as of their discharged effluents. The moderate and major expansions were represented by effluent discharges which increased by 5 and 10 times respectively the existing discharges from the current facilities. A cost analysis was carried out to evaluate the significance of each scenario and of its impact on the MSF performance. The annual operational costs estimated for the three scenarios were estimated at about US $4.58 million, US $23.5 million and US $48.4 million.

It is worth mentioning that such operational costs were based on hydrodynamic simulation results associated with the most frequent wind conditions prevailing in the area, the north-western wind of 5.2 m/s average magnitude. To decide on the optimum intake/outfall configuration, three alternative configurations were considered and evaluated. These alternatives are:

- Alternative 1: Change the intake configuration by extending the intake location offshore to an area receiving cooler and less saline water.
- Alternative 2: Change the outfall configuration by extending the desalination outfall in a location that has a lower impact on the intake location.
- Alternative 3: Change the outfall configuration to discharge the effluent in a deeper zone.

The hydrodynamic simulation results showed little overall reduction in the salinity and temperature achieved with all the tested alternatives and considering the three scenarios, especially the first scenario of existing conditions. This indicates that the current existing intake/outfall configuration has been carefully selected in anticipation of various environmental conditions and different future expansions. The sheltered location of the intake minimized the effect of effluent brine discharged from the existing industrial facilities on the desalination intake.

Alternative 1 did, however, achieve noticeable savings in annual operational costs compared to the existing configurations for moderate and major expansions (1.2 and 2.5 %). Such percentages even though they look small, represent major annual savings estimated at US $291,000 and US $1,193,000 USD respectively.

Studying the effect of the less-frequent reverse wind direction upon the MSF performance reflected a higher operational cost than in the case of dominant wind direction. This was tested for the case of the basic scenario (current effluents) of the existing intake/outfall configuration. Such a result suggests that further savings can be achieved with different alternatives in case of considering the reverse wind direction, especially if future expansion is limited to the western effluents of industrial facilities only. The results indicate that locating the intake near the shoreline in the case of large amounts of discharged effluents is not advisable since the coastal shore water is shallow and is highly affected by any increase in temperature and salinity due to a slow mixing process and low circulation. Although extending the intake configuration requires initial capital costs, the savings attained from the running operational cost can definitely cover such costs in a few years after which savings will be achieved. Finally, the study outcomes strongly recommend greater attention to be paid to studying and investigating the selection of the intake and outfall configurations, especially when there is a plan for a major expansion of the existing facilities. This selection should be made in line with a fuller understanding of the hydrodynamic phenomena prevailing in the area under study as well as an awareness of the relevant environmental conditions, bylaws and regulations.

List of Figures

References

Admiralty Tide Table (ATT) (2001) Hydrographer of the navy, vol 3. United Kingdom
Azam MH, Elshorbagy W, Nakata K (2006) Three dimensional modeling of the Ruwais Coastal Area of United Arab Emirates. ASCE/J Waterway Port Coast Ocean Eng 132(6):487–495
Elshorbagy W, Azam M, Taguchi K (2006) Hydrodynamic characterization and modeling of the Arabian Gulf. ASCE/J Waterway Port Coast Ocean Eng 132(1):47–56
Elshorbagy W, Azam MH, Nakata K, Terasawa T (2008) Temperature-salinity field of the shallow shelf of southern Arabian Gulf. Far East J Ocean Res 1(2–3):99–126
Elshorbagy W, Azam MH, Al-Hakeem A (2013) Temperature-salinity modeling for Ruwais Coast, United Arab Emirates. Mar Pollut Bull 73:170–182
Petroleum Port Authority (2003) Ruwais refinery division, TAKREER Company, ADNOC, Abu Dhabi, United Arab Emirates

Chapter 17
Phytoplankton Species and Associated Bacterial Populations in the Coastal Water of the United Arab Emirates

M.A. Khan, K.G.A. Qalandri, A. Sankaran, L.H. Adnani, and U. AlAlami

Abstract The aim of this study was to observe phytoplankton species and associated bacterial populations in the coastal waters of the United Arab Emirates (UAE). Samples were collected from the Umm al Quwain (UAQ) and the Dibba coastal water area. The samples were analyzed to determine the phytoplankton species and associated bacterial populations. A Fluorescent *in situ* hybridization (FISH) technique, using six sub-groups and a family-specific oligonucleotide probe, was performed on the ethanol and paraformaldehyde fixed samples. The microscope-based analysis of eight samples revealed the presence of both toxic and non-toxic phytoplankton species. The four dominant harmful dinoflagellate species detected in the samples were *Cochlodinium polykrikoides*, *Dinophysis caudata*, *Prorocentrum arenarium* and *Protoperidinium* spp. The FISH analysis showed a diverse bacterial community associated with the phytoplankton species. The overall findings of this study showed the presence of 4 non-toxic and 11 toxic phytoplankton species in the UAE coastal water samples. The FISH analysis indicated the presence of a highly diverse bacterial community belonging to five different sub-groups (Alpha, Beta and Gamma sub-classes of proteobacteria, High and Low G + C sub-groups) and one family (Enterobacteriaceae).

Keywords Bacterial communities • Fluorescent *in situ* hybridization • Harmful algal bloom • Phytoplankton species

---

M.A. Khan (✉) • A. Sankaran • L.H. Adnani • U. AlAlami
Department of Natural Science and Public Health, College of Sustainability Sciences & Humanities, Zayed University, P.O. Box 19282, Dubai, United Arab Emirates
e-mail: Munawwar.Khan@zu.ac.ae; abinaya.sankaran@gmail.com; lata.adnani@gmail.com

K.G.A. Qalandri
Emirates Authority for Standardization & Metrology, P.O. Box 48666, Dubai, United Arab Emirates
e-mail: khadija@esma.ae

M. Baawain et al. (eds.), *Recent Progress in Desalination, Environmental and Marine Outfall Systems*, DOI 10.1007/978-3-319-19123-2_17

17.1 Introduction

Harmful algal blooms (HABs) are caused by single-celled microscopic protists or dinoflagellates, also known as phytoplankton, found in the ocean. They are commonly known as 'red tides' because they change the natural sea water color to red (Yoo 2009). The rapid growth and proliferation of HABs causes fish mortality, seafood contamination and disruptions to microbial ecosystems and to coral reef ecosystems (Glibert 2007; Zhao and Ghedira 2014). Although these algal blooms are thought to occur naturally, they are also attributed to, and exacerbated by, anthropogenic activity (Glibert 2007). There are many types of phytoplankton in the marine environment that change the water color, but all these phytoplankton do not cause harmful algal blooms. The harmful algae form dense, visible patches near the coastal water surface and pose significant adverse impacts on the marine environment. HABs are often linked to significant economic losses through massive fish killings, shellfish harvest closures and the potential threat to humans from shellfish poisoning (Su et al. 2011).

HABs have been reported in several countries and are increasing in frequency and magnitude worldwide as a result of changes in oceanic climate, increased coastal eutrophication and enhanced long-distance dispersal in ballast water (Bauman et al. 2010). However, in recent years, a high incidence rate of red tide has been reported in the United Arab Emirates (UAE) (Fuad 2008; Bauman et al. 2010; Zhao and Ghedira 2014). In Dubai, several man-made coastal lagoons, bays, palm jumeirah, including Dubai Festival City on Dubai Creek were affected by the red tide (Arnold 2009; Landais 2008). The Environmental agency Abu Dhabi (EAD) has reported an increasing occurrence of HABs in Abu Dhabi waters (Al Qubaisi 2006). An extensive algal bloom event affected the Gulf of Oman and the Arabian Gulf from August 2008 to May 2009 and it caused massive fish kills, damaged coral reefs, restricted fishing activities, and forced closure of desalination plants in Oman and in the UAE (Richlen et al. 2010). In order to manage and mitigate the adverse impact of HABs, basic research is required. This could involve understanding the types of algae or phytoplankton (both toxic and non-toxic) present in the seawater samples before or after the HAB event and also their interactions with each other and other biological organisms like bacteria, viruses, protozoan in the seawater affected by red tide. These studies would help in controlling the HAB outbreak and persistence in the UAE in the long run.

As bacteria play an important role in nutrient regeneration and energy transformation in aquatic ecosystems (Azam et al. 1983), algal–bacterial interactions are of particular interest as they are considered to be potentially important regulators of algal growth and toxin production (Doucette et al. 1998). There have been several studies related to bacteria community interaction with the red tide causing algae in other countries (Romalde et al. 1990; Hold et al. 2001; Xin and Yun 2009). A diverse kind of bacteria found associated with dinoflagellates played a role in the accumulation of paralytic shellfish toxins (Hold et al. 2001). Vibrio species were found to be associated with red tide caused by *Mesodinium rubrum* (Romalde et al. 1990). The interaction and diversity of the bacterial population associated with red tide causing phytoplankton species in the UAE coastal waters is not yet known. The aim of this study was to identify the bacterial communities associated with the phytoplankton (both toxic and non-toxic) present in selected coastal areas near Umm Al Quwain in the UAE.

17.2 Methods

17.2.1 Sampling and Monitoring of Phytoplankton Species

Sampling locations were in Umm Al Quwain (UAQ) and Dibba coastal water area. The sampling was performed once per week over a period of 4 weeks. Samples were collected in sterile plastic 1000 ml bottles by vertically towing a plankton net of mesh size 20 μm which is suitable for red tide plankton size of 20 μm. The plankton net was drawn several times to the surface of the water until the water in the sample collector become colored by the algae. The hydrographic data of water were recorded using the equipment hydrolab (Hyrolab DS5) in situ. The following hydrographic parameters were recorded: temperature, depth of the water column, pH, salinity, dissolved oxygen, and chlorophyll concentration. The sample was stored as subsamples and a fresh sample (100–250 ml) in appropriately labelled sample containers and transported to the laboratory in a cold box under dark conditions. Monitoring of the sample was conducted within an hour of sample collection at the Ministry of Environment and Water laboratory in UAQ. A compound light microscope was used to determine the phytoplankton species and the phytoplankton abundance that caused the red tide. The Sedgewick-Rafter cell counting method was used for counting the phytoplankton and the sample was analyzed to identify the red tide causing plankton species in the seawater samples. For qualitative analysis, the plankton samples were prepared as wet and permanent mounts. For accurate identification of diatoms, the samples were cleaned using appropriate methods. The slides were analyzed under compound light microscopes (Olympus model IX2-100 fitted with DP72 Camera) and different magnifications were used, depending on size, to determine the general phytoplankton community composition and to identify it at the genus and species level. Photomicrographs were taken of the different identified species. Plankton genera and species were identified using a variety of bibliographic references (Tomas 1997; Rajan and Al Abdessalaam 2008; Al Kandari et al. 2009).

17.2.2 Monitoring of Bacterial Populations

The fluorescence in situ hybridization technique was performed on the red tide sample using the methods described earlier (Daims et al. 2005). The samples were fixed both in ethanol and 4 % (w/v) paraformaldehyde by procedures described earlier (Schuppler et al. 1998). In order to determine the total area occupied by all the bacteria present in the biomass sample, staining with DAPI (4, 6-diamino-2-phenylindole) dye was employed. DAPI would stain the entire DNA present in the sample and once they were stained, a visual signal would be emitted which would be captured by the imaging system (Daims et al. 2005). Hybridization with the samples was performed by domain, group/class, genus and species specific probes described in earlier studies (Amann et al. 1990; Manz et al. 1992; Schuppler et al. 1998; Daims et al. 1999; Meier et al. 1999). After the washing step, the slides were visualized by epifluorescence microscopy at 1,000× magnification on an Olympus BX-51 (Olympus, Japan) microscope equipped with a 100-W mercury lamp and filter sets for TRITC, FITC and UV for DAPI. Images were captured and analyzed using the Olympus digital camera system (Olympus, Japan). Pictures were processed as tagged-image file format (TIFF) files on a personal computer. All oligonucleotide probes were labelled at their 5′ end by tetramethyl rhodamine isothiocyanate (TRITC).

17.3 Results and Discussion

17.3.1 Monitoring of Phytoplankton Communities

Table 17.1 shows seawater parameters (Hydrographic data) measured by the hydrolab instrument at the time of sample collection.

As indicated in the Table 17.1, the pH of all samples was slightly alkaline and the salinity was normal except in sample 3, collected from Dibba, where the salinity was 3.881 %, which is lower than normal salinity of seawater. In addition, the oxygen level in sample 3 was found to be greater than in other samples. The monitoring results (microscopic-based analysis) of phytoplankton species number and types found in eight water samples collected during the entire sampling period are shown in Table 17.2. Figure 17.1 shows the four dominant harmful dinoflagellate species found in the red tide samples during this study sampling period.

Table 17.1 shows that there was no bloom of algal species in the control sample which was taken from the upper surface water. Sample 1 was found to have seven different kinds of phytoplankton, four of them were diatoms (dominant) harmless bloom species, namely *Asterionellopsis glacialis*, *Chaetoceros*, *Guinardia flaccida* and *Eucampia zodiacus*. Another three types of dinoflagellates were harmful bloom species such as *Dinophysis caudata*, *Prorocentrum micans* and *Prorocentrum baltіcum*. Sample 2 was found to have seven different kinds of phytoplankton and

Table 17.1 Hydrographic data of the samples

Date/sample #	Station location	Depth (m)	Temp (°C)	DO (mg/L)	Salinity (%)	pH	Fluorescence (μg/L)
4/3/2011/ sample 1	UAQ	1.29	24.08	6.99	4.52	8.07	0.51
4/17/2011/ sample 2	UAQ	1.86	24.38	6.71	4.224	7.88	0.72
4/20/2011/ sample 3	Dibba	1.56	25.83	7.52	3.881	8.08	2.86
5/1/2011/ sample 4	UAQ	1.23	27.54	6.54	4.112	7.82	0.72

all seven phytoplankton species were found to be in the harmful category (diatoms: *Pseudonitzschia pungens*, *Pseudonitzschia seriata* and *Skeletonema costatum*). On the other hand, sample 3 showed only two nontoxic phytoplankton species (diatoms "dominant" *Guinardia striata* and *Thalassiosira*). Sample 4 showed the presence of five different kinds of phytoplankton (diatom "dominant": *Guinardia flaccida*, *Guinardia striata* and *Navicula anca*), (dinoflagellate "harmful": *Ceratium fusus* and *Protoperidinium*).

17.3.2 Monitoring of Bacterial Communities

Figure 17.2 shows the monitoring results of bacterial populations by the fluorescent *in situ* hybridization (FISH) technique. *In situ* hybridization of the samples was performed by domain, group/class, and family specific 16S rRNA targeted oligonucleotide probes against Alpha, Beta, Gamma sub-class of proteobacteria, High and Low G+C sub-class of bacteria and the Enterobacteriaceae family. In total, eight samples (fixed with ethanol and paraformaldehyde) were used in the FISH analysis.

This is the first study to analyze bacterial community members associated with the phytoplankton species in the UAE coastal waters. The microscopic-based analysis of eight coastal water samples collected over a period of 4 weeks detected both toxic and non-toxic phytoplankton. A harmful dinoflagellate species *Dinophysis caudata* was observed in some samples and it is known for its ability to cause diarrhetic shellfish poisoning (Nishitani et al. 2008). *Dinophysis caudata* has been implicated earlier in creating red tides with resulted massive fish mortality in Japan (Okaichi 1967). *Dinophysis caudata* was observed in two out of eight samples (samples 1 and 2). This species is common in temperate to tropical neritic waters (Abè 1967). The temperature of water was 24 °C during the collection of samples 1 and 2 (Table 17.1). This shows that water temperature was ideal for the growth of *Dinophysis caudata* species. We also observed two bacterial species (presumptively identified as Vibrio) from these two samples (data not shown). These species were successfully isolated and purified on thiosulfate-citrate-bile-

Table 17.2 Biological parameters: Phytoplankton species number and their abundance

Samples	Phytoplankton species number	Phytoplankton abundance
Control	Nil	Nil
S 1	NC*	Diatoms (dominant):
		Asterionellopsis glacialis, Chaetoceros, Guinardia flaccida, Eucampia zodiacus
		Dinoflagellates:
		Dinophysis caudata, Prorocentrum micans, Prorocentrum balticum
S 2	Diatoms: 3	Diatoms:
	Dinoflagellates (dominant): 4	*Pseudo-nitzschia pungens, Pseudo-nitzschia seriata, Skeletonema costatum*
		Dinoflagellates (dominant):
		Cochlodinium polykrikoides, Dinophysis caudata, Prorocentrum minimum
Sub-2a	Diatoms: 3	Diatoms:
	Dinoflagellates (dominant): 4	*Pseudo-nitzschia pungens, Pseudo-nitzschia seriata, Skeletonema costatum*
		Dinoflagellates (dominant):
		Cochlodinium polykrikoides, Dinophysis caudata, Prorocentrum minimum
Sub-2b	Diatoms: 3	Diatoms:
	Dinoflagellates (dominant): 4	*Pseudo-nitzschia pungens, Pseudo-nitzschia seriata, Skeletonema costatum*
		Dinoflagellates (dominant):
		Cochlodinium polykrikoides, Dinophysis caudata, Prorocentrum minimum
S 3	Diatoms: 2	Diatoms (dominant):
		Guinardia striata, Thalassiosira
S 4	NC*	Diatoms (dominant):
		Guinardia flaccida, Guinardia striata, Navicula
		Dinoflagellates:
		Ceratium fusus, Protoperidinium

*NC** not counted

sucrose (TCBS) agar which is a specific culture media for the isolation of Vibrio spp. The yellow colony on TCBS agar is a characteristic feature of Vibrio species including that of *V. cholera*. Vibrio species have been reported as the dominant genera within a microbial community associated with phytoplankton species (Yang et al. 2012). Culture independent analysis of bacterial populations by fluorescence *in situ* hybridization of the sample-1 and 2 (fixed by ethanol) revealed the presence of members of all six bacterial groups (Alpha, Gamma and Beta sub-class of Proteobacteria, low G+C (LGC), High G+C (HGC) sub-class and Enterobacteriaceae family). However, the same sample-1 fixed by paraformaldehye (PFA) failed to detect any member of LGC group bacteria. This is due to the fact that

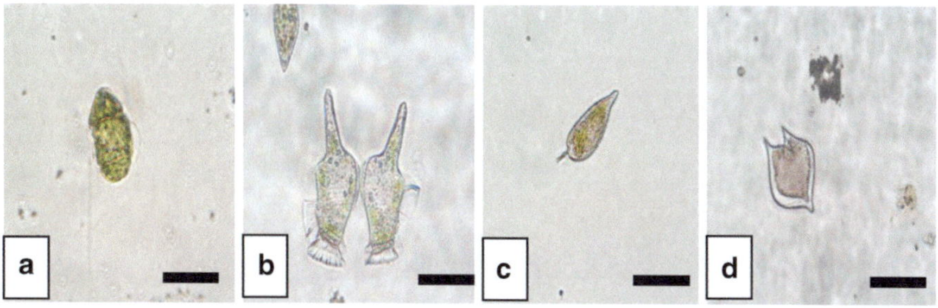

Fig. 17.1 Dominant harmful marine dinoflagellates species found in the samples (**a**) *Cochlodinium polykrikoides* (**b**) *Dinophysis caudata* (**c**) *Prorocentrum arenarium* (**d**) *Protoperidinium* spp.

ethanol is considered as a more suitable fixation method for the detection of gram positive bacteria and LGC group predominantly consists of gram positive bacteria. We have also performed a direct bacterial count of samples using the membrane filtration technique. All the samples, except the control, have shown a large number of uncountable bacteria during the study period (Data not shown).

Another planktonic species *Prorocentrum micans* was found in sample 1, and it is reported to form extensive red tides in many parts of the world (Fukuyo et al. 1990), although it is usually considered as harmless (Graneli et al. 1990). On the contrary, this organism has been shown to be associated with mortality of fish or shellfish, which were led by anoxia condition (Cho et al. 2009). *Cochlodinium polykrikoides* was observed in sample-2 (S2, Sub-2a& Sub-2b), and it is known to cause blooms which are strongly ichthyotoxic and also kill many other marine organisms (Kudela and Globler 2012). *C. polykrikoides* was implicated in one of the significant HAB events that occurred throughout the Arabian Gulf and the Gulf of Oman from 2008 through 2009 when a massive bloom of *C. polykrikoides* caused fish kills (Richlen et al. 2010; Kudela and Globler 2012; Zhao and Ghedira 2014). Another study in the UAE reported *C. polykrikoides* as the primary species during the UAB event (Bauman et al. 2010). This sample also detected a Vibrio species which formed a yellowish colony on TCBS agar. This bacterium was presumptively identified as *Vibrio alginolyticus*.

FISH analysis, using 16S rRNA targeted probes for the six different classes of bacteria, for sample 2, detected tiny rod-shaped bacteria belonging to all six types of bacterial group. The control and sample 4 were collected in the same physical conditions (pH, temperature, location and salinity), except that the control sample was collected from the sea surface and sample 4 from a depth of 1.29 m. Microscopic analysis of the control sample indicated that it was free from planktonic species growth. However, sample 4 was found to have two toxic and three non-toxic planktonic species. The direct filtration of sample 4, using the membrane filtration technique, showed five different types of bacteria on the culture media (data not shown). On the other hand, the isolation attempt for the Vibrio species from sample 4 resulted in the formation of blue/green colonies on the TCBS media. Vibrio

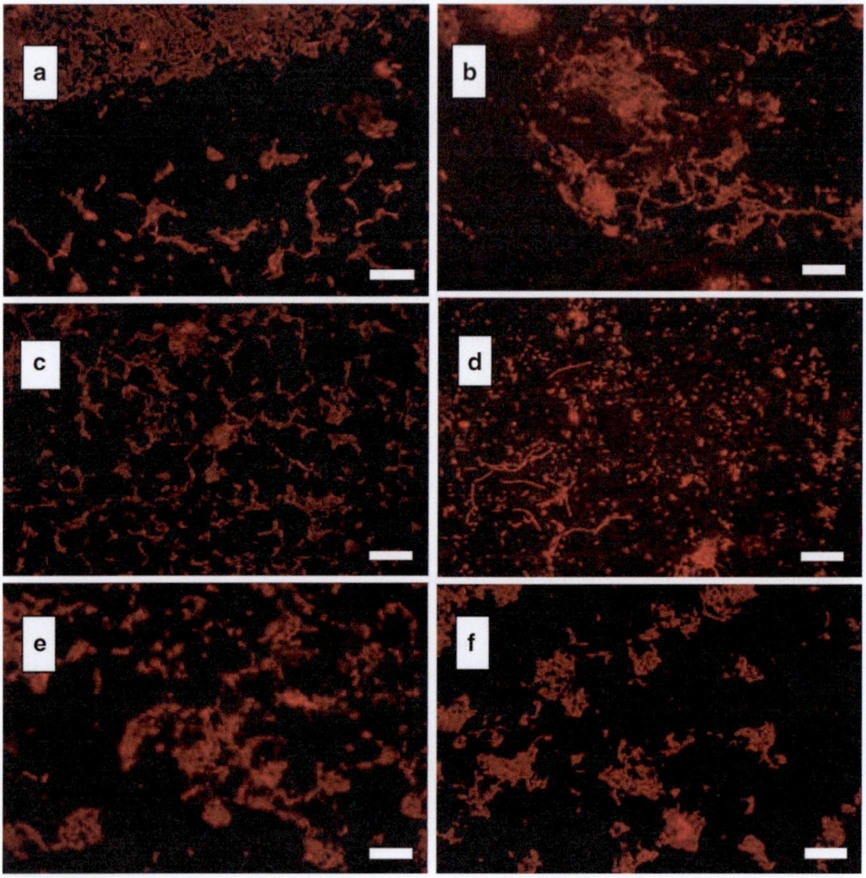

Fig. 17.2 Whole cell rRNA targeted fluorescence *in situ* hybridization of bacterial community members belonging to: (**a**), Alpha sub-class of proteobacteria (**b**) Beta sub-class of proteobacteria (**c**) Gamma sub-class of proteobacteria (**d**) Enterobacteriaceae family (**e**) High G + C sub-class (**f**) Low G + C sub-class. All samples were hybridized by the sub-class specific TRITC–labelled oligonucleotide probe. For each panel, identical field was viewed by epifluorescence microscopy. Bar = 10 μm applies to all photomicrographs. Original magnification: 1000×

species with algicidal effect have been associated with several planktonic species (Su et al. 2011; Yang et al. 2012) Overall, during this study, it was found that there were at least three distinct types of bacteria that were targeted by the GAM42a, HGC69a and Alpha1b probes. However, the bacterial population targeted by GAM42a and HGC69a were always observed in all the samples during the whole period of the study. The algicidal bacteria belonging to the Gamma-subclass of the proteobacteria group have been reported to be dominant during the HAB event in an earlier study (Yang et al. 2012). We have also detected several distinct members of the γ-subclass of proteobacteria and of the enterobacteriaceae family. The role of algicidal bacteria that are harmful to or feed on algae would be an important factor in the termination of algal blooms (Doucette et al. 1998; Su et al. 2011). A recent study reported high bacterial productivity and numbers during the key period of bloom decline (Yang et al. 2012). We too observed a very high bacterial population in most of the samples. The interaction between bacteria and phytoplankton is recognized as an important factor in the physiology and dynamics of harmful algal blooms (Barlaan et al. 2007; Liu et al. 2008). Therefore a long term monitoring study is highly recommended to understand the role and activities of diverse bacterial community members during pre- and post-algal bloom events.

17.4 Conclusions

In this study, 13 harmful dinoflagellates and six harmless dinoflagellates species were observed in a total of eight coastal water samples. To conclude, the findings of this study showed a highly diverse bacterial community belonging to five different sub-groups (Alpha, Beta and Gamma sub-class of proteobacteria, High and Low G + C sub-group) and one family (Enterobacteriaceae) associated with the phytoplankton species present in the UAE coastal waters. However, their specific role, whether algicidal or favourable, against harmful or nontoxic phytoplankton species needs to be established. This study could be used to form the basis of providing possible choices for controlling HABs in the UAE. A few isolated bacteria obtained in this study should be studied further for their algicidal features on both toxic and nontoxic phytoplankton species. Moreover, preliminary data obtained in this study will increase the information available concerning algicidal bacteria in the UAE and so will improve our understanding of phytoplankton–bacterial interactions in the UAE coastal waters.

Acknowledgement We are thankful to the Ministry of Environment and Water (MOEW), Dubai, UAE for permission to carry out a part of this work at the Marine resources research centre, Umm Al Quwain, UAE. We are grateful to Dr. Rashid Mohammed AlShihi and Dr. Jaishinimol Santhamma (Ministry of Environment and Water, UAE) for their assistance during the collection and analysis of samples in the Marine resources research centre laboratory, Umm Al Quwain, UAE. We appreciate Dr. Robert Boldi's valuable comments for the improvement of the manuscript.

List of Figures

Fig. 17.1 Dominant harmful marine dinoflagellates species found in the samples (**a**) *Cochlodinium polykrikoides* (**b**) *Dinophysis caudata* (**c**) *Prorocentrum arenarium* (**d**) *Protoperidinium* spp.

Fig. 17.2 Whole cell rRNA targeted fluorescence *in situ* hybridization of bacterial community members belonging to: (**a**), Alpha sub-class of proteobacteria (**b**) Beta sub-class of proteobacteria (**c**) Gamma sub-class of proteobacteria (**d**) Enterobacteriaceae family (**e**) High G + C sub-class (**f**) Low G + C sub-class. All samples were hybridized by the sub-class specific TRITC–labelled oligonucleotide probe. For each panel, identical field was viewed by epifluorescence microscopy. Bar = 10 μm applies to all photomicrographs. Original magnification: 1000×

References

Abè TH (1967) The armoured Dinoflagellata: II. Prorocentridae and Dinophysidae (A). Publ Seto Mar Biol Lab 14:369–389

Al Kandari M, Al Yamani F, Al Rifaie K (2009) Marine phytoplankton atlas of Kuwait's waters. Kuwait Institute of Scientific Research, Safat 54 pp

Al Qubaisi BS (2006) A study of factors influencing the occurrence of Harmful Algae Bloom (HAB) along the Abu Dhabi Coast, UAE. United Arab Emirates University Abu Dhabi: Environment Agency – Abu Dhabi, UAE, 98 pp

Amann RI, Krumholz L, Stahl DA (1990) Fluorescent-oligonucleotide probing of whole cells for determinative, phylogenetic, and environmental-studies in microbiology. J Bacteriol 172:762–770

Arnold T (2009) Red tide forces closure of Dubai beaches. Retrieved February 10, 2014, from ArabianBusiness.com: http://www.arabianbusiness.com/551844-red-tide-forces-closure-of-dubai-beach

Azam F, Fenchel T, Field JG, Gray JS, Meyer-Reil LA, Thingstad F (1983) The ecological role of water-column microbes in the sea. Mar Ecol Prog Ser 10:257–263

Barlaan EA, Furukawa S, Takeuchi K (2007) Detection of bacteria associated with harmful algal blooms from coastal and microcosm environments using electronic microarrays. Environ Microbiol 9:690–702

Bauman AG, Burt JA, Feary DA, Marquis E, Usseglio P (2010) Tropical harmful algal blooms: an emerging threat to coral reef communities. Mar Pollut Bull 60:2117–2122

Cho SY, Nagai S, Han MS (2009) Development of microsatellite markers in red-tide causative species *Prorocentrum micans* (Dinophyceae). Conserv Genet 10:1151–1153

Daims H, Bruhl A, Amann RI, Schleifer KH, Wagner M (1999) The domain-specific probe EUB338 is insufficient for the detection of all bacteria: development and evaluation of a more comprehensive probe set. Syst Appl Microbiol 22:434–444

Daims H, Stoecker K, Wagner M (2005) Fluorescence in situ hybridization for the detection of prokaryotes. In: Osborn AM, Smith CJ (eds) Advanced methods in molecular microbial ecology. Bios-Garland, Abingdon, pp 213–239

Doucette GJ, Kodama M, Gallacher S (1998) Bacterial interaction with harmful algal bloom species: bloom ecology, toxigenesis and cytology. In: Anderson DM, Cembella AD, Hallegraeff GM (eds) Physiological ecology of harmful algal bloom. Springer, Heidelberg/Berlin, pp 619–647

Fuad A (2008) Weather change 'will end worst red tide in year'. Retrieved February 10, 2014, from Gulfnews.com: http://gulfnews.com/news/gulf/uae/environment/weather-change-will-end-worst-red-tide-in-years-1.143828

Fukuyo Y, Takano H, Chihara M, Matsuoka K (1990) Red tide organisms in Japan. An illustrated taxonomic guide, Uchida Rokakuho, Co., Ltd., Tokyo, 407 pp

Glibert PM (2007) Eutrophication and Harmful Algal Blooms: a complex global issue, examples from the Arabian Seas including Kuwait Bay, and an introduction to the Global Ecology and Oceanography of Harmful Algal Blooms (GEOHAB) programme. Int J Oceans Oceanogr 2:157–169

Graneli E, Sundstrom B, Edler L, Anderson DM (1990) Toxic marine phytoplankton. Elsevier, New York, 554 pp

Hold GL, Smith EA, Rappé MS, Maas EW, Moore ERB, Stroempl C, Stephen JR, Prosser JI, Birkbeck TH, Gallacher S (2001) Characterisation of bacterial communities associated with toxic and non-toxic dinoflagellates: *Alexandrium* spp. and *Scrippsiella trochoidea*. FEMS Microbiol Ecol 37:161–173

Kudela RM, Gobler CJ (2012) Harmful dinoflagellate blooms caused by *Cochlodinium* sp.: global expansion and ecological strategies facilitating bloom formation. Harmful Algae 14:71–86

Landais E (2008) Dead fish floating in Dubai creek. Retrieved February 10, 2014, from Gulf News: http://www.gulfnews.com/Nation/Environment/10227257.html

Liu J, Lewitus AJ, Kempton JW, Wilde SB (2008) The association of algicidal bacteria and raphidophyte blooms in South Carolina brackish detention ponds. Harmful Algae 7:184–193

Manz W, Amann R, Ludwig W, Wagner M, Schleifer KH (1992) Phylogenetic oligodeoxynucleotide probes for the major subclasses of Proteobacteria: problems and solutions. Syst Appl Microbiol 15:593–600

Meier H, Amann RI, Ludwig W, Schleifer KH (1999) Specific oligonucleotide probes for in situ detection of a major group of gram-positive bacteria with low DNA G + C content. Syst Appl Microbiol 22:186–196

Nishitani G, Nagai S, Sakiyama S, Kamiyama T (2008) Successful cultivation of the toxic dinoflagellate *Dinophysis caudata* (Dinophyceae). Plankton Benthos Res 3:78–85

Okaichi T (1967) Red tides found in and around the Seto Inland Sea in 1965. Tech Bull Fac Agric Kagawa Univ 15:181–185

Rajan A, Al Abdessalaam TZ (2008) Guide to common marine phytoplankton in Abu Dhabi waters. Environmental Agency-Abu Dhabi, 127 pp

Richlen ML, Morton SL, Jamali EA, Rajan A, Anderson DM (2010) The catastrophic 2008–2009 red tide in the Arabian gulf region, with observations on the identification and phylogeny of the fish-killing dinoflagellate *Cochlodinium polykrikoides*. Harmful Algae 9:163–172

Romalde J, Barja J, Toranzo AE (1990) Vibrios associated with red tides caused by *Mesodinium rubrum*. Appl Environ Microbiol 56:3615–3619

Schuppler M, Wagner M, Schon G, Gobel UB (1998) In situ identification of nocardioform actinomycetes in activated sludge using fluorescent rRNA-targeted oligonucleotide probes. Microbiology 144:249–259

Su JQ, Yang XR, Zhou Y, Zheng T (2011) Marine bacteria antagonistic to the harmful algal bloom species *Alexandrium tamarense* (Dinophyceae). Biol Control 56:132–138

Tomas CR (1997) Identifying marine phytoplankton. Academic, New York, 858 pp

Xin W, Yun T (2009) Lysis of a red tide causing algae, caused by bacteria from its phycosphere. Biol Control 52:123–130

Yang C, Zhou Yi Li Y, Zheng W, Tian Y, Zheng T (2012) Bacterial community dynamics during a bloom caused by Akashiwo sanguinea in the Xiamen Sea area, China. Harmful Algae 20:132–141

Yoo A (2009) Red tide rising: an investigation of recent increased Harmful Algal Blooms in the United Arab Emirates. Global climate change: the science, social impact and diplomacy of a world environmental crisis. Graduate Research Paper, Harvard University, August 5, 2009. Retrieved February 10, 2014, from: http://www.climate-talks.net/2013-ENVRE130/PDF/YOO_GRADUATE_PROJECT_S-130.pdf

Zhao J, Ghedira H (2014) Monitoring red tide with satellite imagery and numerical models: a case study in the Arabian Gulf. Mar Pollut Bull 79(1–2):305–313

Chapter 18
Environmental Quality Standards for Brine Discharge from Desalination Plants

Anton Purnama

Abstract When many desalination plants are operated closely together along coastal areas, the continuous brine discharges through marine outfall systems from these cluster plants can have a significant impact on the coastal marine environment. The potential impact can be minimized and regulated by treatment and recycling technologies, by limiting the concentration values of brine at the discharge point and also by imposing concentration values within a prescribed circular mixing zone in the coastal waters via the outfall design.

The maximum concentration values for monitoring the coastal water quality standard are formulated using analytical solutions of the far-field mathematical model for single and two outfall discharges based on a flat seabed, developed to incorporate the effect of a tidally oscillating flow. The results for a single outfall model show that a smaller mixing zone is suitable only for a stronger ambient current condition at the discharge site. If a second outfall is allowed to discharge close to an existing outfall, the results show that the radius of the mixing zone should be increased to accommodate the compounded impacts.

Keywords Effluent discharge • Minimum dilution • Mixing zone • Two sea outfalls • Water quality model

18.1 Introduction

Many large scale seawater desalination plants are operated closely together along the coasts of the Arabian Gulf, Red Sea, Mediterranean Sea, and of the Gulf of Oman (Lattemann and Hopner 2008). These clustered plants dispose of their brine waste product mostly by continuous discharge via marine outfall systems. Current technology limits the efficiency of producing desalinated water and up to 50 % is lost via concentrate that is 1.1–1.5 times typical seawater salinity for the evaporation technologies of Multi-Stage Flash and 1.3–1.7 times for Reverse Osmosis

A. Purnama (✉)
Department of Mathematics and Statistics, College of Science, Sultan Qaboos University, P.O. Box 36, Al-Khod PC123, Muscat, Sultanate of Oman
e-mail: antonp@squ.edu.om

M. Baawain et al. (eds.), *Recent Progress in Desalination, Environmental and Marine Outfall Systems*, DOI 10.1007/978-3-319-19123-2_18

(Ahmad and Baddour 2014). As the steadily discharged brine plume continues to drift away with the currents, due to relatively shallow water depth, the elongated brine plumes spread towards the shoreline which may cause an increase in salinity in the coastal waters. A major concern regarding desalination plants is the impact of the concentrate and chemical discharges on the marine environment which may impair coastal water quality and affect marine life (Roberts et al. 2010). Salinity and temperature are controlling factors for the distribution of marine species, and most organisms can adapt to minor deviations from optimal salinity and temperature conditions, and might even tolerate extreme situations temporarily, but not a continuous exposure to unfavorable conditions. The constant discharge of reject streams with high salinity and temperature levels can thus be fatal for marine life, and can cause a lasting change in species composition and abundance at the discharge site (Sheppard et al. 2010).

There are several approaches to mitigate the environmental effects of brine discharge. To avoid impacts from high salinity, for example, the desalination plant reject stream can be pre-diluted with other waste streams where applicable, such as power plant cooling water (Purnama et al. 2011). Mixing and dispersal of the discharge plume can be enhanced by installing a properly designed diffuser system, and by locating the discharge in a favorable oceanographic site, to achieve high dilution (Bleninger and Jirka 2008). Thus, the potential impact can be reduced and regulated, in addition to some treatment and recycling technologies, by limiting concentration values of brine at the discharge point (better known as ELV, the emission limit values); and in addition to the outfall design, by imposing another concentration limit value (known as EQS, the environmental quality standards) within a circular allocated mixing zone at the coastal waters. The size of these mixing zones may vary from 0 to 500 m from the discharge point (Ahmad and Baddour 2014).

Environmental impact assessments should be required to define regulatory strategies on protection and conservation of the coastal marine environment in a sustainable way (Voutchkov 2011). However, in general, there is little information available on the impacts of desalination plants on the marine environment, and even less data is available to quantify such impacts for regulatory and design purposes (Ahmad and Baddour 2014). Since it is practical and easy to administer, many regulators and permitting authorities prefer ELV over EQS for granting a discharge permit. ELV only restricts the brine concentration at the end-of-outfall and is thus easy to monitor, but it does not directly consider the quality response of the coastal water and the long-term carrying capacity of the marine environment. This makes the permit holder unaware of how the discharge is harming the ecosystems and allows the discharger to refuse to take responsibility for damage to the marine environment. The worst-case scenario may occur when there are intense seawater desalination activities in certain (semi-enclosed) sea areas or coastlines, such as the Arabian Gulf, Red Sea, Mediterranean Sea, and the Gulf of Oman, and thus there is an urgent need to accurately assess both the long-term localized environmental impacts from the individual plant, and the cumulative strategic impacts compounded from the neighboring plants. When two or more sea outfalls discharge brine to shallow coastal waters, the adverse long-term mutual impacts are strongly interdependent and are of a capacity limit of the coastal waters. Therefore, EQS should also be taken into consideration in granting a discharge permit, as this will inform the dischargers of their responsibility regarding damage to the environment. Indeed, the new European Community-Water Framework Directive has integrated EQS in addition to ELV to improve the quality for all European waters (Jirka et al. 2004).

The difficulties in implementing EQS are mainly related to setting up the limit values within the allocated circular mixing zones and its monitoring practices. Predictive models are therefore needed to demonstrate compliance with EQS. Many existing numerical models are generally chosen owing to the personal preference of the user, and the model's system dependence in terms of costs to run and implement it (Ragas et al. 1997).

The model prediction is often criticized also because of uncertainty in obtaining the site-specific input data and parameters for model validation. In searching for a unified and reference quality standard for numerical models, an analytical model based on a flat seabed is proposed that incorporates the effect of a coastal tidal current oscillating with a period of half a lunar day (Purnama and Al-Barwani 2006). Using asymptotic long-time analytical solutions of the far-field mathematical models for single and two outfalls discharge, the paper proposes the maximum value of the brine plume concentration in response to the continuous brine discharge from desalination plants. Furthermore, the radius of a circular mixing zone for monitoring the water quality standard centered at the end-of-outfall can be formulated as a function of the maximum concentration.

For the case of many large scale seawater desalination plants operated closely together along a stretch of coastline, the two sea outfalls discharge model extension shows that the maximum value of concentration depends on the separation distance between the outfalls and discharge load of each outfall. For a shorter separation distance, the interaction and merging of two brine plumes can be detected from the graph of the maximum value of the long-term concentration (Al-Barwani and Purnama 2008a). The results can answer questions such as, if a new large scale seawater desalination plant is to be built on a coastline where an existing plant is operated, how can one calculate the compounded impacts? In particular, will a possible scenario ever arise in which the existing plant with a discharge permit based on EQS is no longer meeting the regulatory requirements?

18.2 Single Outfall Discharge Model

Owing to the highly variable nature of the sea, we do not yet have a full understanding or description of the mixing processes of brine discharges from coastal desalination plants. Immediately after release from the diffuser, vigorous and rapid dilution of concentrate brine is governed by the effluent buoyancy, momentum of the discharge and its interaction with the sea currents (Bleninger and Jirka 2008). At the end of this mixing stage, the established steady discharge brine plume then continues to drift away with the currents.

For simplicity, we are only concerned with the effect of oscillating longshore currents on the long-time (far-field) brine plume, and a highly simplified semi-infinite flat seabed is considered, where the shoreline is straight and of a constant water depth (Al-Barwani and Purnama 2008a; Purnama and Al-Barwani 2006). We assume also that the complexities of flow in coastal water, such as temperature and density, will be ignored, except for time-dependence. The coastal current is assumed to be uniform over water depth and remains in the x-direction parallel to the beach. The dispersion processes are represented by the longitudinal diffusivity D_x and lateral diffusivity D_y. As depicted in Fig. 18.1, the brine waste stream is continuously discharged at a rate of Q_1 from the sea outfall at the position $(x = 0,\ y = \alpha)$ where $\alpha > 0$. As the discharge is commonly made via diffusers and utilizes the best available technology to promote rapid initial dilution (Bleninger and Jirka 2008; Purnama et al. 2011), we also assume that the outfall's brine plume is vertically well-mixed over water depth. Note that for shallow coastal waters, the dispersion in the vertical direction occurs much faster than that in the lateral direction.

The appropriate measure for assessing the impact of brine discharges into the sea would be the long-time maximum concentration (Jirka et al. 2004). To avoid repetition, mathematical details can be found in Al-Barwani and Purnama 2008a and in Purnama and Al-Barwani 2006, and the long-time concentration for the brine plume can be approximated as

$$C_1 = \sqrt{\frac{\pi}{\lambda V(X + \cos T)}} \left[\exp\left(-\frac{\lambda V \eta \{Y - \Lambda\}^2}{X + \cos T} \right) + \exp\left(-\frac{\lambda V \eta \{Y + \Lambda\}^2}{X + \cos T} \right) \right], \tag{18.1}$$

where $C_1 = 4\pi c\sqrt{D_x D_y}/Q_1, \lambda = {U_0}^2/4\omega D_x$ is the distance by which the plume is transported and spread over by advection to that by longitudinal diffusion, $X = \omega x/U_0,\ Y = \omega y/U_0,\ \eta = D_x/D_y$ is the ratio of longitudinal to lateral diffusivities, $\Lambda = \omega\alpha/U_0, V = v/U_0$ is the ratio of drift current v to tidal amplitude U_0, and $2\pi/\omega$ is the tidal period.

Due to unpredictable sea conditions, very little information is available on the model parameters. Using typical numerical values of the period $2\pi/\omega = 4.5 \times 10^4$ s and the mean tidal amplitude $U_0 = 0.2$ m/s, we define a length scale U_0/ω of the order of 1.4 km. Assuming that the longitudinal diffusion is a very efficient process (Smith and Scott 1997), and that larger values of λ are mostly due to a larger current with smaller values of D_x, λ in the range of 5–20 are suitable for a moderate current (Al-Barwani and Purnama 2008a). The value of $V = 0.2$ and $\eta = 30$ will be used in all plots, unless stated otherwise.

The standard regulatory guidelines allocated circular mixing zones around the marine outfall and set the concentration values to control the quality standards of

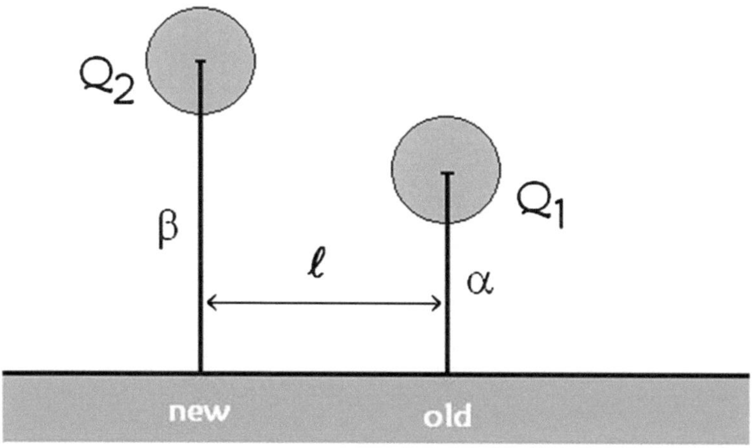

Fig. 18.1 Diagram of two sea outfalls

the receiving water (Jirka et al. 2004; Purnama et al. 2011). The mixing zone is designated as an allocated impact zone within the maximum value of concentration limit that may be exceeded. As we are only interested in a circular mixing zone region centred at the end-of-outfall (with a radius less than the outfall length Λ), the long-time concentration can be approximated further by neglecting the second exponential term in Eq. (18.1), and thus

$$C_{1\infty} = \sqrt{\frac{\pi}{\lambda V(X + \cos T)}} \exp\left(-\frac{\lambda V \eta \{Y - \Lambda\}^2}{X + \cos T} \right). \tag{18.2}$$

Next, by differentiating with respect to X, the maximum value of the long-time concentration is given by

$$C_{1\max} = \frac{1}{\lambda V | \ Y - \Lambda \ |} \sqrt{\frac{\pi}{2\eta e}}, \tag{18.3}$$

which occurs at the position $X_{\max} = 2\lambda V\eta(Y - \Lambda)^2 - \cos T$. From a regulatory viewpoint, $| \ Y - \Lambda \ |$ could be used as a radius for the mixing zone centered at the end-of-outfall, and we will therefore focus our discussions on the numerical values of $| \ Y - \Lambda \ |$ up to 0.4, which correspond to 560 m. Furthermore, the maximum concentration value $C_{1\,\max}$ can also be used as the numerical upper limit for EQS at the end of the mixing zone.

Figure 18.2 shows the variations of the radius of the mixing zone for a single outfall discharge as a function of $C_{1\,\max}$ for three values of the model parameter $\lambda = 5$, 10 and 20 to reflect the uncertainty in sea conditions. For a given value of λ, the radius $| \ Y - \Lambda \ |$ is longer for a smaller value of $C_{1\,\max}$. However, for a specified value of $C_{1\,\max}$, there is a wide range of possible values for the radius of the mixing

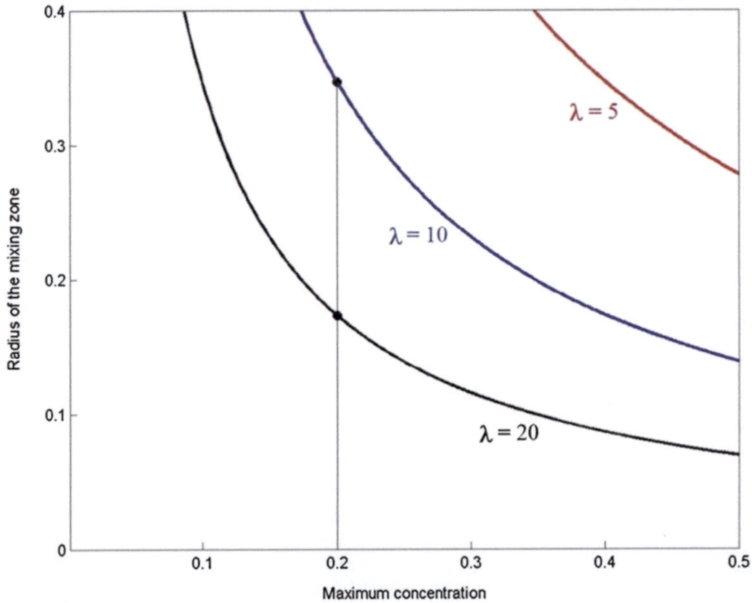

Fig. 18.2 Radius of the mixing zone as a function of $C_{1\,\max}$

zone, and the radius is smaller in a favourable oceanographic mixing condition. For example, for $C_{1\max} = 0.2$, the radius decreases from 0.35 (which corresponds to 490 m) for $\lambda = 10$ to a smaller value of 0.17 (corresponds to 240 m) for $\lambda = 20$. Thus, a larger mixing zone is required for a weaker ambient current condition.

To include the prescribed ELV at the discharge point, a term minimum dilution is commonly used (Ahmad and Baddour 2014), which is defined as the ratio of the emission concentration at the outfall discharge point to the maximum concentration (EQS) at the end of the mixing zone (Jirka et al. 2004). The minimum dilution $1/C_{1\,\max}$ at the end of the mixing zone for a single outfall discharge is illustrated in Fig. 18.3 for three values of $\lambda = 5$, 10 and 20. For a given value of λ, the dilution increases as the radius $|\ Y - \Lambda\ |$ becomes larger. The uncertainty in sea conditions gives rise to a minimum dilution in the range of 1.1–4.3 for a smaller mixing zone with a radius of 0.15 (or 210 m), where a larger dilution value is achieved for a stronger ambient current condition.

18.3 Two Outfalls Discharge Model

As illustrated in Fig. 18.1, we now consider brine discharges from two large scale desalination plants at ℓ distance apart at a rate Q_1 from the first (old) outfall at the position $(x = 0,\ \ y = \alpha)$, and at a different rate Q_2 from the second (new) outfall at the position $(x = 0,\ \ y = \beta)$, where $\beta > 0$. Again, mathematical details can be

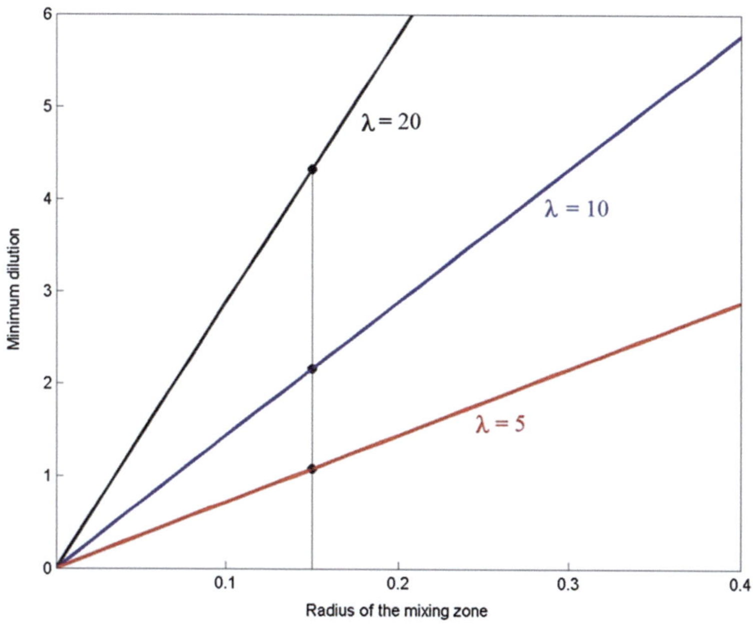

Fig. 18.3 Minimum dilution as a function of the radius of the mixing zone

found in Al-Barwani and Purnama 2008b, and the long-time concentration for the brine plume concentration from the two outfalls discharge can be approximated as

$$C_{2\infty} = C_{1\infty} + q_* \sqrt{\frac{\pi}{\lambda V(X + L + \cos T)}} \exp\left(-\frac{\lambda V \eta \{Y - B\}^2}{X + L + \cos T} \right), \tag{18.4}$$

where $B = \omega\beta/U_0$ and $L = \omega\ell/U_0$. The interaction of two brine plumes is governed by the discharge rate factor $q_* = Q_2/Q_1$ of the second (new) outfall and the separation distance L between two outfalls. It is expected that the longer the separation distance, the smaller the contribution of the brine plume from the second (new) outfall will be.

Since we are only interested in a circular mixing zone centered at the end-of-outfall, we assume that the radius of the mixing zone $|Y - \Lambda| \approx |Y - B|$, and also define the separation distance between two outfalls as a function of the radius of the mixing zone in the form $mL = 2\lambda V\eta(Y - \Lambda)^2$ with $m \geq 1$ (see Table 18.1). The presence of the second (new) outfall will not affect the position of the maximum concentration of the first (old) brine plume X_{max} (Al-Barwani and Purnama 2008b), and thus by substituting X_{max} to Eq. (18.4), the maximum value of the long-time concentration is formulated by

Table 18.1 Separation distance L as a function of $| Y - \Lambda |$

m	L
1	$240\ (Y-\Lambda)^2$
5	$48\ (Y-\Lambda)^2$
10	$24\ (Y-\Lambda)^2$
20	$12\ (Y-\Lambda)^2$
40	$6\ (Y-\Lambda)^2$
80	$3\ (Y-\Lambda)^2$

$$C_{2\max} = C_{1\max}\left[1 + q_* \sqrt{\frac{m}{m+1}} \exp\left(\frac{1}{2m+2}\right)\right]. \tag{18.5}$$

As given in Table 18.1, using a larger value of $\lambda = 20$, it is noted that a larger value of m corresponds to a shorter separation distance L; and thus, ultimately, $C_{2\max} = (1 + q_*)C_{1\max}$.

Figure 18.4 illustrates, for $\lambda = 20$ and $m = 20$, the variations of the radius of the mixing zone for the two outfalls discharge for three values of the discharge rate factor $q_* = 0.5$, 1 and 1.5, where the corresponding values for a single outfall discharge ($q_* = 0$) is shown by the dashed line. To accommodate the long-time compounded impact of the second (new) outfall, the prescribed (old) radius of the mixing zone should be increased. For example, for $C_{2\max} = 0.2$, the (old) radius of 0.17 (or 240 m) should be doubled to 0.35 (490 m) to account for the compounded impact from the new outfall discharging at the same rate ($q_* = 1$).

Next, for $\lambda = 20$ and $m = 20$, the minimum dilution at the end of the mixing zone for two outfalls discharge is shown in Fig. 18.5 for three values of the discharge rate factor $q_* = 0.5$, 1 and 1.5, where the corresponding values for a single outfall discharge ($q_* = 0$) is shown by the dashed line. Larger discharge rates Q_2, of the second (new) outfall will decrease a minimum dilution from 4.3 (single outfall value) to 1.7 ($q_* = 1.5$) for a smaller mixing zone with a radius of 0.15 (or 210 m). Therefore, a possible scenario can occur where the existing (old) plant is no longer meeting the regulatory requirements of achieving the specified maximum concentration value within the prescribed circular mixing zone.

18.4 Conclusion

The long-term impacts of brine discharges can be addressed and regulated by restricting the emission concentration at the discharge point and by imposing the maximum concentration values within the allocated circular mixing zone (Jirka et al. 2004). Monitoring the quality standards of the receiving coastal waters is required to control the mixing capacity of the marine environment.

Using the long-term solutions of a two-dimensional advection-diffusion equation on a flat seabed, the radius of the mixing zone centered at the end-of-outfall can

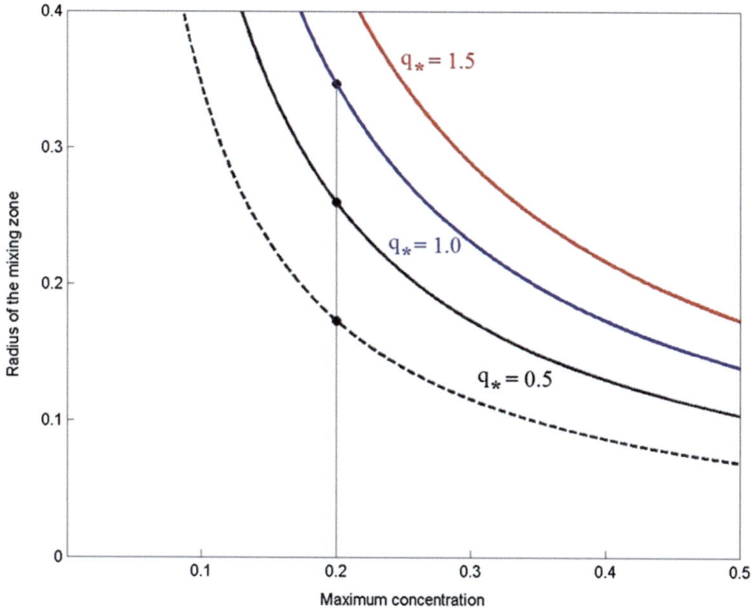

Fig. 18.4 Radius of the mixing zone as a function of $C_{2\,\max}$

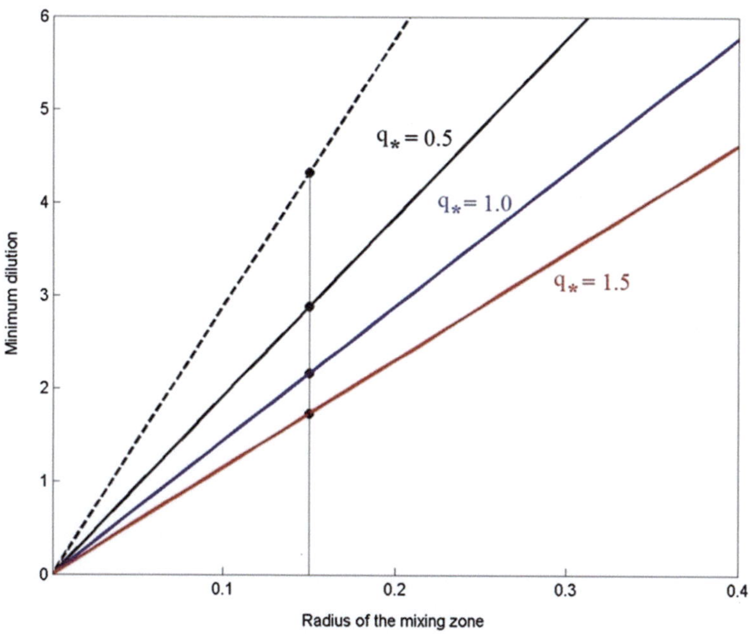

Fig. 18.5 Minimum dilution for the two outfalls discharge model

be formulated as a function of the maximum concentration. Smaller-sized mixing zones are appropriate for regions with a stronger ambient currents condition. For regions with weaker ambient currents, larger mixing zones are required for the longitudinal diffusion processes to take place.

However, when two outfalls are discharging, the additional parameters, such as the discharge rates of each outfall and the separation distance between the outfalls, will influence the radius of the mixing zone. The existing mixing zone for the old outfall should be increased to accommodate the compounded impact of the new outfall discharge. Thus, the initial EQS requirements for the old outfall discharge may no longer be met once the new outfall is in operation.

List of Figures

References

Ahmad N, Baddour RE (2014) A review of sources, effects, disposal methods, and regulations of brine into marine environments. Ocean Coast Manage 87:1–7

Al-Barwani HH, Purnama A (2008a) Simulating brine plumes discharged into the seawaters. Desalination 221:608–613

Al-Barwani HH, Purnama A (2008b) Brine discharges from two coastal desalination plants. In: Delgado DJ, Moreno P (eds) Desalination research progress. Nova Science Publisher, New York, pp 1–10

Bleninger T, Jirka GH (2008) Modelling and environmentally sound management of brine discharges from desalination plants. Desalination 221:585–597

Jirka GH, Bleninger T, Burrows R, Larsen T (2004) Environmental quality standards in the EC-Water Framework Directive: consequences for water pollution control for point sources. European Water Management Online (EWMO). www.ewaonline.de

Lattemann S, Hopner T (2008) Environmental impact and impact assessment of seawater desalination. Desalination 221:1–15

Purnama A, Al-Barwani HH (2006) Spreading of brine waste discharges into the Gulf of Oman. Desalination 195:26–31

Purnama A, Al-Barwani HH, Bleninger T, Doneker RL (2011) CORMIX simulations of brine discharges from Barka plants, Oman. Desalin Water Treat 32:329–338

Ragas AMJ, Haans JLM, Leuven RSEW (1997) Selecting water quality models for discharge permitting. Eur Water Poll Control 7:59–67

Roberts DA, Johnston EL, Knott NA (2010) Impacts of desalination plant discharges on the marine environment: a critical review of published studies. Water Res 44:5117–5128

Sheppard C, Al-Husiani M, Al-Jamali F, Al-Yamani F, Baldwin R, Bishop J, Benzoni F, Dutrieux E, Dulvy NK, Durvasulla SRV, Jones DA, Loughland R, Medio D, Nithyanandan M, Pilling GM, Polykarpov I, Price ARG, Purkis S, Reigl B, Saburova M,

Namin KS, Taylor O, Wilson S, Zainal K (2010) The Gulf: a young sea in decline. Mar Pollut Bull 60:13–34
Smith R, Scott CF (1997) Mixing in the tidal environment. J Hydr Eng ASCE 123:332–340
Voutchkov N (2011) Overview of seawater concentrate disposal alternatives. Desalination 273:205–219

Chapter 19
Hydrodynamic and Thermal Dispersion Modelling of the Effluent in a Coastal Channel

Ahmad Sana

Abstract Thermal dispersion of the effluent from an industrial plant into a tidal channel was studied using Delft3D suite. The potential for morphological changes because of the effluent discharge in the channel was also studied. The simulation was performed for 3 weeks to cover a cycle of neap and spring tides. The heat flux model at the water surface took into account the separate effects of solar and atmospheric radiation, and heat loss due to back radiation, evaporation and convection. The initial and boundary conditions of the temperature were based on field observation data. From the model computations, it was found that the effluent with high temperature starts affecting the lagoon area in front of the outlet structure. The water temperature gradually decreases as the flow continues into the tidal channel to the mouth into the open sea. However, well before it reaches the mouth, the water temperature attains a value close to ambient temperature. The computed bed shear stress was found to be too small to cause sediment movement in the lagoon area for the given bed particle sizes.

Keywords Thermal dispersion modelling • DELFT3D • Bottom shear stress • Sediment movement • Ocean outfall

19.1 Introduction

Various types of industrial plants exist in the vicinity of the coast, such as water desalination units and wastewater treatment plants that discharge effluent into the sea. Generally, an outfall structure is designed to minimise possible impact on the ambient environment. Mostly, the effluent is discharged into the sea through diffusers placed at recommended depths and offshore distance to enhance the mixing process. Local environmental regulations prescribe the allowable effluent water quality parameters after the mixing process has taken place. In certain

A. Sana (✉)
Department of Civil and Architectural Engineering, College of Engineering, Sultan Qaboos University, PO Box 33, Muscat, Postal Code 123, Oman
e-mail: sana@squ.edu.om; sana092@yahoo.com

M. Baawain et al. (eds.), *Recent Progress in Desalination, Environmental and Marine Outfall Systems*, DOI 10.1007/978-3-319-19123-2_19

Fig. 19.1 Lagoon area adjacent to Port Salalah. Transect locations for bathymetric survey are also shown and the outlet is located close to T6 (Google Earth)

situations, the industrial plant is located in the proximity of a natural channel, forming a lagoon-like water body, and only thermal dispersion is the main concern. The disposal of such treated effluent in the channel would be economically feasible, since a submerged outfall will not be required.

One such situation is the topic of the present study. The effluent from the plant will be discharged at a rate of 13,500 m^3/h (3.75 m^3/s) into the natural channel within the lagoon area located on the western side of Salalah Port (Fig. 19.1). The effluent temperature is approximately 40 °C. The excessive discharge of effluent with higher than ambient temperature may cause thermal pollution along-with erosion in the channel. Subsequently, there may be deposition of the eroded sediment in the vicinity of the channel mouth (in the sea).

In order to assess the impact of this effluent, Delft3D suite was used in the present study (Deltares 2003a, b, c), which is one of the modern models capable of predicting the concentration distributions for different pollutants as well as morphologic changes occurring due to high flow rate over greater distances in the water body.

19.2 Methodology

The hydrodynamics of a continuously discharged effluent into a receiving water body can be conceptualized as a mixing process. As the turbulent plume travels further away from the source, the source characteristics become less important. Ambient environmental conditions will control the trajectory and the dilution of the turbulent plume through buoyant spreading motion, passive diffusion due to ambient turbulence, and passive advection by the often time-varying, non-uniform ambient velocity field. The flow is forced by the tide at the open boundaries, wind stress at the free surface, pressure gradients due to free surface gradients (barotropic) or density gradients (baroclinic).

Delft3D-FLOW is the module of the Delft3D suite that provides the hydrodynamic basis for water quality simulations. Delft3D-FLOW calculates the unsteady flow and transport phenomena resulting from tidal and meteorological forcing. The main purpose is the two-dimensional (2D, depth-averaged) and three-dimensional (3D) simulation of tidal and wind-driven flow by solving the unsteady shallow water equations in two (depth-averaged) or three dimensions. In this approach, the vertical momentum equation is reduced to the hydrostatic pressure relation. Vertical accelerations are assumed to be small, compared to the gravitational acceleration and are not taken into account. The equations include velocity gradients, advective terms, barotropic pressure gradients, Coriolis force, bottom shear stress, external forces (wind) and viscosity. The bottom roughness in the Delft3D-FLOW program can be defined in several ways. For a depth-averaged flow (2D), the shear stress on the bed in the x- and y-direction induced by a turbulent flow is given by a quadratic friction law using the Chézy coefficient. The 2D-Chézy coefficient C can be determined with Chézy, Manning or Colebrook formulations.

The main physical phenomena that are accounted for in the Delft3D-FLOW, are:

- Coriolis force.
- Turbulence model to account for the vertical turbulent viscosity and diffusivity based on the eddy viscosity concept. Four options, the k-epsilon, k-L, algebraic and constant models, are provided.
- Shear stress exerted by the turbulent flow on the bottom based on a quadratic Chézy or Manning formula.
- Simulation of the thermal discharge, effluent discharge and the intake of cooling water at any location and any depth in the computational domain (advective-diffusion module).
- Simulation of drying and wetting of inter-tidal flats (moving boundaries) for both 2D and 3D cases.

Besides this, Delft3D-FLOW has the module to model wind stresses on the water surface by a quadratic friction law. However, wind effects are not taken into account in this study. For a complete review of physical phenomena and modules, which are taken into account in their implementation, reference is made to the Delft3D-FLOW user manual (Deltares 2003a).

19.2.1 Grid and Bathymetry

In the horizontal plane, a staggered grid is used. Each cell contains a water level point, a point for the bottom elevation, a point for the velocity in the x-direction (u-velocity) and a point for the velocity in the y-direction (v-velocity). These points are not the same. The water level points are defined in the middle of each cell and the current components are defined on the cell boundaries. An additional program of Delft3D-RGFGRID is used for creating orthogonal curvilinear grids. A curvilinear grid allows a high grid resolution near the area of interest and along the coast and a low grid resolution outside this area. Therefore, the total number of cells is reduced and so is the computational time. The grid allows the boundaries of the grid well to be located outside the area of interest, which reduces errors due to boundary effects. Depending on the available bathymetry data and the desired accuracy in the area of interest, a grid resolution can be chosen for the model (Deltares 2003b).

For generating the bathymetry from measured data, an additional module Delft3D-QUICKIN is used. By using interpolation, the data of sample points is assigned to the grid cells (Deltares 2003c). The bathymetry surveys (echo-soundings) were carried out in the study area.

19.2.2 Initial and Boundary Conditions

A set of differential equations has no unique solution unless appropriate boundary conditions are specified. At the start of the simulation, the currents and water levels are set to zero in the whole area. Boundary conditions are applied on the open boundaries, which are used to keep a limited computational domain.

There are four basic types of open boundaries, each of which is applied in different situations:

- Water level boundaries
- Velocity boundaries
- Flux boundaries
- Riemann boundaries (weakly reflective boundaries)

The Riemann boundary is a weakly reflective boundary. Both the water level and the current have to be known. Usually these boundaries are used to retain a limited computational area. Waves can cross these boundaries unhampered and without reflections. The choice of the type of boundary condition depends on the phenomena to be studied. When modelling tidal flow in a large basin, forcing by prescribing only water levels is generally used. If these boundary conditions dare not satisfactory, combinations of water level and current boundaries can be imposed. The criteria to determine what conditions are satisfactory may be extracted from other models or from measurements.

19.2.3 Numerical Stability

In the Delft3D-FLOW, the Courant number is an indication of numerical stability and accuracy. The derivation of the Courant number is based on experience. For places with large differences in bottom geometry or coastline, the Courant number should not exceed the value of 10. The Courant number gives the relation between the propagation speed and the time step. The magnitude of the time step determines the total computational time. To reduce the total computational time, it is necessary to choose the largest time step possible, without the loss of accuracy and stability. The following parameters are of importance for setting the time step:

- stability
- required accuracy
- size of the smallest grid cell
- water depth
- available computational time

19.3 Model Results and Discussion

The general computation conditions of the hydrodynamic model are shown in Table 19.1.

19.3.1 Model Domain

Figure 19.2 shows the horizontal computational grid and Fig. 19.3 shows the bottom topography derived from the bathymetric and topographic survey. A total of 206 grid cells in M direction and 21 in N direction were used, which means a total of 4326 grid cells were used in the computational domain. The bathymetry contours shown in Fig. 19.3 depict main features of the study area. It is clear that the eastern part of the area is high enough and so would not be inundated by the effluent. However, storm-water coming from the eastern side could inundate the whole lagoon area.

Historical meteorological data (e.g. wind, solar radiation, air temperature, relative humidity) were obtained from Meteorological Department. The simulation was performed for 2 weeks (August 7–21, 2009) to cover a cycle of neap and spring tides. The heat flux model at the water surface took into account the separate effects of solar and atmospheric radiation, and heat loss due to back radiation, evaporation and convection. At open boundaries, the temporal variations of water temperature were set based on the field survey.

No flux condition was applied to the bottom and closed boundaries. Since the area is rather shallow, depth-averaged computation of hydrodynamic parameters

Table 19.1 Computation condition of hydrodynamic model

Item	Condition
Simulation period	August 7–21, 2009
Initial condition	Constant water temperature (25.6 °C)
Computational domain	Horizontal orthogonal curvilinear 206 × 21 with one layer (depth-averaged condition)
Boundary condition	Predicted water level at open boundary (eastern boundary)
Meteorological data	Uniform wind, air temperature, solar radiation and humidity from historical weather data
Bottom stress	Colebrook equation

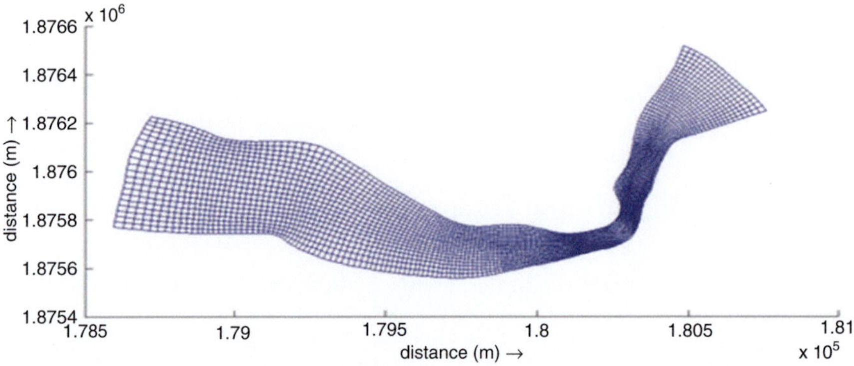

Fig. 19.2 Model grid used in the computations. The axes show UTM coordinates Easting on *horizontal axis* and Northing on *vertical axis*

was considered to be appropriate. The model was calibrated and validated against the observed data from the field measurements.

19.3.2 Hydrodynamic Results

A number of points were selected in the model domain to illustrate main features of the hydrodynamic and thermal dispersion modelling. The locations of these points along with the location of outlet structure are listed in Table 19.2.

Figures 19.4 and 19.5 show the tidal elevations at DPF1 and OP2 points in the model area. The point closest to the outlet structure is DPF1, which clearly shows higher mean elevations due to the effluent discharge. In the absence of effluent discharge, the location DPF1 is usually dry. However, the tidal elevations away from the outlet gradually attain the values imposed at the eastern boundary as depicted by Fig. 19.5.

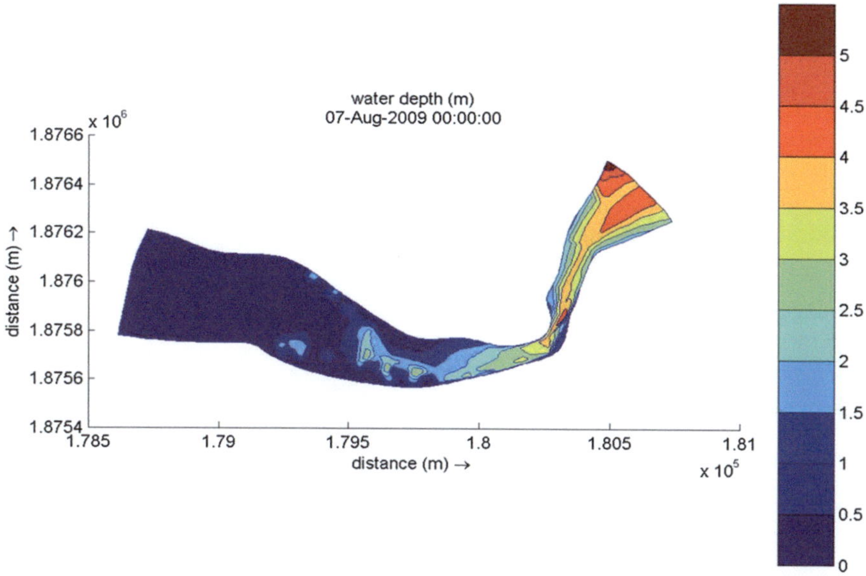

Fig. 19.3 Bathymetry used in the computations. The axes show UTM coordinates, Easting on the *horizontal axis* and Northing on the *vertical axis*

Table 19.2 Location of important points in the model area

Name	Easting (m)	Northing (m)
Outlet Structure (DP)	179,283	1,875,740
DPF1	179,558	1,875,690
DPF2	179,568	1,875,788
DPF3	179,780	1,875,646
OP0	179,981	1,875,662
OP1	180,274	1,875,762
OP2	180,522	1,876,322

It is obvious that the effect of effluent discharge has almost no effect on the water surface elevation in the open sea adjacent to Salalah Port boundary.

19.3.3 Thermal Dispersion Results

The initial and boundary conditions of the temperature were based on field observation data. The temperature just at the exit of the outfall was assumed to be 38 °C. The ambient seawater temperature was set to an average of 25.6 °C as found in the measurements carried out on July 30, 2009. The hydrodynamic data was obtained from the Delft3D-Flow module as mentioned above. From the model computations,

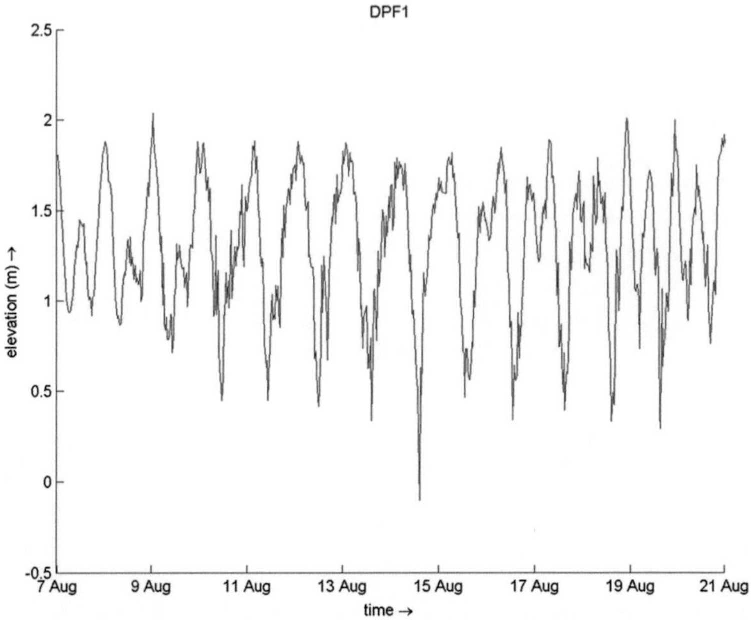

Fig. 19.4 Tidal elevations at the observation station close to the outlet (DPF1)

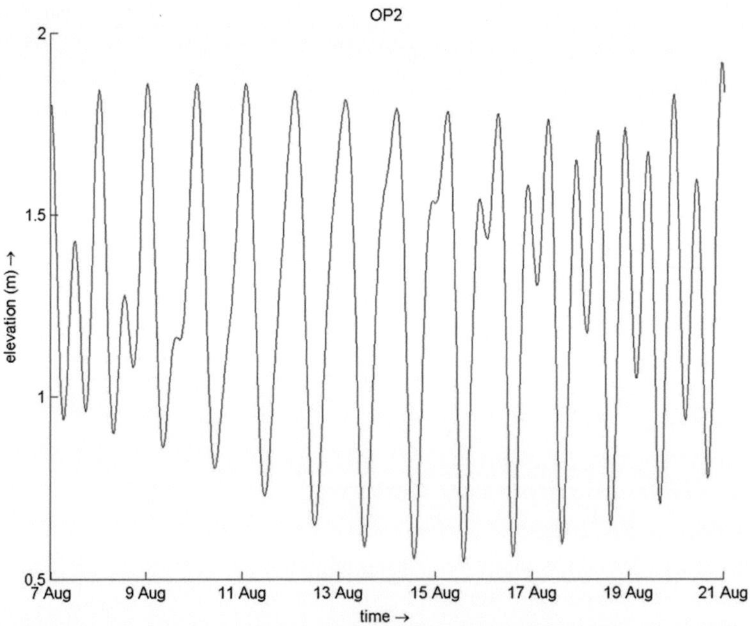

Fig. 19.5 Tidal elevations at the observation station OP2 (Eastern boundary in the sea)

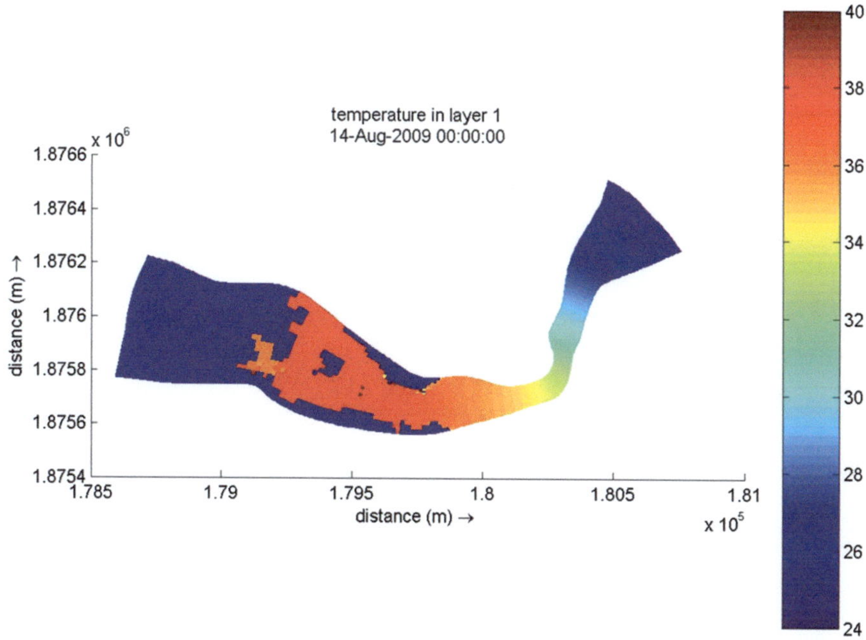

Fig. 19.6 Predicted temperature distribution in the modelled area on August 14, 2009, 00:00 h. Easting (*horizontal axis*) and Northing (*vertical axis*) are shown in meters

it was found that the effluent with high temperature starts affecting the lagoon area in front of the outlet structure (Fig. 19.6).

The western part of the model domain remains unaffected because of higher bottom elevations. The water temperature gradually decreases as the flow continues in the tidal channel to the mouth in the open sea. However, well before it reaches the mouth, the water temperature attains a value close to ambient temperature (25.6 °C). Figures 19.6, 19.7, 19.8, 19.9, 19.10, 19.11, and 19.12 show the distribution of water temperature in the model domain for a complete tidal cycle.

For high tide (Figs. 19.7 and 19.8) the effect of the effluent temperature is confined to a smaller area than for low tide (Figs. 19.9 and 19.10). This is logical because, as the tidal elevations grow larger, the tidal discharge into the lagoon area (opposite to the direction of effluent discharge) is increased, thereby causing the effect of effluent temperatures to remain confined in a smaller area. On the other hand, during low tide, the tidal discharge occurs out of the lagoon area (same direction as the effluent discharge), causing the effect of effluent temperature to extend to a somewhat larger area. However, in both the situations due to the advection process, the seawater temperature at the eastern boundary is unaffected by the effluent discharge.

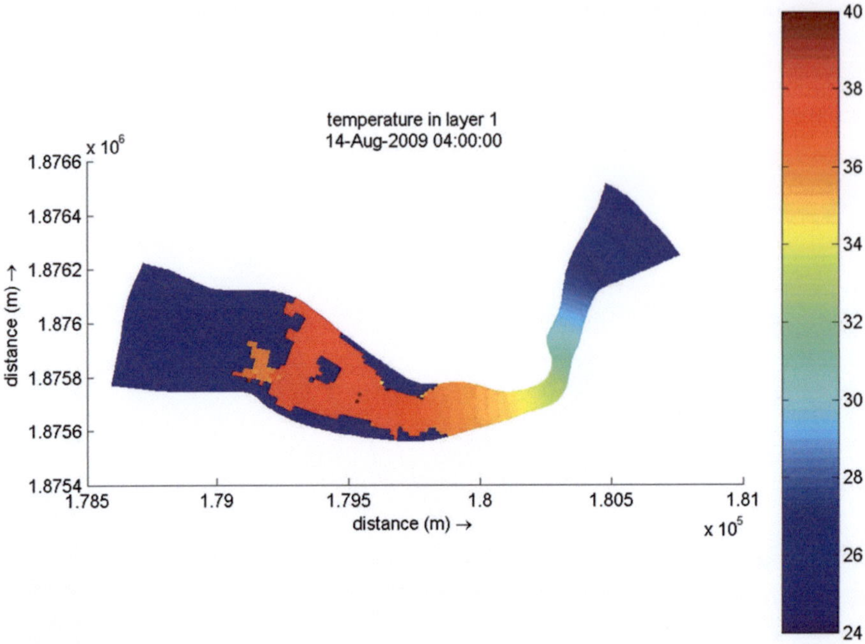

Fig. 19.7 Predicted temperature distribution in the modelled area on August 14, 2009, 04:00 h. Easting (*horizontal axis*) and Northing (*vertical axis*) are shown in meters

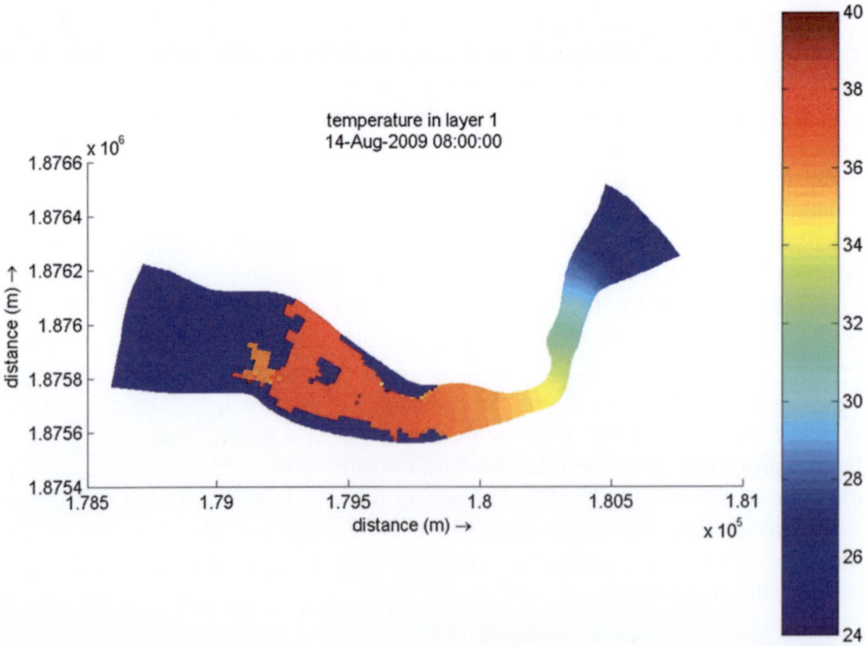

Fig. 19.8 Predicted temperature distribution in the modelled area on August 14, 2009, 08:00 h. Easting (*horizontal axis*) and Northing (*vertical axis*) are shown in meters

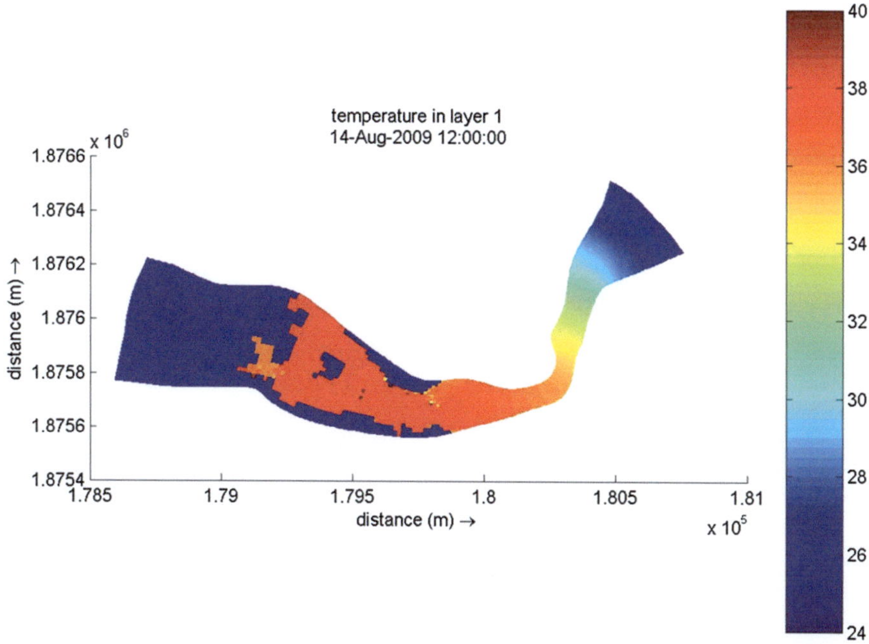

Fig. 19.9 Predicted temperature distribution in the modelled area on August 14, 2009, 12:00 h. Easting (*horizontal axis*) and Northing (*vertical axis*) are shown in meters

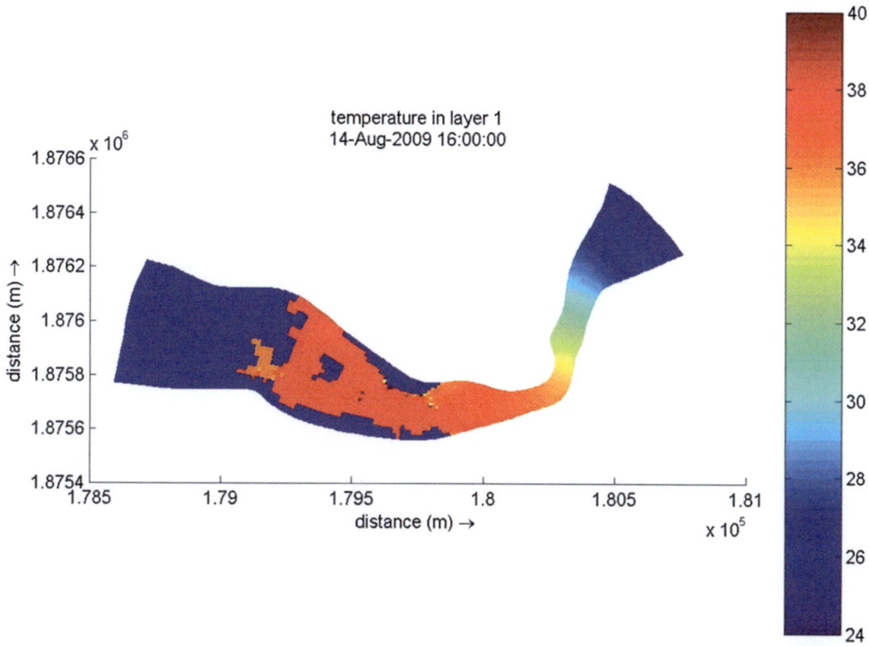

Fig. 19.10 Predicted temperature distribution in the modelled area on August 14, 2009, 16:00 h. Easting (*horizontal axis*) and Northing (*vertical axis*) are shown in meters

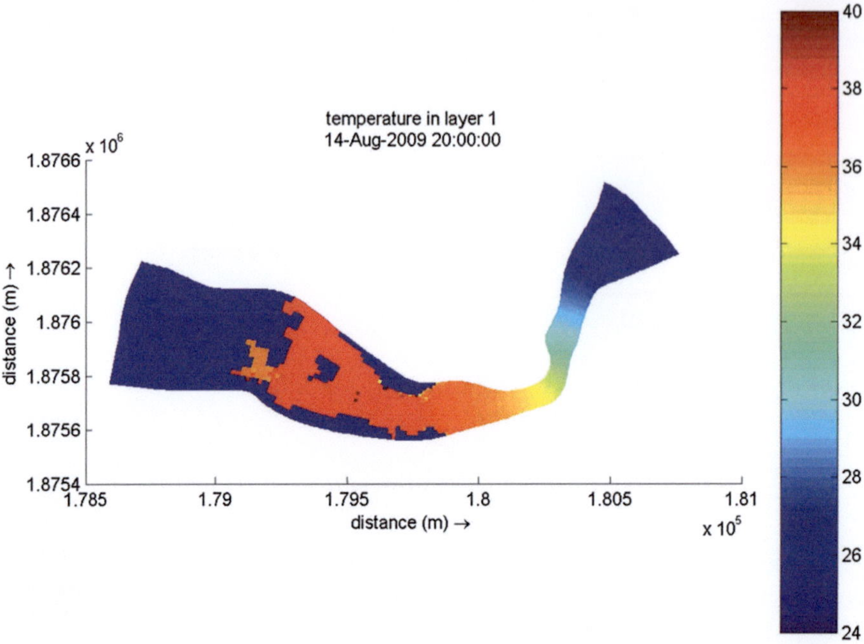

Fig. 19.11 Predicted temperature distribution in the modelled area on August 14, 2009, 20:00 h. Easting (*horizontal axis*) and Northing (*vertical axis*) are shown in meters

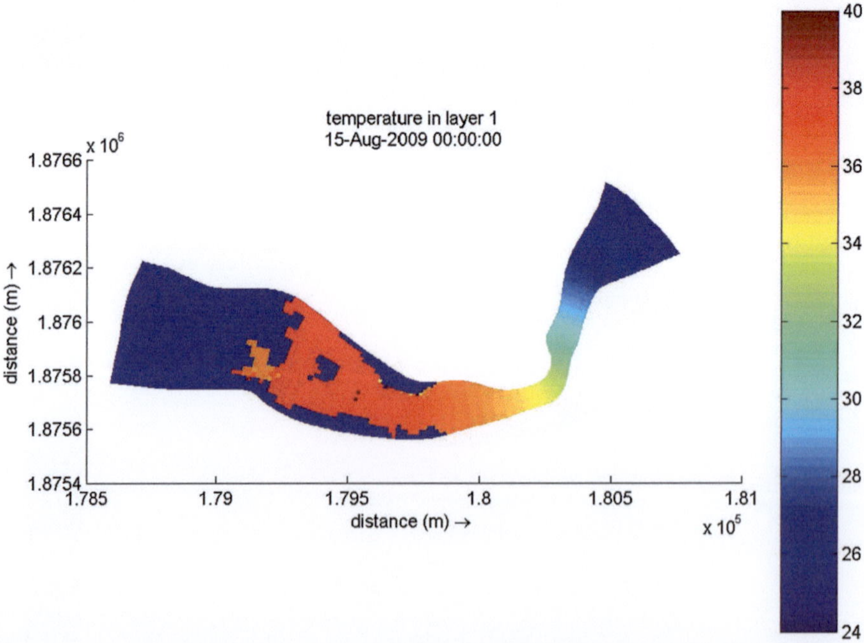

Fig. 19.12 Predicted temperature distribution in the modelled area on August 15, 2009, 00:00 h. Easting (*horizontal axis*) and Northing (*vertical axis*) are shown in meters

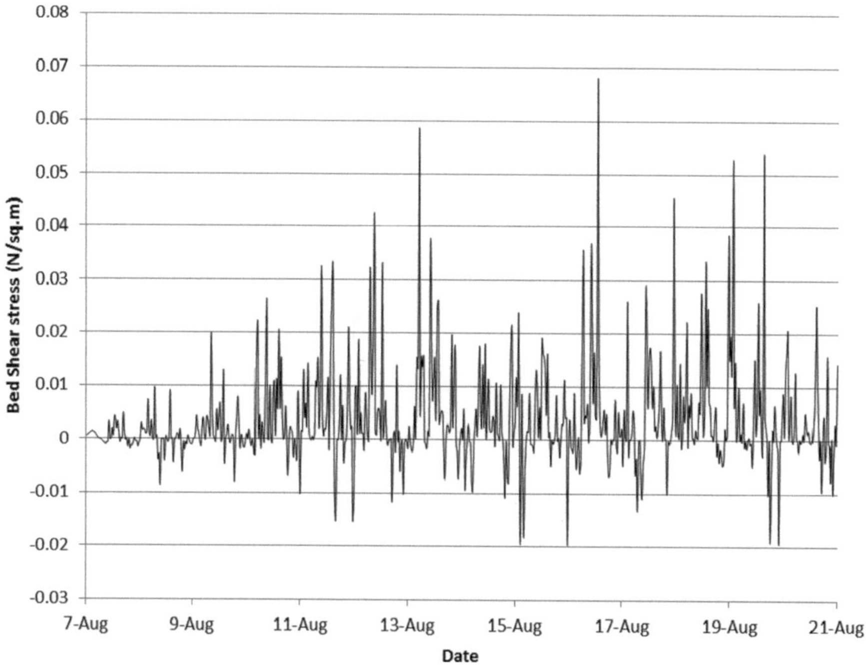

Fig. 19.13 Bed shear stress (Eastward) at OP0

19.3.4 Bed Shear Stress Results

The bed shear stress is one of the most important hydrodynamic parameters in relation to the morphologic changes in the study area because of effluent discharge. The eastward and northward components of bed shear stress at point OP0 are shown in Figs. 19.13 and 19.14. Except for a few spikes, the maximum value of bed shear stress remains less than 0.03 N/m^2. Such a small value of bed shear stress would not cause sediment movement in the lagoon area where the bed material particle sizes vary from 0.075 mm to above 3 cm.

The point OP0 is located to the west of the Salalah Port boundary as evident from the coordinates of this point in Table 19.2. Therefore, the possibility of erosion along the protection wall of the Salalah Port boundary is very little due to the effect of effluent discharge from the plant outlet structure. However, storm-water flow in the lagoon from the east may cause erosion in the tidal channel. However, computation for the effect of storm-water was not in the scope of the present study.

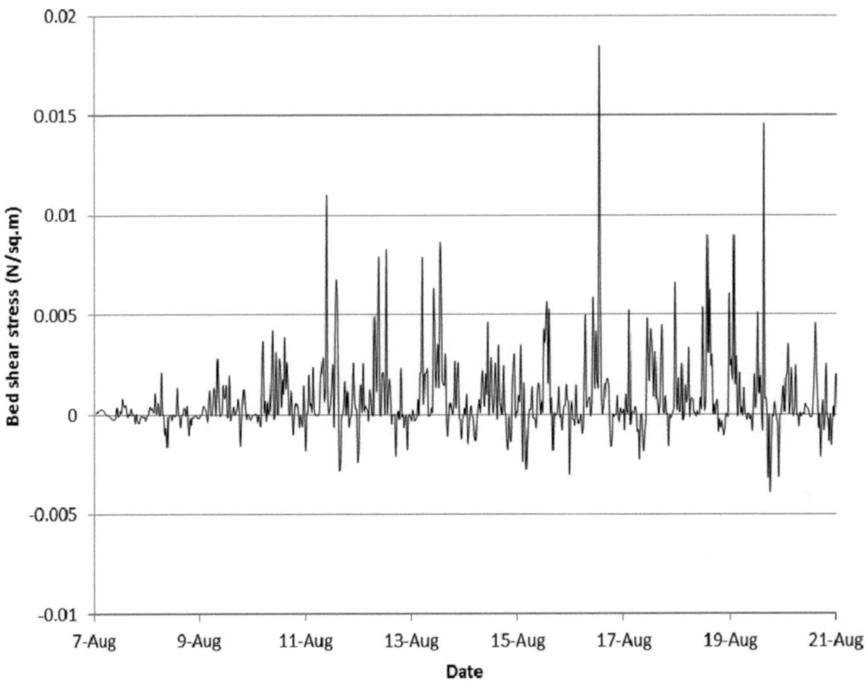

Fig. 19.14 Bed shear stress (Northward) at OP0

19.4 Conclusions

As a result of the hydrodynamic and thermal dispersion modelling of the plant outlet area with the help of the Delft3D model, it can be inferred that:

- During the period of prediction (August 7–21, 2009), the effect of effluent discharge on the seawater temperature remains confined within the lagoon area.
- Detailed information regarding water level, temperature, depth-averaged velocity and bed shear stress is available for the whole model domain.
- Considering the prevailing hydrodynamic properties of the site, the seawater quality in terms of temperature meets the requirements of the Ministry of Environment and Climate Affairs (MECA) Oman, since the impact zone of the effluent is confined within the lagoon area where the discharge point is located.
- Extremely low values of bed shear stress in the model domain suggest that the possibility of erosion and accretion of eroded material is minimal.

List of Figures

Fig. 19.2 Model grid used in the computations. The axes show UTM coordinates Easting on *horizontal axis* and Northing on *vertical axis*
Fig. 19.3 Bathymetry used in the computations. The axes show UTM coordinates, Easting on the *horizontal axis* and Northing on the *vertical axis*
Fig. 19.4 Tidal elevations at the observation station close to the outlet (DPF1)
Fig. 19.5 Tidal elevations at the observation station OP2 (Eastern boundary in the sea)
Fig. 19.6 Predicted temperature distribution in the modelled area on August 14, 2009, 00:00 h. Easting (*horizontal axis*) and Northing (*vertical axis*) are shown in meters
Fig. 19.7 Predicted temperature distribution in the modelled area on August 14, 2009, 04:00 h. Easting (*horizontal axis*) and Northing (*vertical axis*) are shown in meters
Fig. 19.8 Predicted temperature distribution in the modelled area on August 14, 2009, 08:00 h. Easting (*horizontal axis*) and Northing (*vertical axis*) are shown in meters
Fig. 19.9 Predicted temperature distribution in the modelled area on August 14, 2009, 12:00 h. Easting (*horizontal axis*) and Northing (*vertical axis*) are shown in meters
Fig. 19.10 Predicted temperature distribution in the modelled area on August 14, 2009, 16:00 h. Easting (*horizontal axis*) and Northing (*vertical axis*) are shown in meters
Fig. 19.11 Predicted temperature distribution in the modelled area on August 14, 2009, 20:00 h. Easting (*horizontal axis*) and Northing (*vertical axis*) are shown in meters
Fig. 19.12 Predicted temperature distribution in the modelled area on August 15, 2009, 00:00 h. Easting (*horizontal axis*) and Northing (*vertical axis*) are shown in meters
Fig. 19.13 Bed shear stress (Eastward) at OP0
Fig. 19.14 Bed shear stress (Northward) at OP0

References

Deltares (2003a) Delft3D-Flow user manual. Deltares Systems, Delft, The Netherlands
Deltares (2003b) Delft3D-RGFGRID user manual. Deltares Systems, Delft, The Netherlands
Deltares (2003c) Delft3D-QUICKIN user manual. Deltares Systems, Delft, The Netherlands

Chapter 20
Formation of Emerging Disinfection By-products by Chlorination/ Chloramination of Seawater Impacted by Algal Organic Matter

Maolida Nihemaiti, Julien Le Roux, and Jean-Philippe Croué

Abstract The aim of this work was to study the formation of haloacetamides (HAcAms) and other DBPs during chlorination and chloramination of algal organic matter (AlOM). The HAcAms formation potentials of different precursors (amino acids, simulated algal blooms grown in the Red Sea) were evaluated. Experiments with simulated algal blooms were conducted in the presence of bromide ion (synthetic seawater containing 800 μg/L Br^-) to assess the formation of brominated analogues of HAcAms in conditions close to the disinfection of real seawater. Chlorination produced more HAcAms than chloramination from real algae (Synecococcus sp.), thus indicating that the nitrogen of HAcAms comes predominantly from DON through the decarboxylation of amino acids rather than from NH_2Cl. Dibrominated species of DBPs (i.e., DBAcAm, DBAA and DBAN) were the dominant species formed by both chlorination and chloramination of algal bloom samples. Chloramination of the amino acid asparagine produced an important amount of DCAcAm as compared to chlorination, indicating the existence of a specific reaction pathway.

Keywords Chlorine • Chloramines • Disinfection by-products • Algal organic matter • Haloacetamides

M. Nihemaiti • J. Le Roux
Water Desalination and Reuse Center, King Abdullah University of Science and Technology (KAUST), 4700, Thuwal 23955-6900, Saudi Arabia
e-mail: maolida.nihemaiti@kaust.edu.sa; julien.leroux5@gmail.com

J.-P. Croué (✉)
Water Desalination and Reuse Center, King Abdullah University of Science and Technology (KAUST), 4700, Thuwal 23955-6900, Saudi Arabia

Curtin Water Quality Research Centre, Curtin University, Bentley, Perth, Western Australia
e-mail: jean-philippe.croue@curtin.edu.au

M. Baawain et al. (eds.), *Recent Progress in Desalination, Environmental and Marine Outfall Systems*, DOI 10.1007/978-3-319-19123-2_20

20.1 Introduction

Prechlorination is used in the seawater desalination processes to control biofouling in both thermal plants and membrane plants. Important amounts of disinfectant can be required to maintain a residual during algal bloom events. Many disinfection by-products (DBPs) are formed during seawater pre-treatment processes using chlorination, including the regulated trihalomethanes (THMs) and haloacetic acids (HAAs). Nitrogenous DBPs (N-DBPs) generally form in much smaller amounts than regulated DBPs, but have been a growing concern because of their greater health risk (Plewa et al. 2004; Muellner et al. 2007). *In vitro* mammalian cell tests demonstrated that haloacetonitriles (HANs), halonitromethanes (HNMs) and haloacetamides (HAcAms) are more cytotoxic and genotoxic (up to two orders of magnitude) than non-nitrogenous THMs and HAAs (Plewa et al. 2008).

As an alternative disinfectant, monochloramine (NH_2Cl) can be used in cooling systems to get a stable disinfectant residual. However, the use of NH_2Cl can be a source of nitrogen which may lead to the formation of N-DBPs. Moreover, the disinfection of seawater can produce significant amounts of N-DBPs, especially during algal blooms, enriched in nitrogen-containing compounds (e.g., amino acids). Arid regions produce water by desalination of seawater or brackish water, characterized by elevated bromide and iodide ion contents. The presence of high concentrations of bromide (60 mg Br^-/L in the Red Sea) and iodide (0.05 mg I^-/L) ions in seawater favors the formation of brominated and iodinated by-products that are often more toxic than their chlorinated analogues (Richardson et al. 2007, 2008).

Haloacetamides are N-DBPs that produce high levels of toxicity compared to other DBPs, but their formation mechanisms are unclear. They are known to be products of the hydrolysis of HANs (Reckhow et al. 2001), however they were reported to be formed independently by chlorination and chloramination of natural organic matter (NOM) (Huang et al. 2012).

Two reaction pathways have been proposed for the formation of haloacetamides during water disinfection. The first one is the decarboxylation pathway, producing haloacetonitriles that are then hydrolyzed into haloacetamides. In this case, the nitrogen atom of haloacetamides comes from the precursor compound (i.e., amino acids, organic nitrogen) (Huang et al. 2012). The second reaction pathway is the aldehyde pathway, where NH_2Cl reacts with aldehydes to produce HANs and HAcAms (Kimura et al. 2013). In this case, the nitrogen of HAcAms is incorporated from NH_2Cl.

Literature about the formation mechanism of HAcAms and their precursors in NOM, in wastewater, in algal organic matter (AlOM) or in seawater organic matter is limited. The HAcAm formation potential of some amino acids (Chu et al. 2010; Huang et al. 2012), humic acids, wastewater effluent and algal EPS (Huang et al. 2012) and culture bacteria (Huang et al. 2013) were reported but no studies were focused on seawater algal organic matter.

This study investigates the formation mechanisms of haloacetamides during chlorination and chloramination of algal organic matter (AlOM). The aim was to evaluate the importance of the decarboxylation and the aldehyde pathway from different precursors (amino acids, simulated algal blooms). Experiments with simulated algal blooms were conducted in the presence of bromide (synthetic seawater containing 60–65 mg/L Br^-) to assess the formation of brominated analogues of haloacetamides.

20.2 Materials and Methods

All experiments were conducted using deionized water (Milli-Q, Millipore) buffered with a mixture of sodium phosphate monobasic and sodium phosphate dibasic. Solution pH values were adjusted as needed, using sodium hydroxide or sulfuric acid (0.1 N, Fisher Scientific). Fisher Scientific MTBE and ethyl acetate (>99 %) were used without further purification. Amino acids (L-asparagine >98 %, L-tyrosine >99 %, L-tryptophan >99 % and phenylalanine >99 %) were used without further purification and were supplied through Sigma-Aldrich (Fig. 20.1). Sodium hypochlorite (NaOCl, 5.65–6 %, Fisher Scientific) and ammonium chloride (Acros Organics, 99.6 %) were used as chlorination reagents. Ammonium-^{15}N-chloride was purchased from Sigma-Aldrich. Sodium thiosulfate (Fisher Scientific) was used to quench residual oxidant. A mixed standard containing haloacetonitriles (HANs), trichloronitromethane (TCNM) and haloketones (HKs) (EPA 551B Halogenated Volatiles Mix), and another mixed standard containing 9 HAAs (EPA 552.2 Methyl Ester Calibration Mix) and surrogate standard decafluorobiphenyl (99 %), were supplied from Supelco (Sigma-Aldrich). Chloro-, bromo- dichloro-

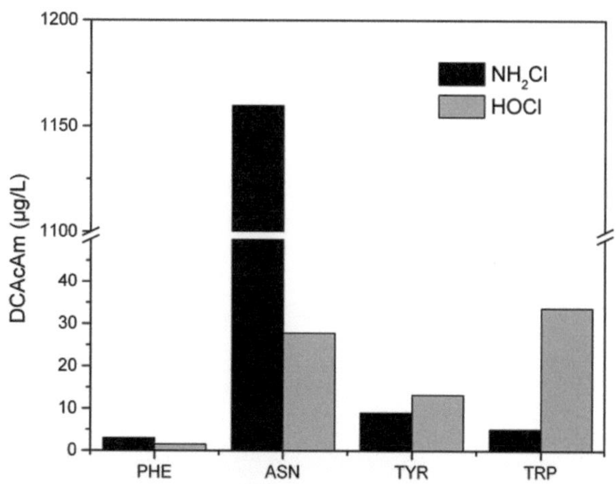

Fig. 20.1 DCAcAm formation potential from chlorination and chloramination (15 mg Cl_2/L) of amino acids (50 μM) after 72 h contact time at pH8 (10 mM carbonate buffer)

Table 20.1 Molecular structures of investigated amino acids

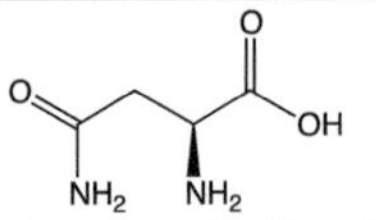	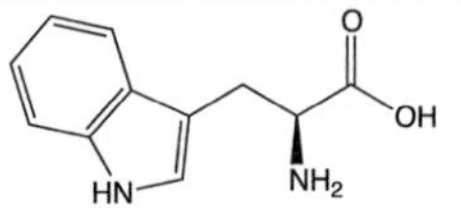	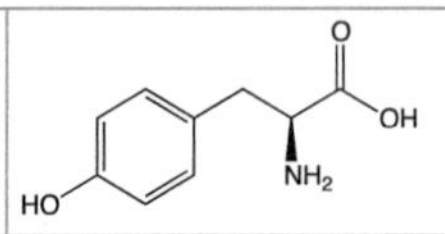	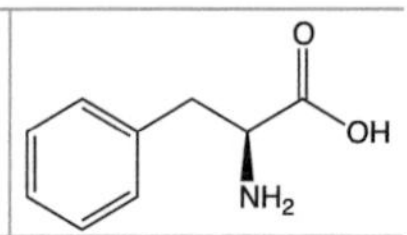
L-asparagine	L-tryptophan	L-tyrosine	Phenylalanine

and tri- chloroacetamides were obtained from Sigma-Aldrich. Other haloacetamides (HAcAms) were purchased from Cansyn Chem. Corp. (Table 20.1).

20.2.1 *Preparation and Analysis of Chlorine and Chloramine*

Monochloramine (NH_2Cl) stock solutions (50 mM) were prepared fresh daily by dissolving ammonium chloride (NH_4Cl) in deionized water adjusted to pH = 8.5 with sodium hydroxide. Sodium hypochlorite (NaOCl) was then added slowly to the rapidly stirred solution, at a Cl:N molar ratio of at least 1:1.2 to avoid breakpoint chlorination resulting from local excess of hypochlorite. Adjusting the pH at 8.5 minimizes the disproportionation of NH_2Cl to dichloramine ($NHCl_2$), since $NHCl_2$ forms at pH < 8 according to equilibrium 1:

$$2NH_2Cl + H^+ = NHCl_2 + NH_4{}^+ \quad (20.1)$$

NH_2Cl and $NHCl_2$ were quantified in stock solutions by monitoring absorbance at their respective λ_{max} ($\lambda_{NH2Cl} = 245$ nm; $\lambda_{NHCl2} = 295$ nm). Residual oxidant in samples after experiments was analysed iodometrically (Standards Methods for the Examination of Water and Wastewater 1995).

20.2.2 *Experimental Methods*

All glassware used during these experiments was washed and baked at 500 °C for at least 5 h prior to use. Reactions were conducted in sealed 65 mL amber glass bottles at room temperature (20 °C). Chloramination experiments were conducted following the approach of Krasner et al. (2004), using an excess of oxidant (15 mg/L as Cl_2) and a reaction time of 72 h at pH 8.

Simulated algal blooms grown directly in the Red Sea (mesocosms) were employed as the algal source (main species Synecococcus sp.). Aqueous solutions were prepared by the dilution of seawater samples (simulated algal blooms) with synthetic seawater (Grasshoff et al. 1976) to the desired DOC concentration (1.5 mg C/L). Amino acids solutions were prepared by dissolving a pre-determined amount of amino acids in Milli-Q water containing 10 mM carbonate buffer. 275 μL of preformed monochloramine were then added. Each series of experiments included a blank (buffered Milli-Q water) and each sample was triplicated. After 72 h of contact time, 5 mL of the samples were used for residual chlorine analysis and 3 × 20 mL were used for DBP analysis. Molar yields were calculated based on the molar concentrations of amino acids (mol DBP formed × 100/mol initial compound).

20.2.3 DBP Analysis

4 THMs, 4 haloacetonitriles (HANs), 2 haloketones (HKs) and chloropicrin were extracted following the EPA method 551, which consists of a liquid-liquid extraction using MTBE, followed by detection using an Agilent 7890A gas chromatograph coupled with an electron capture detector (GC-ECD) (Munch and Hautman 1995). 9 HAAs were extracted and analyzed following the EPA method 552.2, which is based on a liquid-liquid extraction with MTBE in acidic conditions followed by derivatization to methyl esters using acidic methanol (Munch et al. 1995). Extracts were analysed using an Agilent 7890A GC system coupled with a 5975C mass spectrometer (GC-MS). HAcAms were analysed following the same protocol as the EPA method 551, but using ethyl acetate for the liquid-liquid extraction step. All DBPs analysed are described in Table 20.2.

20.3 Results and Discussion

20.3.1 Formation of Haloacetamides from Amino Acids

The formation of haloacetamides was investigated after 72 h of chlorination and chloramination (15 mg/L as Cl_2) of 4 amino acids (50 μM each) at pH 8 (Fig. 20.1). Asparagine was found to be a major dichloracetamide (DCAcAm) precursor during chloramination, producing up to 1160 μg/L DCAcAm. This can be attributed to the amide group present in the asparagine structure (Table 20.1). However, this does not explain why chloramination formed much more DCAcAm than chlorination. Other amino acids were much lower precursors of DCAcAm. In the case of tyrosine and tryptophan, more DCAcAm was formed by chlorination. These results indicate that different reaction pathways occur during chlorination and chloramination of amino acids. A specific pathway seems to occur specially during the chloramination of asparagine.

Huang et al. (2012) proposed a formation mechanism to explain the formation of DCAcAm from asparagine, but did not describe a specific pathway for NH_2Cl, because the formation of DCAcAm was similar by chlorination and by

Table 20.2 Description of monitored DBPs

Method of determination	Class of compounds and structures	Abbreviation	Full name
EPA 551	Trihalomethanes (THMs)	TCM	Trichloromethane (chloroform)
	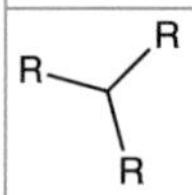	DCBM	Dichlorobromomethane
		DBCM	Dibromochloromethane
		TBM	Tribromomethane (bromoform)
	Haloacetonitriles (HANs)	DCAN	Dichloroacetonitrile
	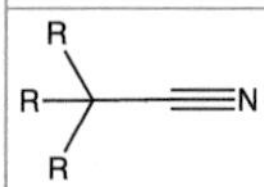	TCAN	Trichloroacetonitrile
		BCAN	Bromochloroacetonitrile
		DBAN	Dibromoacetonitrile
	Haloketones (HKs)	1,1-DCP	1,1-dichloropropanone
	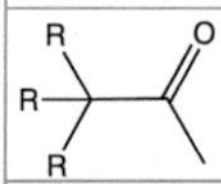	1,1,1-TCP	1,1,1-trichloropropanone
	Halonitromethane	TCNM	Trichloronitromethane
	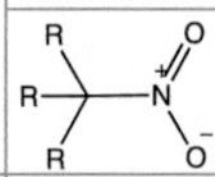		
EPA 551 – EA	Haloacetamides (HAcAms)	CAcAm	Chloroacetamide
	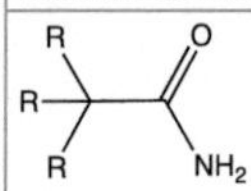	BAcAm	Bromoacetamide
		DCAcAm	Dichloroacetamide
		TCAcAm	Trichloroacetamide
		DBAcAm	Dibromoacetamide
		TBAcAm	Tribromoacetamide
		BCAcAm	Bromochloroacetamide
		CIAcAm	Chloroiodoacetamide
		BIAcAm	Bromoiodoacetamide
		DIAcAm	Diiodoacetamide
EPA 552.2	Haloacetic acids (HAAs)	MCAA	Monochloroacetic acid
	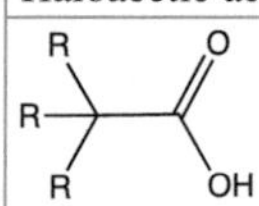	MBAA	Monobromoacetic acid
		DCAA	Dichloroacetic acid
		TCAA	Trichloroacetic acid
		BCAA	Bromochloroacetic acid
		DCBAA	Dichlorobromoacetic acid
		DBAA	Dibromoacetic acid
		DBCAA	Dibromochloroacetic acid
		TBAA	Tribromoacetic acid

chloramination in their experimental conditions (100 μM asparagine, 10 mg/L as Cl_2, $t = 2$ h, pH 6.9).

TCAcAm formation was more important than DCAcAm formation during the chlorination of amino acids, indicating a higher incorporation of chlorine atoms

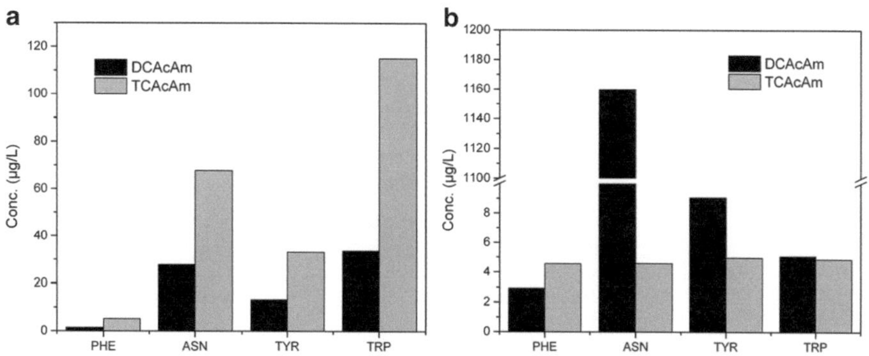

Fig. 20.2 Formation of DCAcAm and TCAcAm from (**a**) chlorination and (**b**) chloramination of amino acids ([amino acids] = 50 μM; [oxidant] = 15 mg/L as Cl_2; t = 72 h; pH = 8)

because of higher chlorine transfer from HOCl to haloacetamides (Fig. 20.2a). Tryptophan was the most important precursor of TCAcAm from chlorination, with up to 115 μg/L of TCAcAm formed. During chloramination, levels of TCAcAm formed were very low and similar for all studied amino acids (i.e., 4.5–5 μg/L), indicating lower chlorine incorporation during chloramination.

20.3.2 Formation of Haloacetamides from Simulated Algal Blooms

Simulated algal blooms grown directly in the Red Sea (8000 L mesocosms) were employed as an algal source (main species Synecococcus sp.) for DBPs formation potential tests. Four different mesocosms were studied, each existing in a different set of nutrient conditions: (a) nitrates and phosphates added every day; (b) nitrates and phosphates added once; (c) nitrates, phosphates and silica added every day, and (d) nitrates, phosphates and silica added once. All four samples were diluted with synthetic seawater in equal concentrations of DOC (1.5 mg C/L) and were chlorinated and chloraminated (5 mg Cl_2/L) in the lab for 72 h. Results showed a higher formation of haloacetamides by chlorination than by chloramination (Fig. 20.3a). The first mesocosm formed the highest concentration of HAcAms (240 μg/L), probably because of the presence of a higher concentration of dissolved organic nitrogen (DON = 0.19 mg N/L in samples from mesocosm 1 and 3; DON = 0.12 mg N/L in samples from mesocosms 2 and 4) due to the continuous addition of nitrates. These results indicate that the nitrogen atom of haloacetamides would predominantly originate from organic nitrogen contained in algal cells, and not from NH_2Cl. This implies that the predominant mechanism of HAcAms formation from algae cells is the decarboxylation pathway, where the nitrogen atom of HAcAms originates from DON. This is also suggested by a similar trend in HANs formation which exhibits higher concentrations from chlorination than

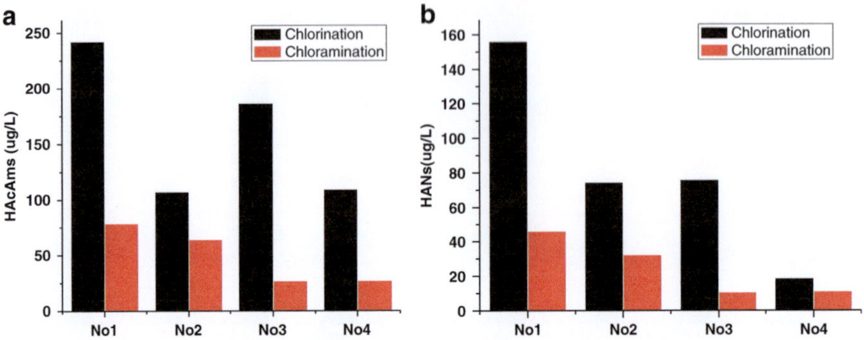

Fig. 20.3 Formation of (**a**) total HAcAms and (**b**) total HANs from chloramination and chlorination (5 mg/L as Cl_2) of four simulated algal blooms (mesocosms) (1.5 mg C/L) during 72 h

from chloramination (Fig. 20.3b). In the decarboxylation pathway, HAcAms are produced through HANs hydrolysis. Hence our results suggest a similar pathway for HANs and HAcAms formation, favoured by the presence of free chlorine.

As could be expected with the presence of high bromide ion content in the seawater samples (60–65 mg/L as Br^-), the proportion of brominated DBPs formed after chlorination and chloramination was more important than the amount of chlorinated DBPs. TBM, DBAcAm, DBAA, and DBAN were the predominant species among THMs, HAcAms, HAAs and HANs, respectively. This is explained by the fast reaction between chlorine and bromide ions leading to the production of hypobromous acid (HOBr) (Eq. 20.2), which is an oxidizing species even more reactive than hypochlorous acid (HOCl) (Heller-Grossman et al. 1993).

$$\mathrm{HOCl} + \mathrm{Br}^- \rightarrow \mathrm{HOBr} + \mathrm{Cl}^- \tag{20.2}$$

20.4 Conclusions

The aim of our work was to study the formation of HAcAms from chlorination and chloramination of marine algal organic matter. DBP formation potential tests were performed from model compounds (amino acids) and from simulated algal blooms in real seawater conditions. Results show that chlorination produces more HAcAms than chloramination from real algae (Synecococcus sp.), thus indicating that the nitrogen of HAcAms comes predominantly from DON through the decarboxylation of amino acids rather than from NH_2Cl. Results of HAcAm formation potentials from amino acids are in agreement with this pathway, where tyrosine and tryptophan exhibit higher DCAcAm formation by chlorination than by chloramination. However, asparagine was found to be a major DCAcAm precursor during chloramination, thus indicating a specific reaction pathway. While the amide group of asparagine is probably the main reason of its important DCAcAm formation potential, the specific role of NH_2Cl as compared to HOCl needs to be further investigated. The use of isotopically labelled monochloramine($^{15}NH_2Cl$) could help in understanding the incorporation of nitrogen from asparagine or from NH_2Cl.

Dihalogenated species of DBPs (i.e., DBAcAm, DBAA, DBAN) were the dominant species formed by both the chlorination and chloramination of mesocosms samples, while the chlorination of amino acids formed more TCAcAm than DCAcAm, indicating a higher incorporation of chlorine.

Future work will focus on the formation of haloacetamides by chloramination of different marine AlOM obtained from algae monocultures grown in the laboratory to understand the specificities of haloacetamide precursors from real algae. Reactions will be conducted in the presence of isotopically labelled $^{15}NH_2Cl$ to assess the proportion of nitrogen coming from organic nitrogen sources and from monochloramine.

List of Figures

References

Chu W-H, Gao N-Y, Deng Y, Krasner SW (2010) Precursors of dichloroacetamide, an emerging nitrogenous DBP formed during chlorination or chloramination. Environ Sci Tech 44 (10):3908–3912

Grasshoff K, Ehrhardt M, Kremling K, Anderson LG (1976) Methods of seawater analysis. Wiley

Heller-Grossman L, Manka J, Limoni-Relis B, Rebhun M (1993) Formation and distribution of haloacetic acids, THM and TOX in chlorination of bromide-rich lake water. Water Res 27:1323–1331

Huang H, Wu Q-Y, Hu H-Y, Mitch WA (2012) Dichloroacetonitrile and dichloroacetamide can form independently during chlorination and chloramination of drinking waters, model organic matters, and wastewater effluents. Environ Sci Tech 46(19):10624–10631

Huang H, Wu Q-Y, Tang X, Jiang R, Hu H-Y (2013) Formation of haloacetonitriles and haloacetamides during chlorination of pure culture bacteria. Chemosphere 92(4):375–381

Kimura SY, Komaki Y, Plewa MJ, Mariñas BJ (2013) Chloroacetonitrile and N,2-dichloroacetamide formation from the reaction of chloroacetaldehyde and monochloramine in water. Environ Sci Tech 47(21):12382–12390

Krasner SW, Sclimenti MJ, Guo YC, Hwang CJ, Westerhoff P (2004) Development of DBP and nitrosamine formation potential tests for treated wastewater, reclaimed water, and drinking water. In: Proceedings of the 2004 AWWA water quality technology conference. AWWA, Denver

Muellner MG, Wagner ED, Mccalla K, Richardson SD, Woo Y-T, Plewa MJ (2007) Haloacetonitriles vs. regulated haloacetic acids: are nitrogen-containing DBPs more toxic? Environ Sci Tech 41:645–651

Munch DJ, Hautman DP (1995) EPA Method 551.1. Determination of chlorination disinfection by-products, chlorinated solvents, and halogenated pesticides/herbicides in drinking water by liquid-liquid extraction and gas chromatography with electron-capture detection. Revision 1

Munch DJ, Munch JW, Pawlecki AM (1995) EPA Method 552.2. Determination of haloacetic acids and dalapon in drinking water by liquid-liquid extraction, derivatization and gas chromatography with electron capture detection, Revision 1

Plewa MJ, Wagner ED, Jazwierska P, Richardson SD, Chen PH, McKague AB (2004) Halonitromethane drinking water disinfection by-products: chemical characterization and mammalian cell cytotoxicity and genotoxicity. Environ Sci Tech 38:62–68

Plewa MJ, Wagner ED, Muellner MG, Hsu K-M, Richardson SD (2008) Comparative mammalian cell toxicity of N-DBPs and C-DBPs. In: Karanfil T, Krasner SW, Westerhoff P, Xie Y (eds) Disinfection byproducts in drinking water, occurence, formation, health effects and control, Symposium series no. 995. American Chemical Society, Washington, DC, pp 36–50

Reckhow DA, MacNeill AL, Platt TL, MacNeill AL, McClellan JN (2001) Formation and degradation of dichloroacetonitrile in drinking waters. J Water Supply Res Technol AQUA 50(1):1–13

Richardson SD, Plewa MJ, Wagner ED, Schoeny R, DeMarini DM (2007) Occurrence, genotoxicity, and carcinogenicity of regulated and emerging disinfection by-products in drinking water: a review and roadmap for research. Mutat Res Rev Mutat Res 636:178–242

Richardson SD, Fasano F, Ellington JJ, Crumley FG, Buettner KM, Evans JJ, Blount BC, Silva LK, Waite TJ, Luther GW, Mckague AB, Miltner RJ, Wagner ED, Plewa MJ (2008) Occurrence and mammalian cell toxicity of iodinated disinfection by-products in drinking water. Environ Sci Tech 42:8330–8338

Standard Methods for the Examination of Water and Wastewater (1995) 19th ed. American Public Health Association/American Water Works Association/Water Environment Federation, Washington DC

Chapter 21
Evaluation of Marine Outfall at Mumbai for Domestic Waste Discharges Through Tracer Technique

Shivani Dhage, Ritesh Vijay, and Prakash Kelkar

Abstract Many coastal cities release liquid waste into marine water bodies through point and non-point discharges. Marine outfalls with multi-port diffusers are one of the modern systems adopted by mega cities along the coast for disposal of large volume of domestic wastewaters. Once the marine outfall is operational, it is cumbersome to find its functional and operational behavior in the coastal region. The evaluation of the extent of dilution achieved is complicated due to many reasons like variation in direction and buoyancy of upward flow, density, currents, tidal effects and also the submergence of the physical components of the diffuser. All these limitations create difficulty in establishing the fulfillment of the rationale behind outfall installation because accurate information on the compliance of the coastal water standards is not readily computable.

Hydrodynamic simulation through mathematical models and through the experimental application of tracers are the commonly adopted techniques for finding the effectiveness of the outfall for the mixing behavior, dilution and dispersion patterns of waste plumes released through diffusors. Simulation through mathematical models provides an expected dilution and dispersion pattern of the waste plume but these models need precise calibration for hydrodynamic and water quality parameters. A tracer investigation is a valuable tool to obtain factual information about the dispersion of wastewater. Natural, fluorescent and radio tracers are used during the studies conducted for the city of Mumbai, India. The paper presents the comparison of natural and fluorescent dye tracers to find out the behavior and the

S. Dhage (✉)
National Environmental Engineering Research Institute, Mumbai Zonal Centre, Worli, Mumbai 400 018, India
e-mail: shivanid25@hotmail.com; shivanid25@gmail.com

R. Vijay
National Environmental Engineering Research Institute, Nehru Marg, Nagpur 440 020, India
e-mail: r_vijay@neeri.res.in

P. Kelkar
Formaly, Advisor, Rajiv Gandhi Science and Technology Commission, Government of Maharashtra, Nagpur, India
e-mail: ps_kelkar@rediffmail.com

M. Baawain et al. (eds.), *Recent Progress in Desalination, Environmental and Marine Outfall Systems*, DOI 10.1007/978-3-319-19123-2_21

travel path of pollutant concentration through the distribution of the tracers in both longitudinal and transverse directions from an existing ocean outfall in Mumbai.

Keywords Domestic wastewater disposal • Marine outfalls • Simulation through mathematical models • DO • BOD • Fluorescent dye tracer

21.1 Introduction

The joining of seven islands has formed the metropolitan city of Mumbai, the commercial capital of India which has a population of 12.4 million (2011). The Municipal Corporation of Greater Mumbai (MCGM) is intensifying essential infrastructure services like water supply, wastewater collection, treatment, disposal, solid waste management etc. to fulfill the basic civic needs of present and future inhabitants. Due to the scarcity of land, instead of conventional wastewater treatment schemes, preliminarily treated domestic waste was earlier discharged in the near-shore shallow west coast. This has resulted in the deterioration of the marine environment and the spoiling of recreational beaches. To comply with stipulated stringent marine water standards, urgent action was taken to minimize pollution by adopting alternative treatment options for the disposal of domestic waste. Preliminary treated domestic wastewater is now released through "marine outfalls", one of the favored and cheapest options for coastal cities.

In Mumbai, the organized sewerage system collects about 60–70 % domestic wastewater and the remaining quantity reaches the near-shore coast via non-point sources (Environmental Status Report 2006–2007). About 900 mld of preliminary treated effluent is discharged through two marine outfalls at Worli and Bandra located on the city's west coast. The details of the wastewater discharges and BOD load through point discharges are presented in Table 21.1.

Table 21.1 Pollution load through organized wastewater discharge[a]

Service zone	Organized wastewater flow (m^3/d)	Method of disposal	Avg. BOD (mg/L)	Pollution load (kg/day)
Colaba	35	Marine outfall	312	10,920
Worli	393	Marine outfall	316	124,188
Bandra	473	Marine outfall	170	80,410
Ghatkopar	120	Aerated lagoon	134	16,080
Bhandup	121	Aerated lagoon	67	8107
Versova	190	Aerated lagoon	85	16,150
Malad	101	Marine outfall (Proposed)	220	22,220

[a]Total flow = 1433 MLD (2009) Total pollution load = 278,075 kg/day

21.2 Description Wastewater Treatment Facility with Marine Outfall

The marine outfall system comprises a shaft above ground level receiving treated wastewater by gravity, a specially designed underground tunnel 60 m below ground level to convey the wastewater 3.2 km from the shoreline inside the sea, and a diffuser section of 270 m with 10 ports located at the bottom of the sea bed. Each port has a number of openings with a facility to adjust the degree and angle of aperture so as to achieve even distribution of waste flows which offers an environmentally-sound engineering solution for proper release of waste flows. Typical layout of the treatment system at the wastewater treatment facility is presented in Fig. 21.1.

21.3 Methods and Experimental Programme

The marine environment is a very complex and dynamic system and hence the process of dilution and dispersion is quite multifaceted. The plume of wastewater released through the outfall's diffusers moves in the upward direction without merging. The waste field generated at the surface of the water gets diluted and is dispersed into the tidally-influenced coastal region. The mixing behavior, dilution and dispersion of released wastewater all depend on the interplay of ambient conditions of the receiving water body (Gupta et al. 2006). The main goal of the present study was to establish the changes in water quality of the coastal region due to the impact of the marine outfall, compliance with the stipulated coastal standards and conservation of the natural ecosystem (Kumta and Kumar 1998). At the near-field, the wastewater comes out to the surface through the risers of the diffuser

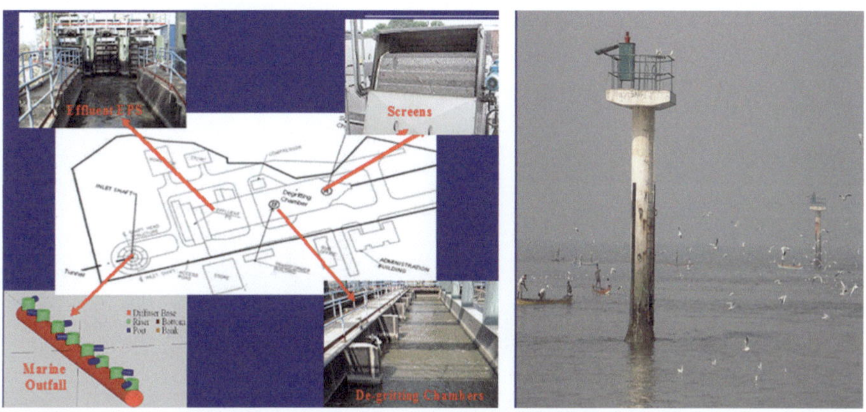

Fig. 21.1 Layout of preliminary wastewater treatment facility and poles of marine outfall, Mumbai

without merging (i.e. non-merging and surface plume) which is a characteristic of a shallow coastal marine outfall (Lee and Cheung 1990). The flooding and receding tidal cycles of 12 h each occur every day and the strength of the tide is governed by the lunar cycle. While planning and finalising the monitoring program, a large number of factors like the quantity and quality of the waste discharge, currents, tides, approachability, relevant laws, stipulated standards for the receiving water body and minimum disturbance to fishing activities were given due consideration.

The extent of dilution depends on the variation in the direction and buoyancy of the upward flow, density, currents, tidal effects and also the submergence of the physical components of the diffuser. Tracer studies and mathematical simulation provide valuable tools to test the performance and adequacy of the diffusor systems. It is essential to compare the reliability of the findings obtained with different tracers for an accurate system evaluation. The experiments conducted for deciding the evaluation procedure for the outfall performance are described below.

(a) Pollutants like BOD, FC, NH3 existing in wastewater were tried as natural tracers. The effluent had BOD of 150 mg/L, NH3-N: 14.0 mg/L and FC count in the range 10 E+07. Water samples were collected for about 3 h after the waste field appeared at the surface near the diffuser and the travel corridor of the study area. The results of the identified parameters and dilution factors are presented in Figs. 21.2 and 21.3.

(b) One square km area in the near-field diffuser region was monitored for flooding and receding tides of 12 h duration for DO and BOD to find out the compliance status of DO and BOD standards and the counters are presented in Fig. 21.4.

(c) Studies using chemical and radioactive tracers were carried out to investigate the dilution and dispersion pattern of the waste plume in the marine environment. The fluorescent tracer rhodamine is one of the most useful tracers for quantitative studies due to its minimum detectability, photochemical and

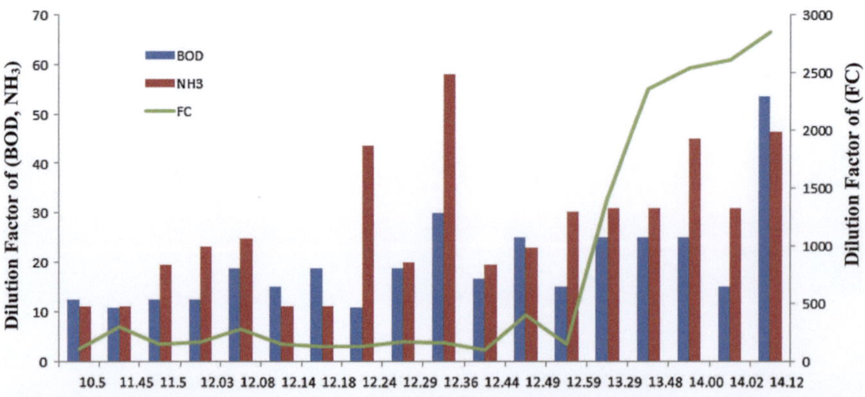

Fig. 21.2 Dilution factor during flooding tide with BOD, FC and NH3-N. Travel time of plume (hours). Tide conditions: LT 9:50 a.m., 1.42 m, HT 3:35 p.m., 3:30 m. Time of travel through outfall tunnel for waste plume appearance 1 h 35 min

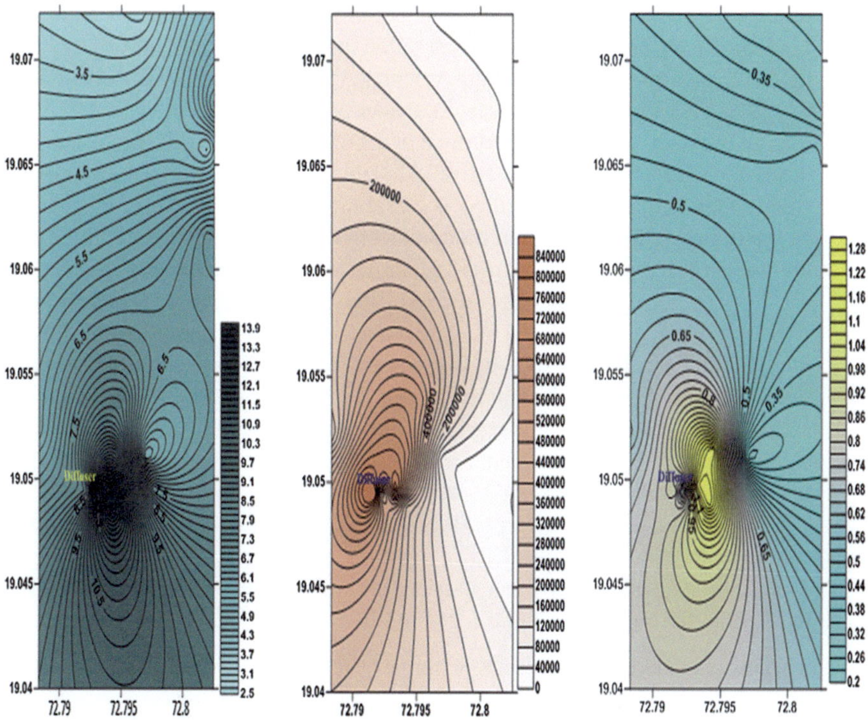

Fig. 21.3 Analytical results for dilution experiment conducted during flooding tide. Tide Conditions: LT 9.50 am, 1.42 m, HT 3.35 pm, 3.30 m. Time of travel through outfall tunnel for waste plume appearance at surface: 1 h 35 min

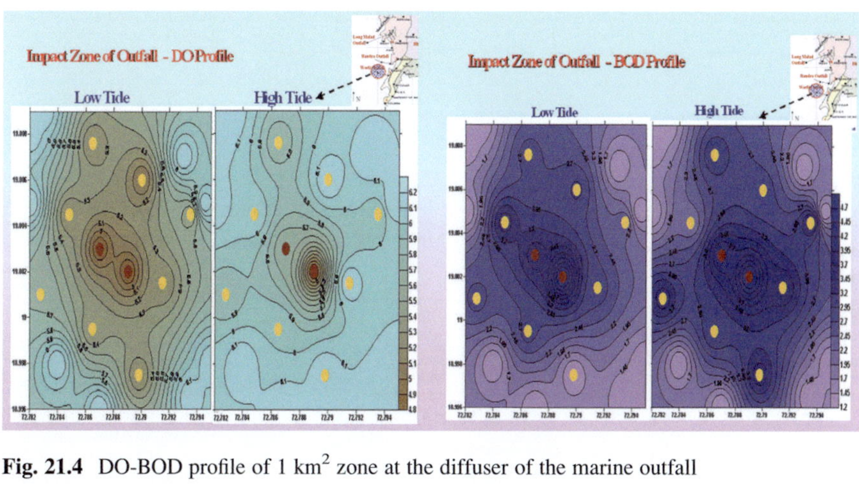

Fig. 21.4 DO-BOD profile of 1 km^2 zone at the diffuser of the marine outfall

biological decay rates, good aqueous solubility, and adsorption (Smart and Laidlaw 1977). Hubbard et al. (1982) stated that it is the most conservative dye tried for streams or karst tracing. Aulenbach and Cleseeri (1980) also found this dye successful as a tracer in a sandy medium. Toxicological data, established with an oral dose to mice as LD 50: 887 mg/kg, indicate that the dose of dye applied will not be harmful to aquatic life. Laboratory experiments were conducted to establish impacts of salinity, turbidity, organic matter etc. on the fluorescent properties of rhodamine. The fluorescent dye rhodamine and the radiotracer bromine-82 provide comparable results. These studies were carried out during the spring receding tide with the assistance of Bhabha Atomic Research Centre (BARC), Mumbai. Boats were employed to collect samples in the expected zone of dispersion. The comparative results for both tracers and mathematical modeling using the Brooks model for the near-field and DIVAST for the far-field region are presented in Table 21.2 and Fig. 21.5.

(d) Organizing the field experiments with radiotracers have become difficult due to its non-availability and necessary safety requirements since it has been established that the radiotracer (Dhage et al. 2001) and fluorescent dye provide comparable results within ±15 % variation in the near-field region. An evaluation of the outfall was carried out using only the fluorescent dye rhodamine during flooding and receding tides.

The fluorescent dye solution of rhodamine was injected for 1 h in the effluent channel after the grit chamber (Fig. 21.6).

Table 21.2 Comparison of percentage difference of dilutions obtained from radiotracer, dye, DIVAST and Brooks-model

	Percent difference					
Distance from pole (m)	Radiotracer-dye	Radiotracer-Brooks	Radiotracer-DIVAST	Dye-Brooks	Dye-DIVAST	Brooks-DIVAST
225	4.5	1.6	1.6	6.1	3.0	3.1
625	12.2	22.2	16.7	11.4	26.8	35.2
930	24.8	23.0	8.9	2.3	31.5	29.8

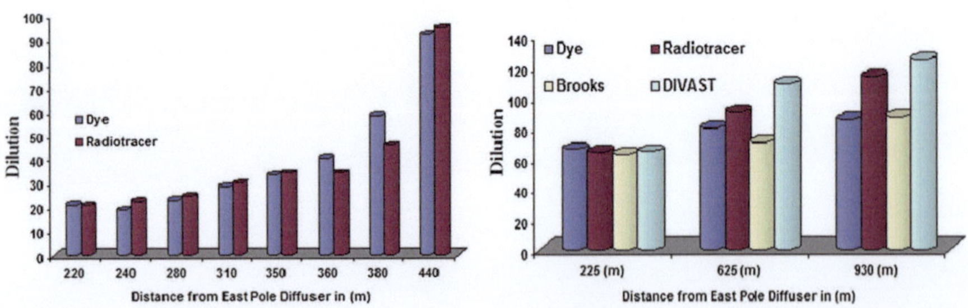

Fig. 21.5 Performance of radiotracer, fluorescent dye rhodamine and mathematical models for estimation of dilutions

Fig. 21.6 Rhodamine application after grit chamber and near diffuser pole

Table 21.3 Results of dilution at various locations during two tidal conditions

	Location with diffuser	Distance (m)	Dye conc. (ppb)	Dilution factor (DF)	Av. dilution factor (DF)	Time[a] (min)
1	Center	500	15–40	25–60	38	15–45
2	East pole	500–1300	12–26	37–82	54	40–70
3	Center	1000	14–38	26–62	40	45–60
4	West	400–900	8–16	24–129	87	35–70
5	East pole	1100–2950	11–20	51–93	72	100–130
6	Center	1100	17–20	51–58	54	100–150
7	West	850–2500	9–17	109–133	121	120–150
8	East pole	1000–2350	11–12	80–86	83	150–180
9	Center	1200	10–12	100–105	102	130–170
10	West	1150–2500	8–16	109–133	121	120–150

[a]Time after appearance of the dye

The input concentration of the dye was 1000 ppb. The dispersion was monitored by moving the boats along the travel path of the dye for 3 h after the first appearance of the dye until the tide had almost approached a steady slag condition. The S4 current meter (Inter Ocean) was deployed to measure the temporal variation of ambient current velocities. Water samples were collected from the expected impact zone of the waste field in the ocean to determine the movement of the waste plume and the extent of dilution achieved. The fluorescence was measured with the Shimadzu RF 5000 spectrofluorometer. The results are presented in Table 21.3.

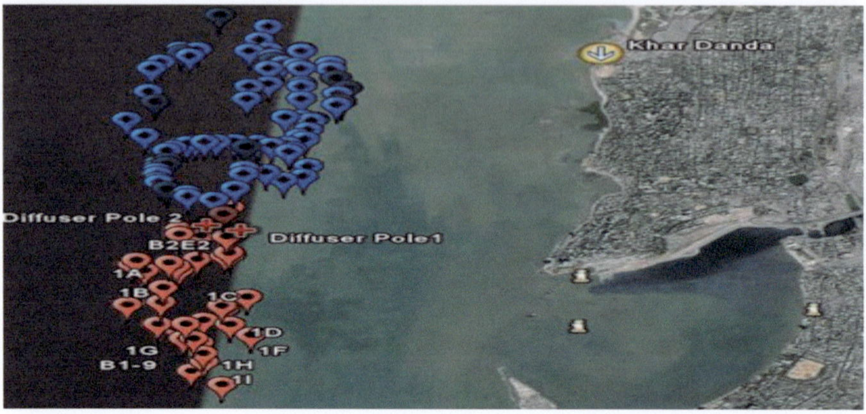

Fig. 21.7 Dispersion pattern at Bandra outfall during flooding and reseeding tide (based on GPS readings with boat movement)

The dispersion pattern is depicted in Fig. 21.7 based on the dye tracer test during low and high tide.

(e) The mathematical model Mike21 model was used to carry out hydrodynamic and water quality simulation for solute transport processes, dilution and dispersion in the study area under variable tidal conditions and time intervals. This model simulates two dimensional distributions of current, water surface elevations and various water quality parameters within the modeling domain as a function of time, taking into account the hydraulic characteristics governed by bed topography, surface wind effects and boundary conditions. Simulation results were calibrated and validated with the experimental observed data of DO, BOD, FC and dye. The hydrodynamic and water quality validation of the model was carried out to reproduce the agreement between the predicted water surface elevations, current velocities and directions with that observed for hydrodynamic and environmental parameters for water quality. The findings of the Mike21 simulation are presented in Fig. 21.8.

21.4 Results and Discussion

The laboratory experiments on the impact of salinity, organic matter, and turbidity on the fluorescent property of Rhodamine dye indicated a marginal reduction of fluorescence (5–7 %). Corrective action was applied during the measurement of the dye concentration by preparing a calibration curve in artificially prepared water with salinity, turbidity and organic content similar to natural coastal water by adding sodium chloride and domestic wastewater to match the field conditions.

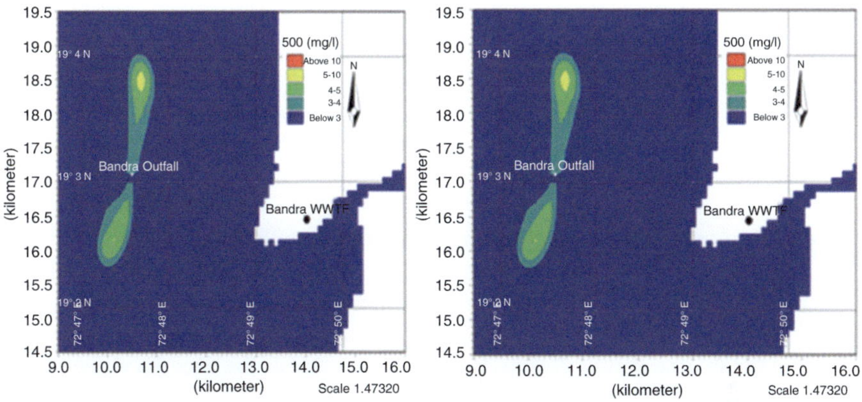

Fig. 21.8 Conceptualization and validation of hydrodynamic simulation and water quality modelling for BOD and fluorescent dye

(a) The analytical results of chemical and microbial parameters did not provide information on the dilution of wastewater discharge either at the initial or far-field locations. The reasons can be first, that the diluted wastewater available from the sea was already contaminated with varying concentrations of parameters selected for analysis and secondly, because of the natural decay of these conservative parameters.

(b) The observations indicated non-compliance of BOD near the diffuser location in the mixing zone. However, the exact dilution could not be established within the natural parameters.

(c) The dilution obtained through radiotracer and fluorescent dye tracer in the near-field region, (500 m from the diffuser) was found to be comparable. The results of the dilution factor obtained with tracer values when compared with the predicted output of the models used for the near- and far-field regions showed a good correlation up to 625 m away from the diffuser pole. The variation in the far-field area was due to constraints and limitations of the applicability of the models for the near- and far-field areas.

(d) The appearance of the dye was first noticed after 1 h 40 min after injection which is comparable with the observations when natural tracers were used.

(e) The calibration of the model was done by comparing the simulated BOD with the observed values during high and low tide with correlation coefficients of 0.9 and 0.8 respectively. The simulation results that were used to establish the movement of the waste plumes represented by BOD and dye follow the same distribution and dispersion trends as observed in actual dye concentrations recorded at the locations established with the help of the GPS as in Fig. 21.7 during high and low tides.

From these experiments, it was observed that the wastewater released through the diffusers was diluted to the extent of 30–60 times in the near-field areas up to a distance of 500–1000 m from the centre of the diffuser. With the wastewater BOD of 150 mg/L, the dilution achieved is not adequate to comply with the designated receiving water quality standard of 3 mg/L BOD at a distance of 3 km from the diffuser location with the currently discharged wastewater quantity.

Based on the results of the dye concentration and dilution factors obtained at the east and west poles, it is concluded that the ten risers and ports along the length of the diffuser had uneven discharge rates. This finding was attributed to the number and extent of openings of each port throughout the length of the diffuser.

The waste plume generated from the outfall during the tidal cycle moved parallel to the coast (Fig. 21.7) thereby indicating minimal risk of wastewater reaching the coast. Thus the purpose of minimizing the pollution on beaches is achieved by adopting the marine outfall system. Before the reversal of the tide, the waste plume traveled between 2.5 and 3.0 km, the current speed not being very strong. The findings also indicated that the outfall impact zone for the deterioration of water quality extended up to 5 km seawards. In the monsoon with high currents, strong winds and turbulence, the waste field is likely to travel a longer distance. This experiment has also pointed out that all the diffusers are not functioning evenly.

The movement of the waste plume based on simulation results through the mathematical model follows observed pattern of dye appearance monitored with the help of GPS locations as in Fig. 21.8 during high and low tides.

21.5 Summary and Conclusions

The experimental and simulation results indicated a dilution and dispersion pattern of dye and water quality parameters under different tidal conditions. Based on the spatial and temporal spread of the waste plume, movement was observed parallel to the coast in both the types of results. This indicates a positive aspect in the prevention of pollutants reaching the beaches, which are used for various recreational purposes.

The experimental study of dye movement, water quality simulation and the advection–dispersion modeling of the dye showed the complementary character of the two methods and a good correlation was also found in the results of both the approaches to evaluate the dispersion pattern of wastewater in the marine environment. The study indicated the need for reorientation and alteration of the degree of opening of ports at each riser in order to improve the initial uniform mixing.

The coast of Mumbai at many places does not meet the requirements of healthy water quality, ecological diversity or even aesthetic quality. The ecosystem of the Mumbai coast in the near-shore region may be at risk due to an increased load of pollutants through non-point discharge. If sewage collection and treatment do not match growing needs, sewage toxicity can lead to aberration in micro- anatomical structures of some fish species and the marine ecosystem will suffer ecological disturbance. Hence major efforts for achieving compliance with water quality standards in the coastal regions are needed, and a number of corrective steps are required to be undertaken simultaneously to attain improvement.

Acknowledgements The authors wish to thank the Director, NEERI, for his constant encouragement and support during the study. The officials from the Municipal Corporation of Greater Mumbai (MCGM) and BARC, Mumbai provided support for the organization of the experimental work, which is gratefully acknowledged. Thanks are due to the supportive project staff from NEERI for their contribution in the field and analytical work.

List of Figures

References

Aulenbach DB, Cleseeri NL (1980) Monitoring for land application of wastewater. Water Air Soil Pollut 14:81–94

Dhage SS, Chandorkar AA, Kumar R (2001) Application of tracers in effluent dispersion studies at Bombay Coast. IANCAS Bull 16(4):49–55

Environmental Status Report (2006–2007) Municipal Corporation of Brihan Mumbai

Gupta I, Dhage SS, Jacob N, Navada SV, Kumar R (2006) Calibration and validation of far field dilution models for outfall at Worli, Mumbai. Environ Model Assess 114:199–209

Hubbard EF, Kilpatrick FA, Martens LA, Wilson JF (1982) Measurement of time of travel and dispersion of streams by dye tracing. In: Techniques of water resources investigations of the U.S. Geological Survey. U.S. Geological Survey, Washington, DC

Kumta A, Kumar R (1998) Aberration in micro anatomical structures of different organs due to impact of chronic exposure to sewage in Oreochromismossambica. Poll Res 17:335–339

Lee JHW, Cheung V (1990) Generalized Lagrangian model of buoyant JETS in current. J Environ Eng ASCE 116(6):1085–1106

MIKE 21 DHI, Water & Environment, ArgenAlle 5, Horsholm, Denmark

Smart PL, Laidlaw IMS (1977) An evolution of some fluorescent dyes for water tracing. Water Resour Res 13(1):15–33

Standard Methods for Examination of Water and Wastewater (2005) 21st ed. APHA, AWWA, WEF

Chapter 22
Hydrodynamic Modeling: Application of Delft3D-FLOW in Santos Bay, São Paulo State, Brazil

Silene Cristina Baptistelli

Abstract This work aims to evaluate the hydrodynamic behavior of Santos Bay through mathematical modeling using the three-dimensional hydrodynamic model Delft3D. The study region is situated in the south coastal area of São Paulo state. The state of São Paulo is located in the southeastern region of the Brazilian Atlantic coast and is the most populous and developed state in Brazil. The Delft3D hydrodynamic model was developed with a set of programs capable of simulating flows in surface water bodies. For modeling implementation, field measurement data were compiled and analyzed and were used to start the model Delft3D in the process of initialization, calibration, validation and evaluation of the modeling results. The results of the simulations demonstrated that the approach was appropriate for the study region. It was also verified that the forcing that determines the hydrodynamics in Santos Bay is the tide, and near the imaginary line connecting Ponta de Itaipu and Ponta Grossa, there is a strong influence of coastal circulation which dominates the NE and SW direction, parallel to Praia Grande. In conclusion, this study contributed to obtain better knowledge about hydrodynamic mathematical modeling and about the studied area.

Keywords Hydrodynamic modeling • Mathematical modeling • Submarine outfall

22.1 Introduction

Increasing environmental awareness has led scientists and engineers to focus their attention on the problem of predicting the hydrodynamic circulation of water and dispersion of contaminants in water bodies. This problem has been aggravated in coastal regions, where the majority of the population lives. For example, the Baixada Santista region has harbor activities in Santos harbor and the Cubatão

S.C. Baptistelli (✉)
Companhia de Saneamento Básico do Estado de São Paulo (SABESP), São Paulo, Brazil
e-mail: sbaptistelli@uol.com.br; sbaptistelli@sabesp.com.br

M. Baawain et al. (eds.), *Recent Progress in Desalination, Environmental and Marine Outfall Systems*, DOI 10.1007/978-3-319-19123-2_22

industrial area, besides tourism and leisure functions. These municipalities have low rates of sewage collection and treatment. Seawater is frequently unsuitable for bathing, mainly because of sewage disposal in the streams of the region.

It is understood that the dispersion of effluent discharge in coastal waters is extremely complex, and to understand its behavior, it is necessary to take into account at the very least, factors such as the type of effluent, the discharge points, the ocean currents, the tidal effects, the meteorological effects, the bathymetry of the seabed and the land contours. Furthermore, the atmospheric and ocean systems interact in a very complicated way and the combination of meteorological and oceanic effects strongly influence coastal areas, as strong winds and storms can cause an increase or decrease in sea level and even affect the behavior of coastal currents for a distance of hundreds of kilometers and for several days.

From these studies, reliable information of the hydrodynamic circulation, water quality, and sediment transport can be obtained by appropriate mathematical modeling tools. The mathematical models represent natural phenomena by means of fundamental differential equations. These mathematical equations require the use of coefficients which are obtained by measurements in the field or in simulated physical models. Data from field measurement are extremely important in the startup, calibration and validation of mathematical models stage. Thus, this work seeks to evaluate the hydrodynamic behavior of Santos Bay through mathematical modeling using the three-dimensional hydrodynamic model Delft3D.

22.1.1 Study Area

Brazil has a coastline of over 8500 km. It has 17 states and harbors by which commodities are exported and imported into the Brazilian economy. The SãoPaulo state is located in the southeastern region of the Brazilian Atlantic coast and is the most populous and developed state in Brazil. The study region is situated in the coastal area, on the south coast of São Paulo state, called Baixada Santista, which is 160 km long and 2,886 km^2 in area, and composed of nine municipalities.

The highlighted study area includes the Santos and São Vicente estuarine systems. The area consists of São Vicente Island, Santos estuary, São Vicente estuary and Santos Bay. Administratively, São Vicente Island is divided into two municipalities: Santos and São Vicente. This area is influenced by oceanic and continental waters.

Geographically, Santos Bay is bordered to the north by the beaches of Santos and São Vicente, on the south by an imaginary line joining Ponta de Itaipu and Ponta da Munduba, to the east by the Santos harbor bar and to the west by São Vicente bar and Ponta de Itaipu (Fig. 22.1). Its NS axis is about 6.7 km long and about 6.8 km in the EW axis. The slope of the bottom is smooth and the depths range from 0 to 13 m, along with its NS axis. The total area of the estuarine region of Santos, including the bay, is approximately 100 km^2 (Hidroconsult 1974).

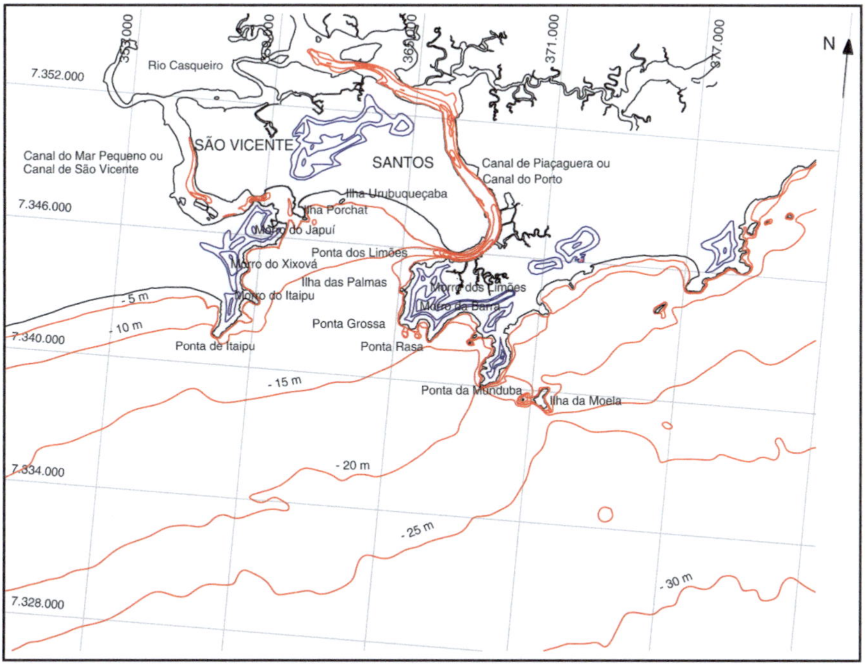

Fig. 22.1 Study area

The bay of Santos is affected by ocean water through the section that unites Ponta de Itaipu and Ponta da Munduba, and by freshwater through the estuaries of Santos and São Vicente. The continental freshwater comes from the estuarine system through the channels Piaçaguera to the east and Mar Pequeno (São Vicente bar) to the west. The estuarine system Santos/São Vicente is powered by a set of watersheds from the slopes of Serra do Mar and the source of their coastal region. Rivers originating in Serra do Mar are torrential due to the small slope of the marshland, making water flow difficult. This phenomenon produces a labyrinth of channels and streams which are characteristic of this region (SABESP 2006).

The sewage system of the island is integrated and designed to accommodate the sewage collected in the urban areas of both municipalities (Santos and São Vicente) and to treat it in the wastewater treatment plant, with its final disposal through the Santos submarine outfall which is located in Santos Bay with the discharge point at the coordinate 7,344,809 N and 362,446 E. It has a total length of 4,425 m, and has 4,000 m of outfall pipeline and 425 m of diffuser stretch. The diffuser stretch has 79 risers spaced 5.38 m apart, each having a diameter of 0.24 m. Each riser has two diametrically opposed ports with diameters of 0.11 m. The depth of discharge is approximately 9.50 m and the local depth is 10.70 m. The average flow rate of discharge of the effluent is $Q_m = 3.34\ m^3/s$ and the maximum projected discharge rate is $Q_{max} = 5.3\ m^3/s$ (SABESP 2005).

22.1.2 Conceptual Description of Hydrodynamic Mathematical Modeling

The use of computer models as a tool for simulations of the hydrodynamic pattern of flow and water quality is of fundamental importance in coastal engineering studies, particularly with regards to environmental studies. Mathematical models are associated with computational tools and appropriate techniques for optimizing the search for solutions to complex programming problems. These models enable the viability of the research as the studies consist of a large number of variables (Costa 2002).

Mathematical models represent natural phenomena nature by the use of differential equations which give the flow pattern of water bodies, defined by currents and surface elevation along with the modeled domain, depending on the condition of the bathymetry and forcing. However, this is not a simple task due to the complexity of the processes involved. The utilization of numerical methods is necessary because the analytical solutions of the equations, usually non-homogeneous differential equations, involved in the process are not known. With advances in information technology, there has been significant progress in the use of numerical methods, enabling them to produce the most comprehensive and complex mathematical models. According to Harari (1989), numerical modeling allows not only the solution of equations that describe physical phenomena, but is also a tool for experiments related to processes that both compose and interact with the characteristics observed in the oceans.

The application of two-dimensional (2D, depth-averaged) and three-dimensional (3D) hydrodynamic models to study the impact of engineering solutions have several obstacles, such as the difficulty to start (initial configurations), the processing time, the number of processes to be implemented, besides the difficulty of calibration. Moreover, one of the main problems is the specification of appropriate boundary conditions at the open borders.

In this study, the hydrodynamics model Delft3D, developed by Delft Hydraulics, was used. It is a unique, fully-integrated computer software suite for a multidisciplinary approach for 3D computations for coastal, river and estuarine areas. It can carry out simulations of flows, sediment transport, waves, water quality, morphological developments and ecology. The Delft3D suite is composed of several modules, grouped around a mutual interface, capable of interacting with each another (Delft3D-FLOW 2006).

The hydrodynamic module Delft3D-FLOW simulates 2D or 3D unsteady flow and transport phenomena resulting from tidal and/or meteorological forcing, including the effect of density differences due to a non-uniform temperature and salinity distribution (density-driven flow). The flow model can be used to predict the flow in shallow seas, coastal areas, estuaries, lagoons, rivers and lakes. It aims to model flow phenomena of which the horizontal length and time scales are significantly larger than the vertical scales. If the fluid is vertically homogeneous, a depth-averaged approach is appropriate. Delft3D-FLOW is able to run in 2D mode (one computational layer), which corresponds to solving the depth-averaged equations. The 3D model is of particular interest in transport problems where the horizontal flow field shows significant variation in the vertical direction. This variation may be generated by wind forcing, bed stress, Coriolis force, bed topography or density differences. Examples are dispersion of waste or cooling water in lakes and coastal areas, upwelling and down-welling of nutrients, salt intrusion in estuaries, fresh water river discharges in bays and thermal stratification in lakes and seas (Delft3D-FLOW 2006).

Delft3D-FLOW solves the Navier-Stokes equations for an incompressible fluid under the shallow water, and also uses the Boussinesq assumptions. In the vertical momentum equation, the vertical accelerations are neglected which leads to the hydrostatic pressure equation. In the 3D model, the vertical velocities are computed from the continuity equation. The set of partial differential equations in combination with an appropriate set of initial and boundary conditions is solved on a finite difference grid (Delft3D-FLOW 2006). For the implementation of the modeling, field measurements data were compiled and analyzed and then used to start the model Delft3D in the process of initialization, calibration, validation and evaluation of the modeling results.

22.2 Methods

The methodology consists of the usage of mathematical models in order to assess the behavior of currents circulating in the study area. In addition, field data measurements have been used in order to calibrate the models and to acquire understanding of hydrodynamic processes involved in the behavior of maritime currents in the Santo and São Vicente estuarine system.

Field data measurements were compiled and analyzed and their utilization was verified. Spreadsheets and data charts that allowed temporal and spatial visualization were prepared. The data analyzed were used as input in the mathematical model, as well as used for the comparison and analysis of compliance with the results of mathematical modeling. Table 22.1 shows the initial and boundary conditions and the physical and numerical parameters used in the modeling.

The calibration of the model is a recursive process which performs known and trusted data entry, for example, bathymetry, variation of the tide or the wind, etc., by using the calibration parameters which are adjustable in the course of various processes The model is considered calibrated after a series of processes and the results achieved were the nearest possible values measured in the field.

In this study, the model was calibrated by comparing with the winter 2005 measurement data and validated for the other periods of simulation (summer 2002 and winter 2002). The calibration parameters that were used are the wind friction coefficients and the Manning coefficient.

Table 22.1 Initial and boundary conditions, physical and numerical parameters used in Delft3D

Parameter	Value			
Grid:	Grid-points in M-direction: 102			
	Grid-points in N-direction: 112			
	Grid spacing: 300 m			
	Number of layers:			
	Summer 2002: 5 layers			
	Winter 2002: 10 layers			
	Winter 2005: 3 layers			
Period of simulation:	Summer 2002: 15-Feb to 09-Mar-2002			
	Winter 2002: 20-Jul to 18-Aug-2002			
	Winter 2005: 23-Jul to 14-Aug-2005			
Time step:	Summer 2002: 2 min			
	Winter 2002: 2 min			
	Winter 2005: 1 min			
Physical processes:	Wind, salinity and temperature			
Boundary conditions:	Flow conditions	South	East	West
	Type of open border	Elevation surface	Neumann	Neumann
	Type of forcing	Astronomical	Time series	Time series
Astronomical conditions:	Constituent-Name	Amplitude (m)	Phase (degree)	
	Q1	0.027	62.21	
	O1	0.106	80.87	
	P1	0.026	150.62	
	K1	0.069	163.88	
	N2	0.043	101.02	
	M2	0.290	89.13	
	S2	0.205	87.60	
	K2	0.058	87.55	
	M3	0.052	233.41	
Turbulent viscosity:	1.0 m^2/s			
Wind friction coefficients:	0.009 – first and second "breakpoints"			
Roughness formula:	Manning: uniform: U = V = 0.03			
Wind:	Uniform – Interpolation linear – Time series (NCEP/NCAR Reanalysis Project and local wind)			

Briefly, the tasks carried out for implementation of the model are

- choosing the area to be modeled;
- generating the bathymetry and the grid from ASCII files using Delft3D-RGFGRID and Delft3D-QuickIN modules;
- specifying open borders and boundary conditions;
- specifying surface elevation with the adoption of the constituent tide (astronomical conditions);
- specifying the physical processes involved (wind, temperature and salinity);
- choosing simulation periods; and
- choosing parameters and coefficients that served as the calibration parameters of the model.

The Delft3D-QuickIN, which is the module for generation, interpolation and data manipulation of the spatial variation of the bathymetry, boundary condition or field parameters, will be used by Delft3D-FLOW. The "Sample" file that was the basis for the development of the grid used in the processing of simulations is shown in Fig. 22.2. In Fig. 22.3, the rectangular grid, with a spacing of 300 m drawn from the file "Sample", is displayed.

The problem of the specification of the boundary conditions in the side edges is that a certain water level or velocity distribution will develop across the border due to combining processes in the model domain. For a correct specification of the boundary conditions, its distribution must be known beforehand, otherwise disturbances in the edges can be developed. There are two ways to solve this problem.

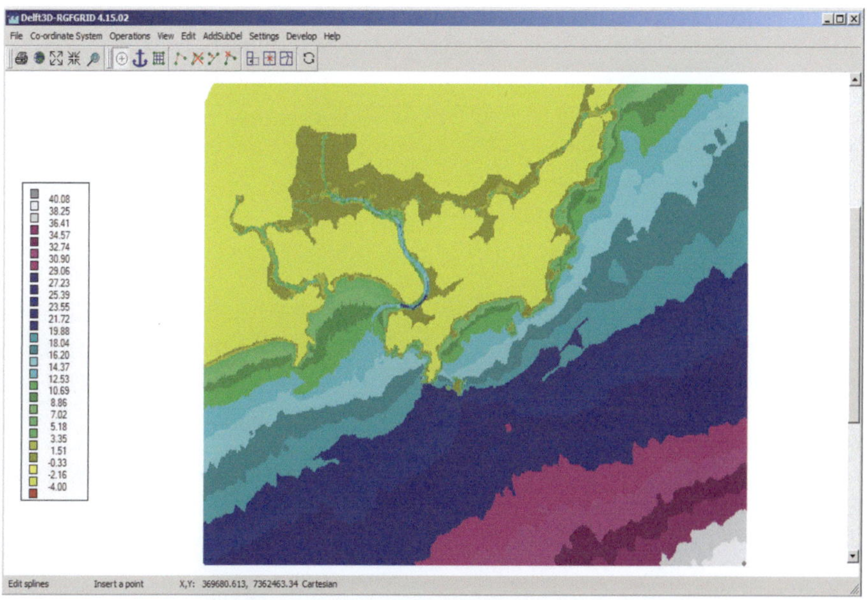

Fig. 22.2 Bathymetry of the estuary and bay of Santos generated from Delft3D-RGFGRID

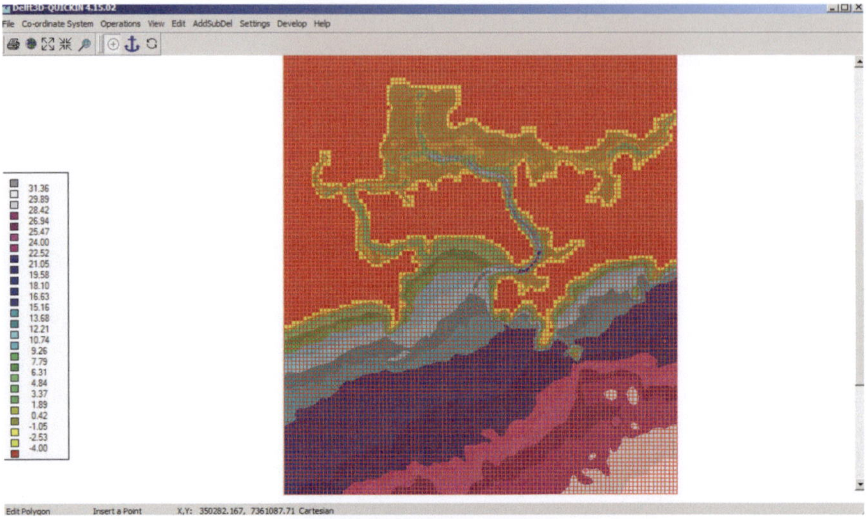

Fig. 22.3 Grid of the estuary and bay of Santos generated from Delft3D-QuickIN

The first one is to try to predict the configuration of the water level or flow velocity along the open side edge through solving a 1D or 2D model, and then to impose this condition. For simple cases this is possible but for more complex combinations of conditions forcing that is complicated and slow.

According to Roelvink and Walstra (2004), the best option is to let the model determine the correct solution to the open side edge through the imposition of a gradient for the water level along the edge (Neumann boundary condition) instead of fixing the water level or the speed of the current. The present study considers this to be the boundary condition, as shown in Table 22.1.

22.2.1 Database

The field data used in the study is presented in Table 22.2 and the location of field measurements points is presented in Fig. 22.4. The graphic representations of these field measurements are as follows: Figure 22.5 for the wind rose at Point 1 for summer 2002 and winter 2002; Fig. 22.6 for the wind time series and Fig. 22.7 for the wind rose at Point 2 for winter 2005. In addition, wind data from the National Oceanic and Atmospherics Administration (NOAA 2006), through the National Center for Environment Prediction Reanalysis Project (NCEP/NCAR), were used to input in the model, and simulations were carried out in specific periods of time in summer 2002, winter 2002, and winter 2005.

The surface elevation time series data measured at Point 3 for summer 2002 and winter 2002 is given in Fig. 22.8. The currents time series data measured at Point

Table 22.2 Location of field measurements points

Data description – (Point)	Project/ Location	Coordinates				Period
		Geodesics		UTM		
		Latitude	Longitude	Latitude	Longitude	
Anemograph – Wind (Point 1)	CODESP (2002) Guarujá – Cabras Island	24° 0.5′ S	46° 13.1′ W	7,344,279 N	376,078 E	7-Feb to 4-Mar-2002 and 19-Jul to 27-Sep-2002
Anemograph – Wind (Point 2)	SABESP (2006) Praia Grande	24° 1.46′ S	46° 27.6′ W	7,342,274 N	351,568 E	22-Jul to 27-Oct-2005
Surface elevation – Tide (Point 3)	CODESP (2002) Palmas Island	24° 0.6′ S	46° 19.6′ W	7,343,995 N	365,060 E	9-Feb to 27-Mar-2002 and 18-Jul to 13-Sep-2002
ADCP – Currents (Point 4)	CODESP (2002) Santos	24° 5.2′ S	46° 17.8′ W	7,334,187 N	362,441 E	9-Feb to 27-Mar-2002 and 18-Jul to 13-Sep-2002
ADCP – Currents (Point 5)	SABESP (2006) Praia Grande	24° 2.95′ S	46° 26.5′ W	7,339,555 N	353,476 E	11-Jul to 15-Aug-2005

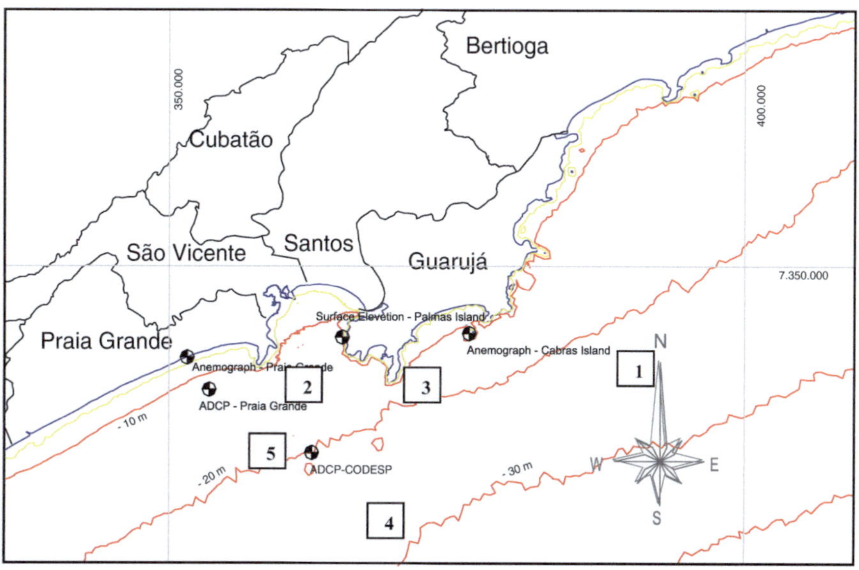

Fig. 22.4 Location of the field measurements points

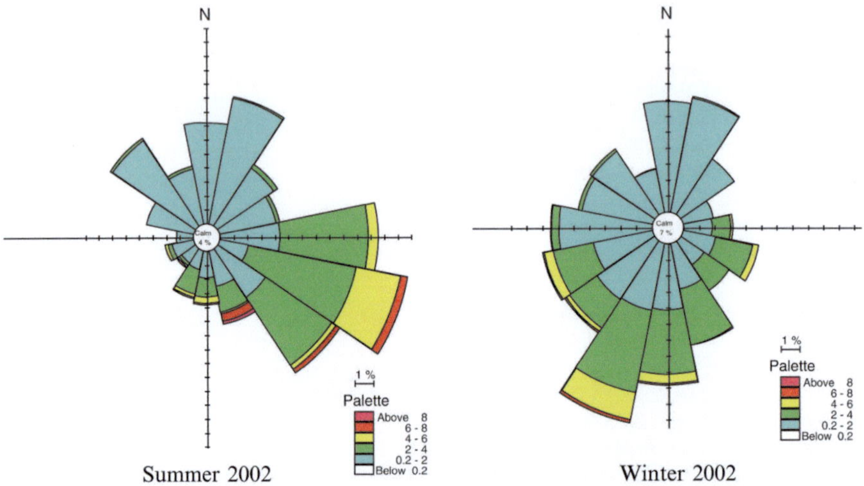

Fig. 22.5 Wind rose for data measured at Point 1 in Guarujá–Cabras Island (CODESP 2002)

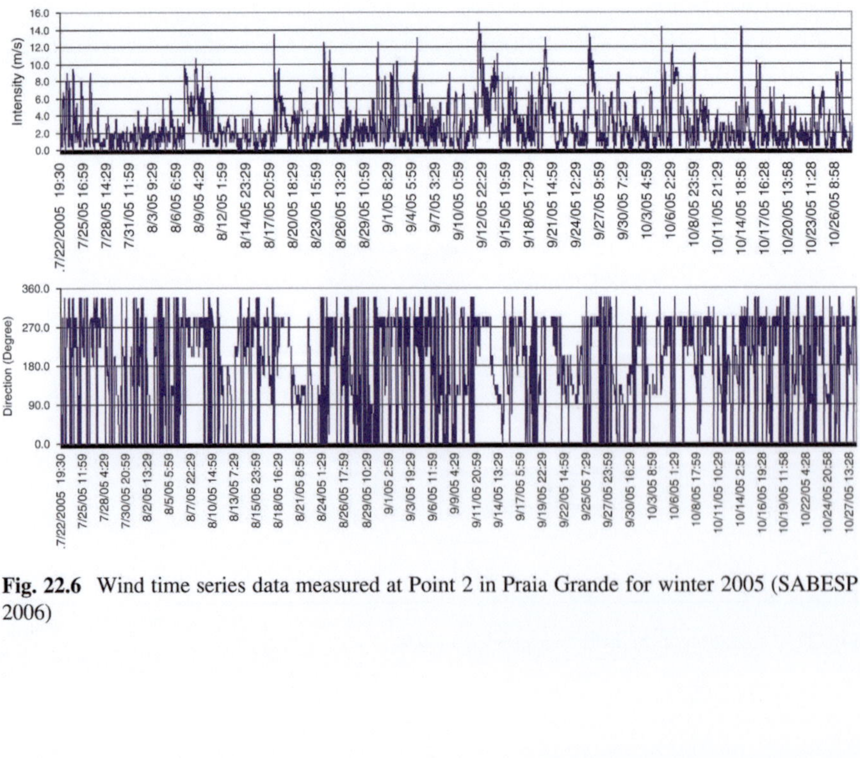

Fig. 22.6 Wind time series data measured at Point 2 in Praia Grande for winter 2005 (SABESP 2006)

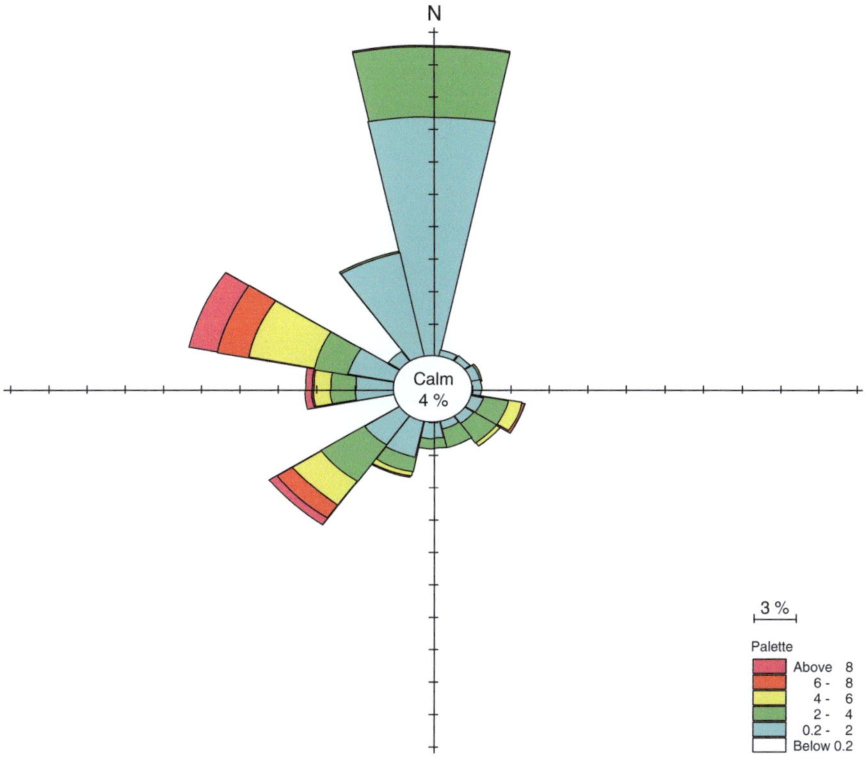

Fig. 22.7 Wind rose for data measured in Praia Grande (Point 2)

4 for summer 2002 is plotted in Fig. 22.9 and for winter 2002 in Fig. 22.10. The current rose for data at Point 4 is given in Fig. 22.11. Finally, the currents time series data measured at Point 5 for winter 2005 is plotted in Fig. 22.12, and its current rose is given in Fig. 22.13.

22.3 Results and Discussions

In this section, the results of the simulations and discussions are presented. Figure 22.14 presents the bathymetry generated by modules Delft3D-RGFGRID (Fig. 22.2) and Delft3D-QuickIN (Fig. 22.3).

In Fig. 22.15, the graphics of the comparison between the results extracted from Delft3D simulation and field measurements for surface elevation (tide), for the three simulated periods, are presented. For these comparisons, a time series from the simulation results of surface elevation at the grid point near Point 3 (Table 22.2) was extracted. So, the time series was compared with the temporal series already shown in Fig. 22.9.

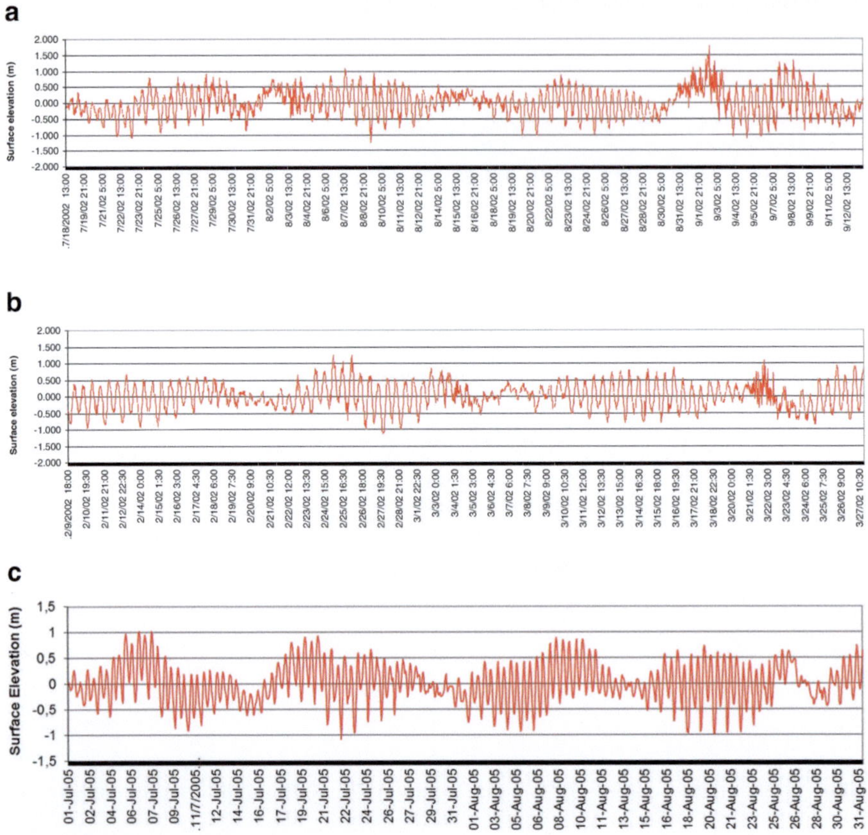

Fig. 22.8 Surface elevation time series data measured at Point 3 in Palmas Island for (**a**) winter 2002, (**b**) summer 2002 (CODESP 2002); and (**c**) for winter 2005 (SABESP 2006)

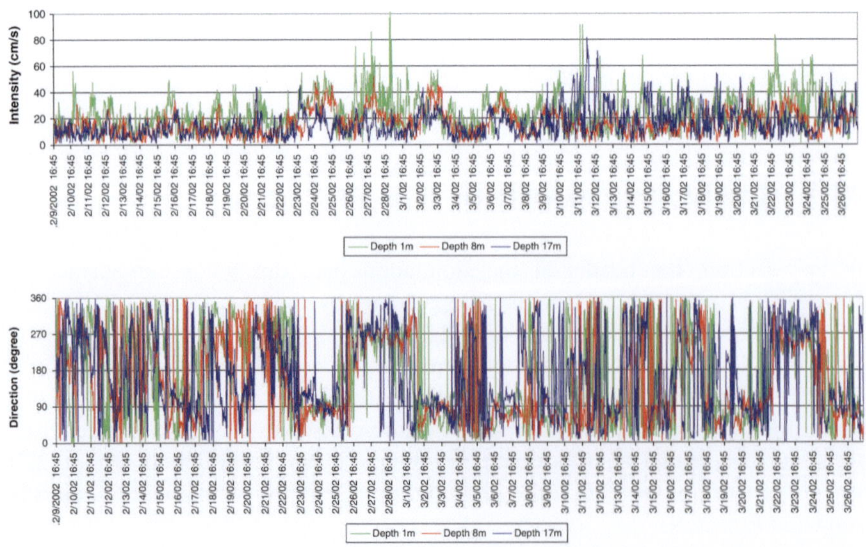

Fig. 22.9 Currents time series data measured at Point 4 in Santos for summer 2002 (CODESP 2002)

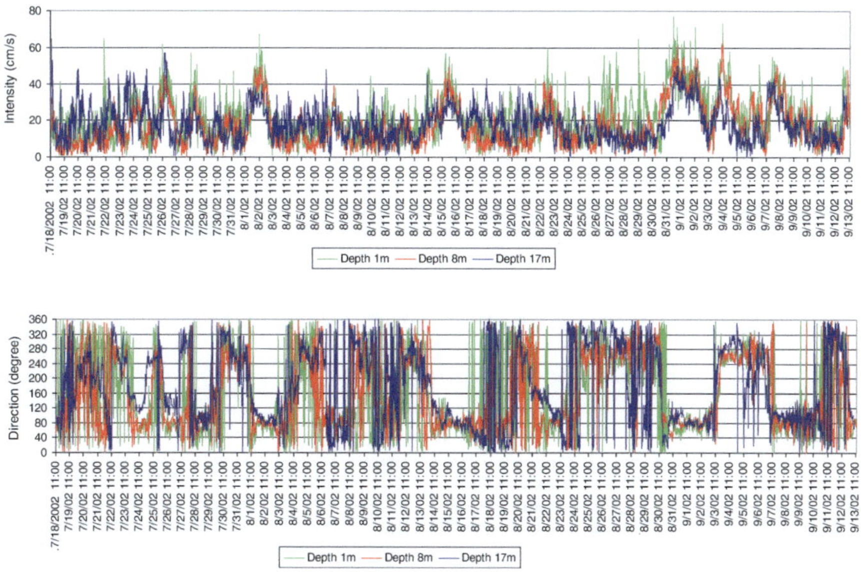

Fig. 22.10 Currents time series data measured at Point 4 in Santos for winter 2002 (CODESP 2002)

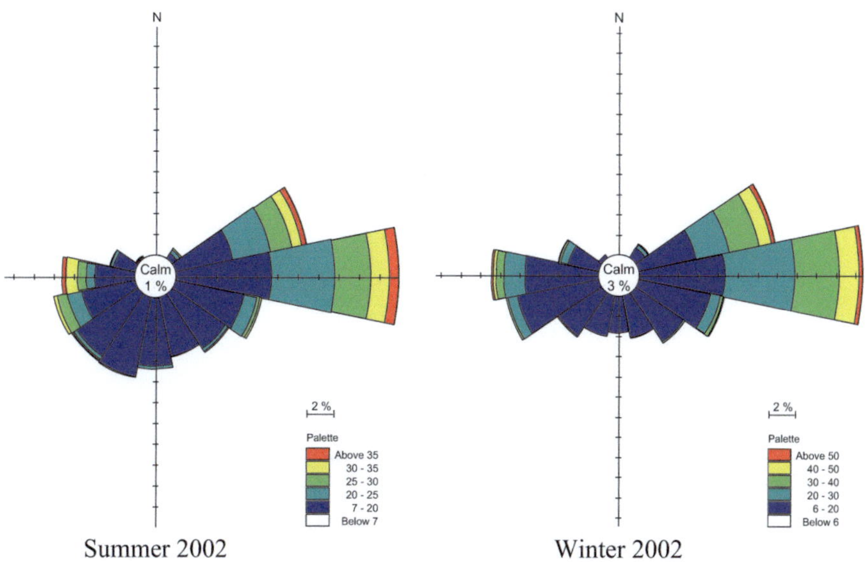

Fig. 22.11 Current rose for data measured at Point 4 in Santos

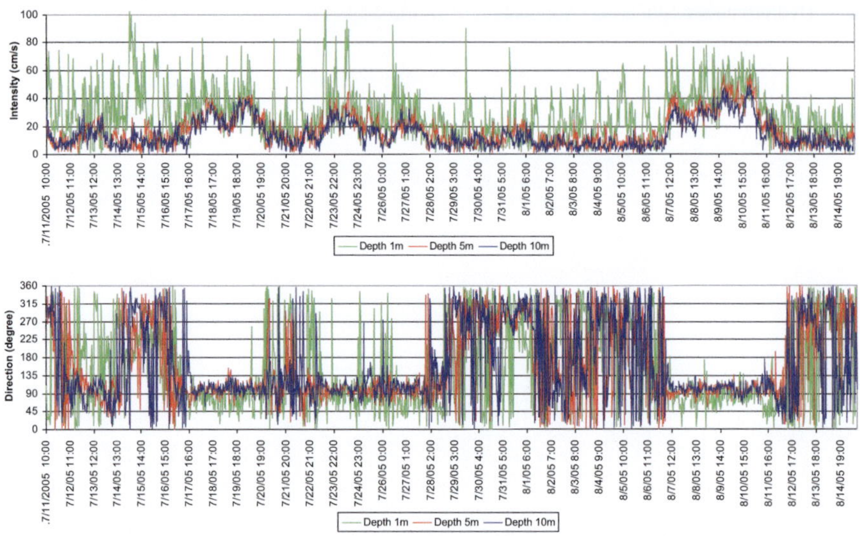

Fig. 22.12 Currents time series data measured at Point 5 in Praia Grande for winter 2005 (SABESP 2006)

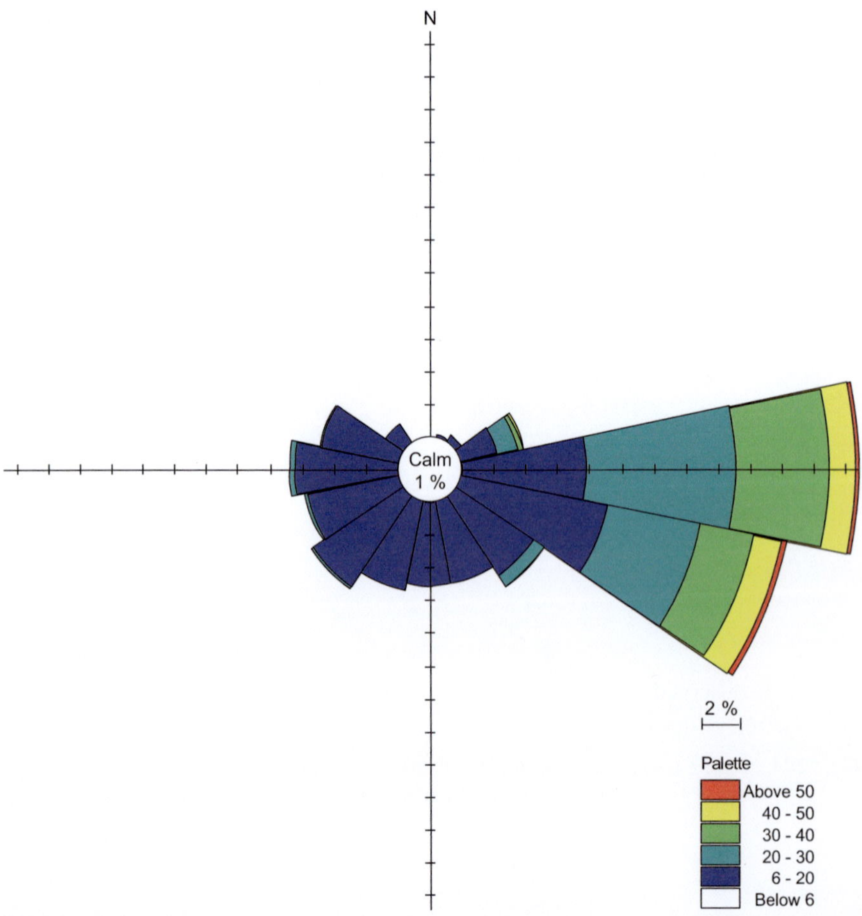

Fig. 22.13 Current rose for data measured in Praia Grande (Point 5)

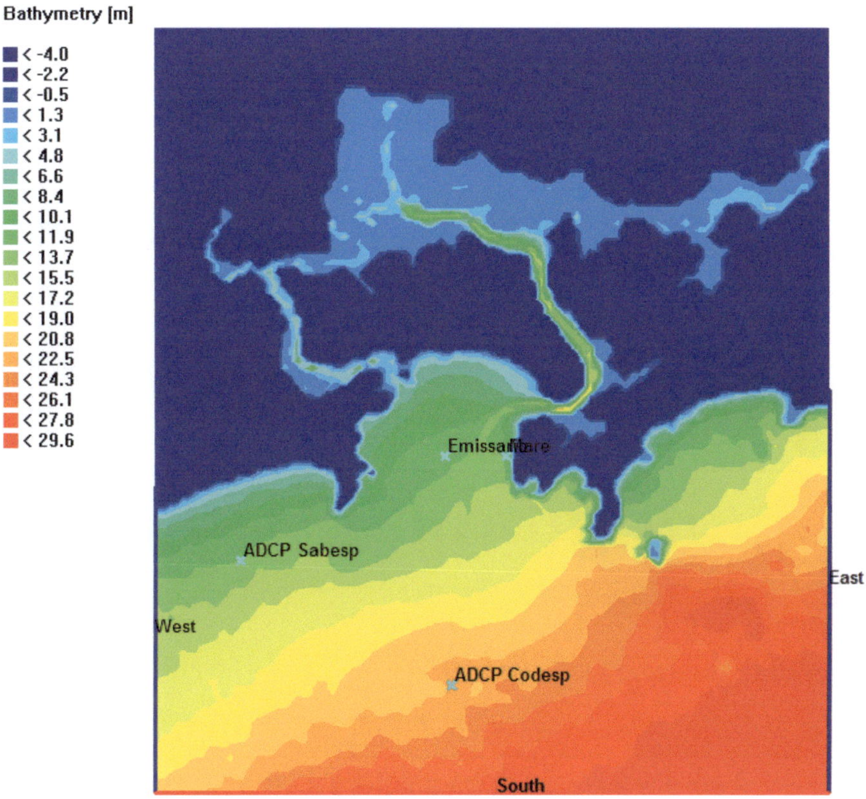

Fig. 22.14 Bathymetry generated by Delft3D using a horizontal spacing of 300 m with 102 × 112 grid points

In Fig. 22.16, scatter plots between the values of surface elevation of Delft3D simulation results and field measurements at Point 3 are presented. A good correlation is obtained, specially for the results for summer 2002.

In Fig. 22.17, graphs comparing current results of Delft3D simulations and field measurements are presented. The graphs are divided into speed, direction component-x (component "u" of the velocity vector – EW) and component-y (component "v" of the velocity vector – NS). The graphs show the values of average velocities in depth. For these comparisons in the temporal data field measurement series, a filter moving average (harmonic mean) equivalent to 5 h was applied. This "smoothing" of the data aimed to reduce the high frequency variability observed in the data.

The point of discharge from Santos outfall into the bay of Santos is an important point to be analyzed. Therefore, the time series of the predicted currents from

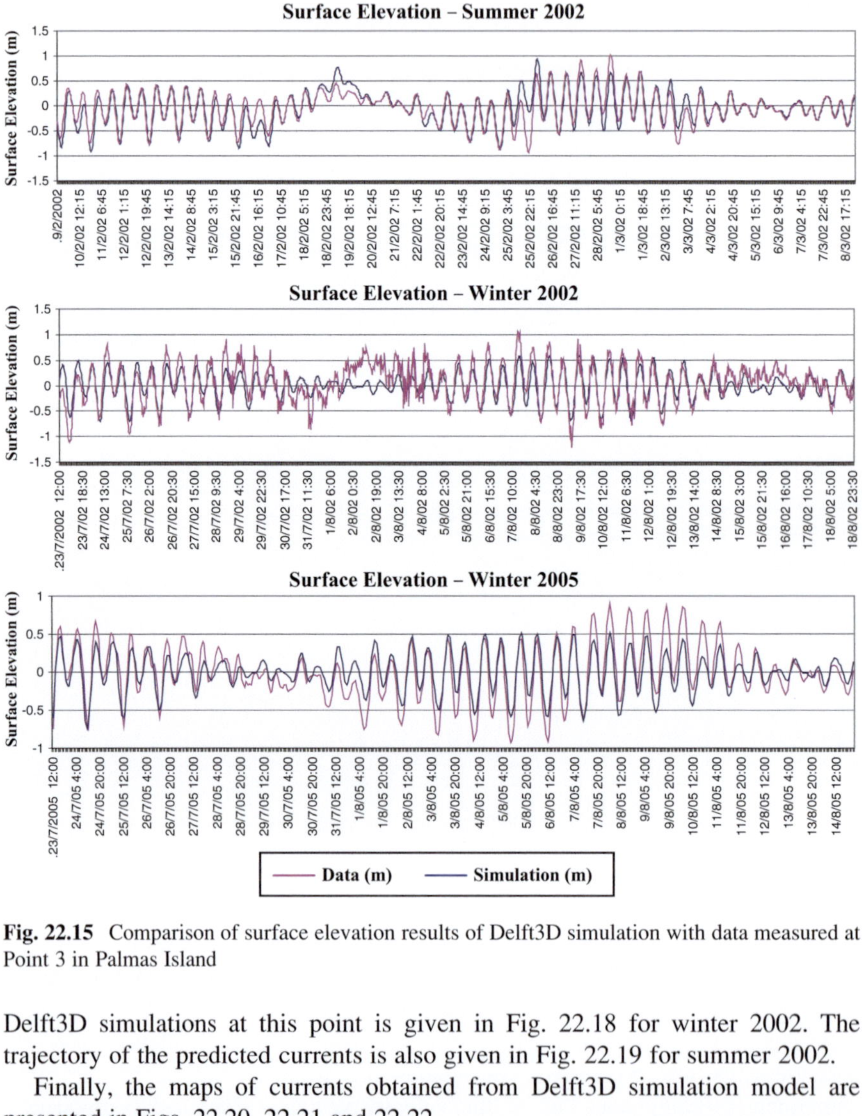

Fig. 22.15 Comparison of surface elevation results of Delft3D simulation with data measured at Point 3 in Palmas Island

Delft3D simulations at this point is given in Fig. 22.18 for winter 2002. The trajectory of the predicted currents is also given in Fig. 22.19 for summer 2002.

Finally, the maps of currents obtained from Delft3D simulation model are presented in Figs. 22.20, 22.21 and 22.22.

Based on the results of Deflt3D simulations and data obtained from field measurements, it is possible to characterize the hydrodynamics of local circulation, very often corroborating the results of previous studies (Harari and Camargo 1998; Baptistelli 2003; Cunha et al. 2006; SABESP 2006). These characteristics are itemized below:

- Regarding the hydrodynamics of Santos Bay, the tide is basically what determines the dynamics of the currents. This characteristic can be observed in the results extracted from modeling with Delft3D and field measurement data.

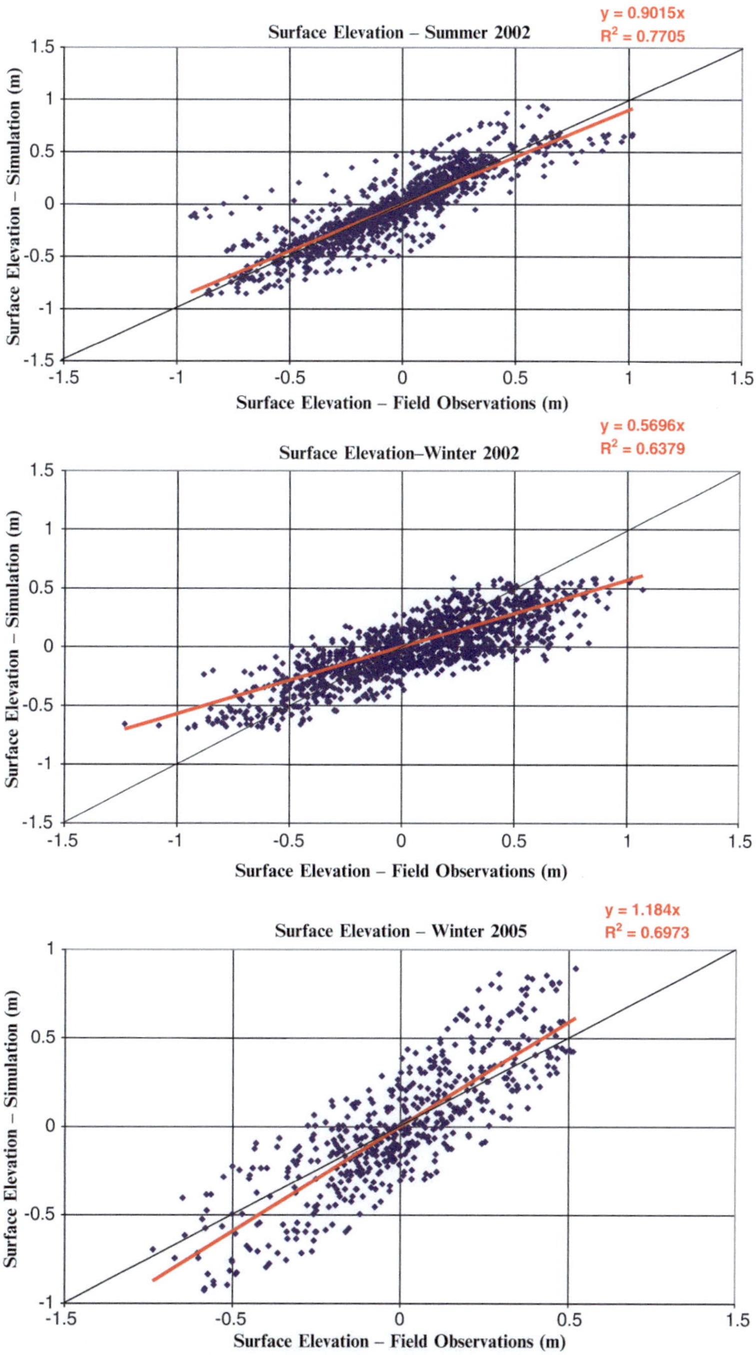

Fig. 22.16 Scatter plots of surface elevation of Delft3D simulation results and data measured at Point 3 in Palmas Island

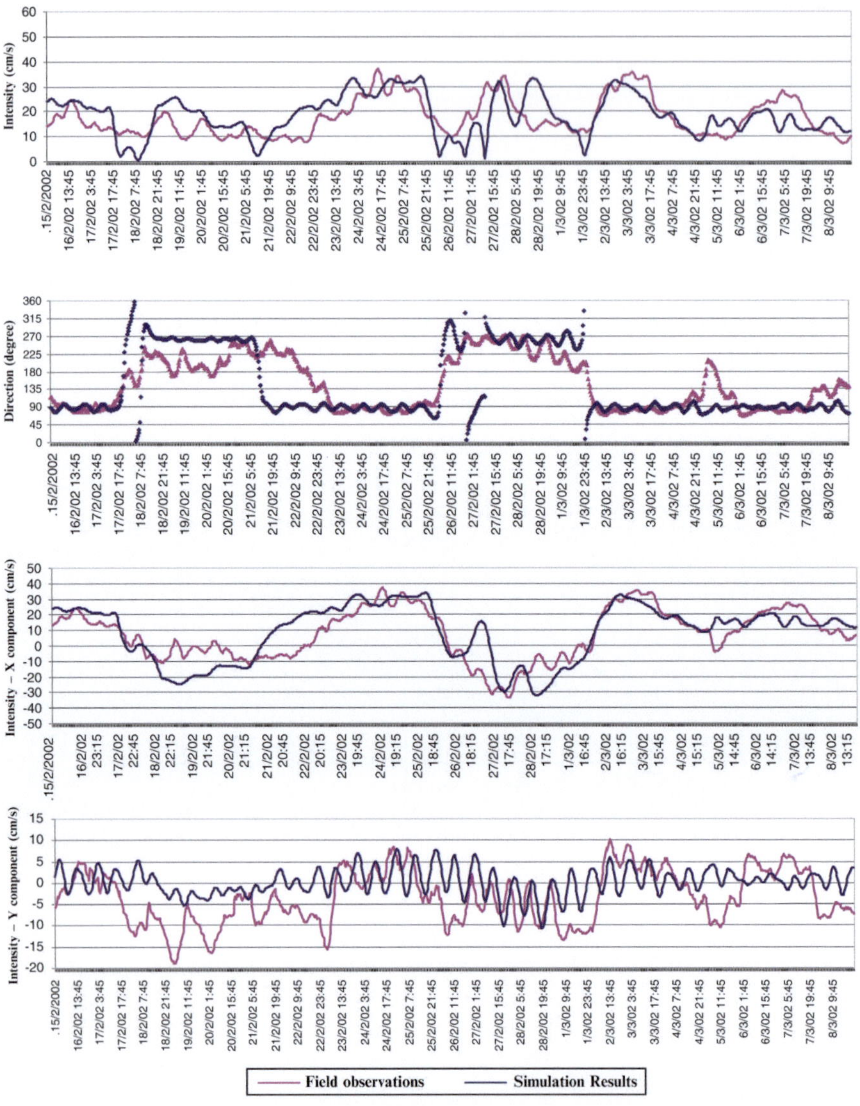

Fig. 22.17 Comparison of current results of Delft3D simulation with data measured at Point 4 in Santos for summer 2002

Differences between periods of syzygy and quadrature can also be observed (Fig. 22.22). Although the speeds further offshore are influenced by meteorological effects, the center of the bay and natural channels are dominated by tidal effects.

- At the entrance to the bay area, near the imaginary line connecting Ponta de Itaipu and Ponta Grossa, there is a strong influence of coastal circulation which dominates the NE and SW direction, in other words parallel to Praia Grande.

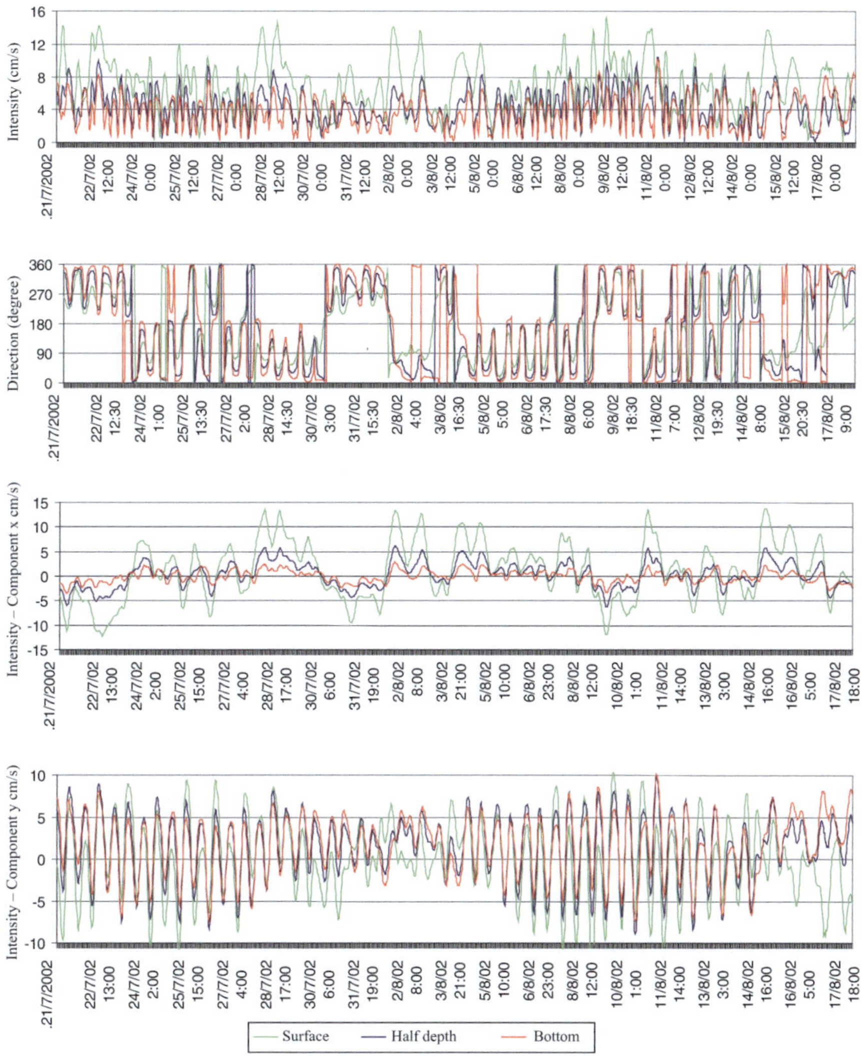

Fig. 22.18 Delft3D simulation results of speed and direction of currents at the Santos/São Vicente submarine outfall point of release for winter 2002

This characteristic is shown on the maps of currents presented in Figs. 22.20, 22.21, and 22.22.

- It is observed that the rotation characteristic of the currents is anti-clockwise.
- For the region outside Santos Bay, in Praia Grande, weather conditions have a great influence on coastal circulation. In some periods of higher current velocities, there were periods that were followed by higher wind speeds, and the highest speeds occurred on the surface.

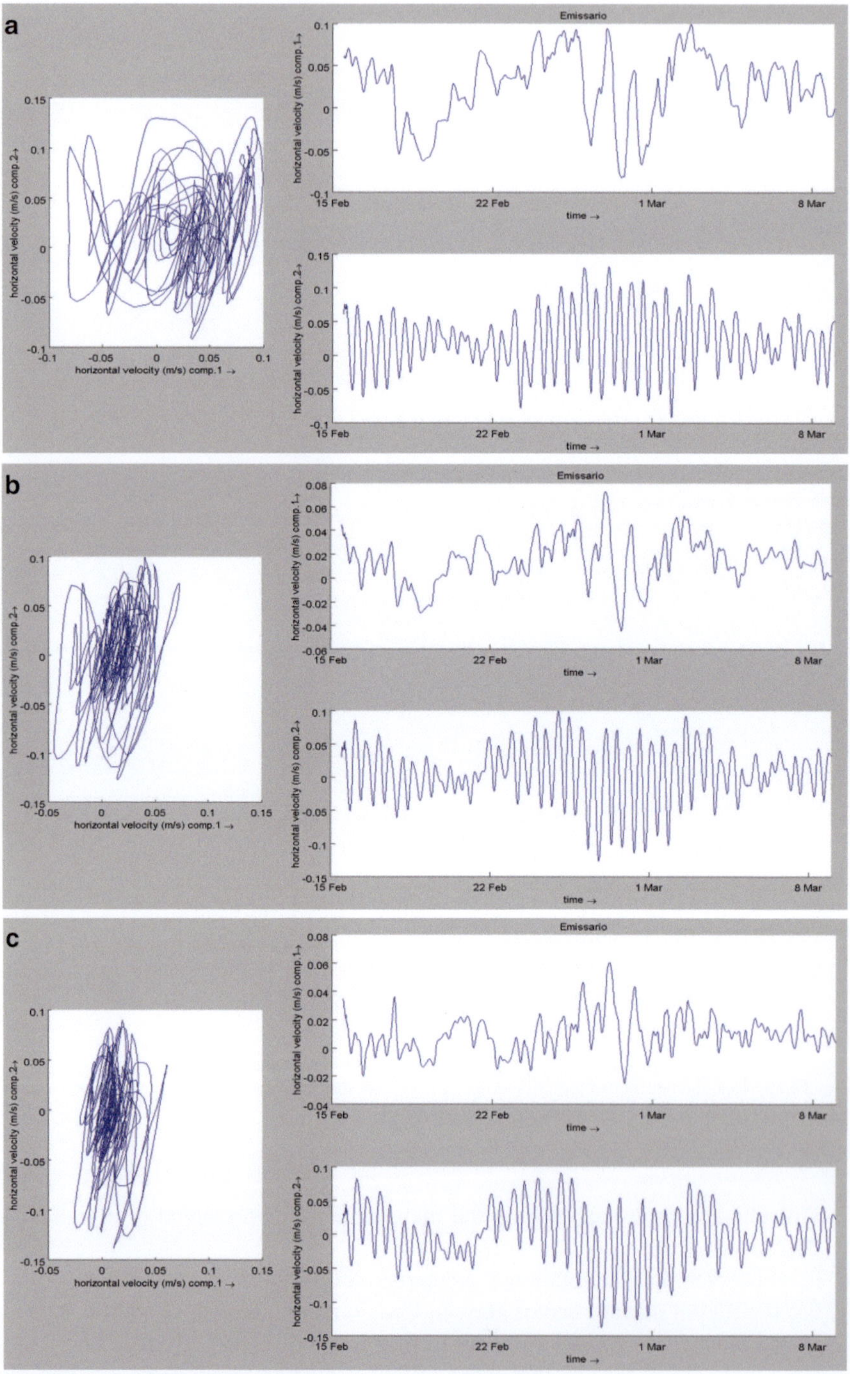

Fig. 22.19 Delft3D simulation results of trajectory and velocity (x and y component) of currents at the point of discharge from the Santos/São Vicente outfall for summer 2002 (**a**) surface, (**b**) half depth, and (**c**) bottom

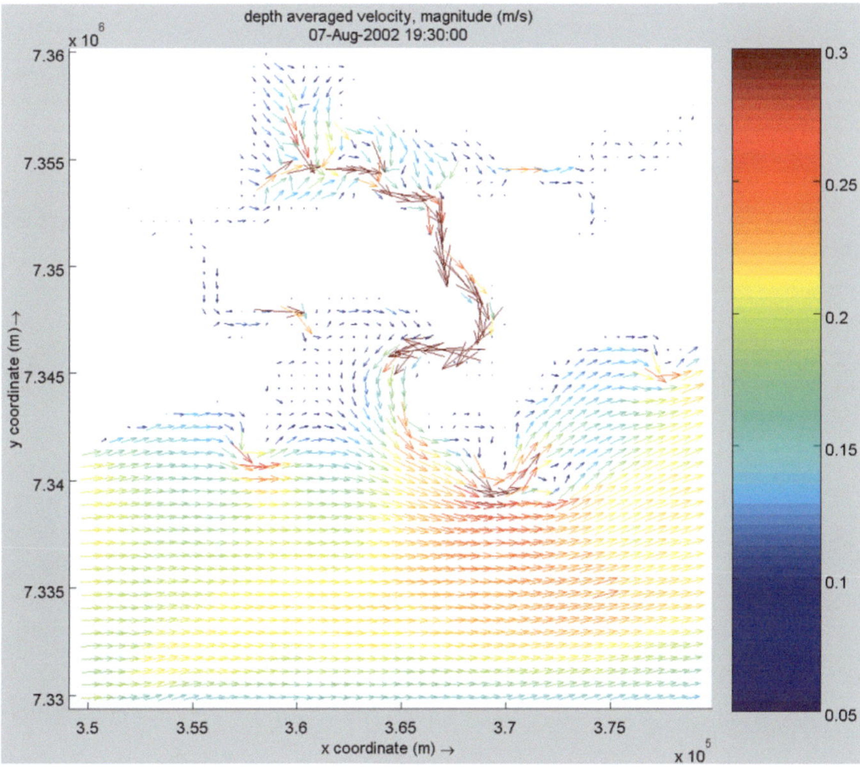

Fig. 22.20 Map of current results of Delft3D simulation for 08/07/2002 (syzygy – Ebb)

In addition, this study evaluated the magnitude of the currents at the discharge point of the Santos outfall for winter 2002 (Fig. 22.18). For comparison, a summary of the field measurements, taken in 2005 and 2006, for 13 consecutive hours at quadrature periods, in four different seasons, is presented in Table 22.3.

From the results of Delft3D simulations, the predicted maximum and minimum (depth average) values at the point of discharge from Santos outfall are

- (summer 2002) for the period 15 February to 8 March 2002, there was a maximum speed of 0.12 m/s and a minimum of 0.02 m/s (values less than 0.02 m/s are excluded, which correspond to 19 % of the total). The maximum speed at the surface is 0.16 m/s.
- (winter 2002) for the period of 21 July to 18 August 2002, there was a maximum speed of 0.10 m/s and a minimum of 0.02 m/s (values less than 0.02 m/s are excluded, which correspond to 13 % of the total). The maximum speed at the surface is 0.15 m/s.

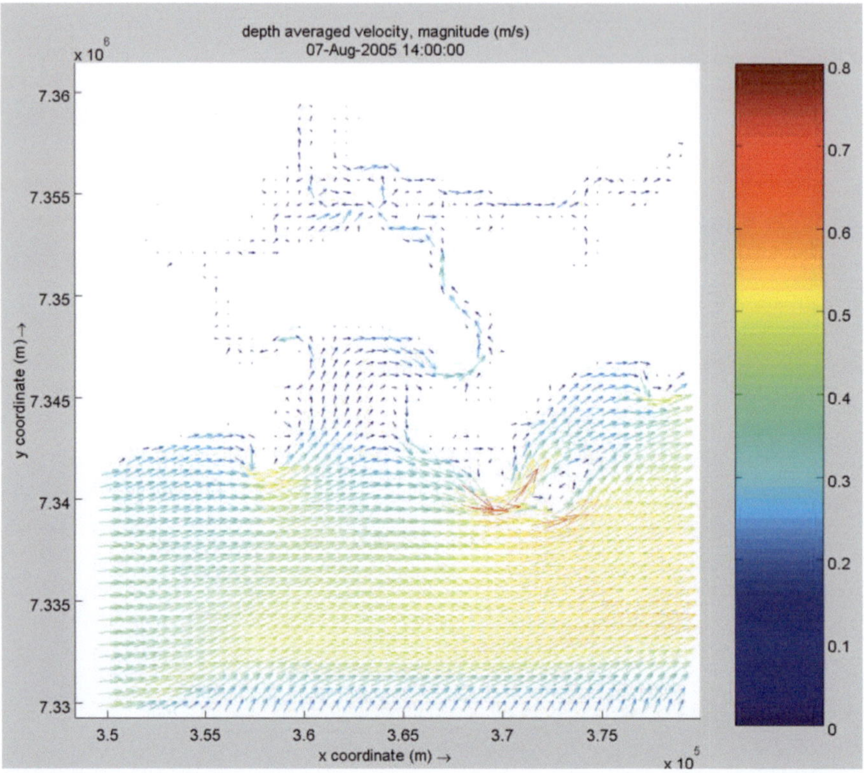

Fig. 22.21 Map of current results of Delft3D simulation for 08/07/2005 (syzygy – Flood)

- (winter 2005) for the period of 23 July to 15 August 2005, there was a maximum speed of 0.14 m/s and a minimum of 0.02 m/s (values less than 0.02 m/s are excluded, which correspond to 30 % of the total). The maximum speed at the surface is 0.16 m/s.

Comparing Delft3D simulated values and those in Table 22.3, it can be observed that magnitude of the currents at the discharge point of the Santos outfall is quite low, and the minimum values vary (depth averaged) from 0.02 and 0.16 m/s, and the maximum values (at the surface) vary from 0.15 and 0.36 m/s, and maximum values (depth averaged) vary from 0.14 and 0.27 m/s.

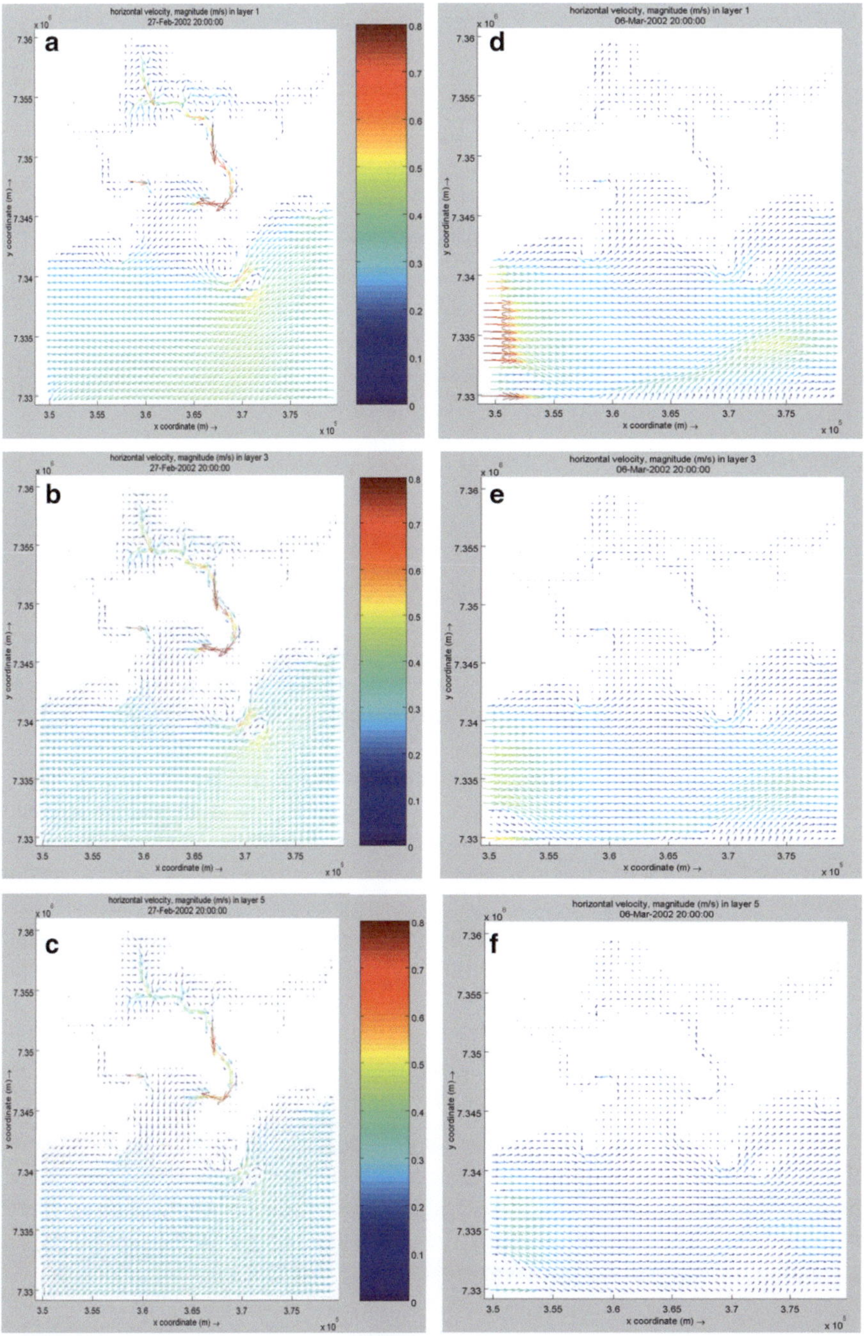

Fig. 22.22 Maps of currents results of Delft3D simulation for: 27-Feb-2002(syzygy) (**a**) surface, (**b**) half depth, and (**c**) bottom; and 6-Mar-2002 (quadrature) (**d**) surface, (**e**) half depth, and (**f**) bottom

Table 22.3 Field observations of maximum and minimum velocities at the Santos outfall

Velocity	13-Jul-2005 (13 h of measurement)	10-Aug-2005 (13 h of measurement)	01-Jun-2006 (13 h of measurement)	03-Jun-2006 (12 h of measurement)
Average	0.11 m/s	0.20 m/s	0.12 m/s	0.19 m/s
Maximum	0.20 m/s (surface, half depth, & bottom)	0.31 m/s (surface)	0.22 m/s (surface)	0.36 m/s (surface)
Minimum	0.04 m/s (bottom)	0.12 m/s (bottom)	0.07 m/s (half depth)	0.13 m/s (bottom)
Maximum (average depth)	0.14 m/s	0.25 m/s	0.17 m/s	0.27 m/s
Minimum (average depth)	0.08 m/s	0.16 m/s	0.09 m/s	0.14 m/s

22.4 Conclusions

The main conclusions are:

- the results of Delft3D simulations are in satisfactory agreement with the field measurement data, demonstrating that the approach was appropriate for the study region.
- the tide is the forcing mechanism that determines the hydrodynamics in Santos Bay. There are significant differences between ebb and flood periods and between syzygy and quadrature. Also, there are clear differences in the magnitude and direction of the surface currents and the currents in the other layers in depth.
- at the entrance of Santos Bay, near the imaginary line connecting Ponta de Itaipu and Ponta Grossa, there is a strong influence of coastal circulation which dominates the NE and SW direction, parallel to Praia Grande.
- at the discharge point of the Santos submarine outfall, minimum values of the current vary from 0.02 to 0.16 m/s and the maximum values (at the surface) from 0.15 to 0.36 m/s.

This work is based on the results and conclusions of Baptistelli (2008). Despite the Santos submarine outfall compliance with the discharge conditions of the current legislation, SABESP has been studying a proposal for improving their performance. The proposal assesses the increasing length of the outfall in an attempt to identify a discharge point outside the center bay.

List of Figures

References

Baptistelli SC (2003) Modelação numérica da circulação de correntes de maré e induzidas pelo vento aplicada a estudos de disposição oceânica de efluentes na Praia Grande – São Paulo. 168p. Dissertação (Mestrado). Escola Politécnica, Universidade de São Paulo. São Paulo, Brazil (in Portuguese)

Baptistelli SC (2008) Análise crítica da utilização de modelagem matemática na avaliação da dispersão de efluentes leves no litoral da Baixada Santista (Estado de São Paulo). Doctoral Thesis. University of São Paulo, São Paulo, Brazil (in Portuguese)

CODESP – Companhia DOCAS do Estado de São Paulo (2002) Pesquisa de Determinação de áreas de Descarte de Material de Dragagem na Zona Oceânica Exterior à Baía de Santos.

Relatório Final, Trabalho realizado pela Universidade Santa Cecília – UNISANTA, São Paulo, Brazil (in Portuguese)
Costa FP (2002) Solução Numérica de Equações Diferenciais. Mini curso. Universidade Estadual de Santa Cruz, Santa Catarina, Brazil (in Portuguese)
Cunha DLN, Rosman PCC, Ferreira AP, Monteiro TCN (2006) Hydrodynamics and water quality models applied to Sepetiba Bay. Continental Shelf Res 26:1940–1953
Delft3D-FLOW (2006) Simulation of multi-dimensional hydrodynamic flows and transport phenomena, including sediments – user manual. The Netherlands
Harari J (1989) Modelos numéricos aplicados a processos costeiros e estuarinos. Notas de aulas do curso de Pós-graduação em Oceanografia Física no Instituto Oceanográfico da Universidade de São Paulo, São Paulo, Brazil (in Portuguese)
Harari J, Camargo (1998) Modelagem numérica da região costeira de Santos (SP): Circulação de maré. Revista Brasileira Oceanografia 46(2):135–156 (in Portuguese)
Hidroconsult (1974) Investigações e Estudos Realizados Para o Projeto do Sistema de Disposição Oceânica dos Esgotos de Santos e São Vicente. São Paulo, Brazil (in Portuguese)
NOAA (2006) NCEP/NCAR Reanalysis Project (CDAS). http://www.cpc.ncep.noaa.gov/prod ucts/wesley/reanalysis.html. Accessed 2006
Roelvink D, Walstra DJ (2004) Keeping it simple by using complex models. In: Altinakar MS, Wang SSY, Holz KP, Kawahara M (eds). Advances in hydro-science and engineering, vol VI. The University of Mississippi
SABESP (2005) Estudo de Otimização do Sistema Difusor do Emissário Submarino de Santos/São Vicente.Relatório SABESP– Departamento de Gestão e Desenvolvimento Operacional da Baixada Santista – RSO.São Paulo, Brazil (in Portuguese)
SABESP (2006) Programa de Monitoramento Ambiental da área sob Influencia dos Emissários Submarinos de Esgotos de Santos/São Vicente e Praia Grande Subsistemas 1, 2 e 3.Relatório Final. São Paulo, Brazil (in Portuguese)

Chapter 23
Mathematical Model Study of the Effluent Disposal from a Desalination Plant in the Marine Environment at Tuticorin, India

D.R. Danish, B.V. Mudgal, G. Dhinesh, and M.V. Ramanamurthy

Abstract Consequent to rapid urbanization and industrialization, desalination of sea water is considered as a potential source of water along coastal areas. One of the challenges in putting up a desalination plant along coastal areas is the safe disposal of brine discharge and its dispersion in seawater, as it can have a profound impact on the marine environment. This study has been carried out to ensure better disposal and dispersion of the effluent from a proposed desalination plant at Tuticorin. Taking a point source as the outfall point, the brine solution is discharged at a rate of 1.58 m^3/s into the marine environment. This was done at a distance of 2–2.5 km from the shore at a depth of 7–9 m from the mean sea level. Since the depth is relatively small and initial vertical momentum is high, the discharge surfaces initially and then sinks in. The fate of the brine discharged from the outfall of the desalination plant is predicted using a commercial numerical model, MIKE-21, by considering the local bathymetry and assessing the hydrodynamic parameters such as waves, tides and currents of the area as well as wind and the other local environmental parameters included in the study.

Keywords Brine discharge • Computational model • Dispersion • Hydrodynamic parameters • MIKE-21 • Outfalls

---

D.R. Danish (✉)
Institute for Ocean Management, Department of Civil Engineering, Anna University, Chennai, Tamil Nadu, India
e-mail: civildanish@gmail.com

B.V. Mudgal
Centre for Water Resources, Department of Civil Engineering, Anna University, Chennai, Tamil Nadu, India
e-mail: bvmudgal@gmail.com

G. Dhinesh • M.V. Ramanamurthy
National Institute for Ocean Technology, Chennai, Tamil Nadu, India
e-mail: dhinesh@niot.res.in; mvr@niot.res.in

M. Baawain et al. (eds.), *Recent Progress in Desalination, Environmental and Marine Outfall Systems*, DOI 10.1007/978-3-319-19123-2_23

23.1 Introduction

Desalination is any of the several processes involved in removing dissolved minerals from sea water (Palomar and Losada 2011). Due to increase in population, rapid urbanization and industrialization, scarcity of water has increased and desalination of sea water is considered as a potential solution for the water problem along coastal areas (Fernandez-Torquemada et al. 2009). Recently, there have been many proposals for a desalination plant along the eastern coast of Tamil Nadu, especially in Chennai, Ramanathapuram and Tuticorin of 100 million litres per day (MLD) capacity.

One of the challenges in putting up a desalination plant along the coastal area is the brine discharge and dispersion in the sea water as it can have a profound impact on the marine environment (Safrai and Zask 2006). This study mainly concentrates on the Tuticorin sea water reverse osmosis (SWRO) plant of 100 MLD capacity. The Tuticorin coast in Tamil Nadu, south India, is well known for its major port. The SWRO by-product is at ambient temperature compared with those of other technologies but salinity might reach more than 100 % than that of sea water salinity (El-Gamal and Abdrabbo 2004).

Thus desalination of sea water leaves brine waste containing high salt concentration to be disposed of into the environment. For coastal desalination plants, there is no better alternative than to continuously discharge brine into the sea water through an outfall (Purnama et al. 2004). Over 90 % of large seawater desalination plants in operation dispose of concentrate through a new ocean outfall designed specifically for this purpose (WHO 2007).

Discharges of brine in to the coastal waters have to be adapted to be site specific and optimum. Most important in the brine dilution is the combination of two physical processes such as primary dilution in the near field and the natural dilution happening in the far field due to diffusion and mixing produced by the prevailing currents in the sea. Thus the spreading of an effluent released in an aquatic environment is governed by advection by large- scale water movement and its diffusion is caused by comparatively small-scale random and irregular movements without causing any net transport of water (Kumar et al. 2014).

Discharge outfalls should be located in a less sensitive coastal region with good transport and flushing characteristics to avoid accumulation and allow further mixing (Bleninger and Jirka 2010). The fate of the effluent discharge can be predicted using an explicit scheme for advection-diffusion modelling in two dimensions using MIKE-21 (Ekebjaerg and Justesen 1991).

23.1.1 Study Area

Tuticorin is located at 8°40′N-8°55′N latitude and 78°0′E-78°15′Elongitude. It is also known as Pearl City. It is one of the industrial cities in the southeastern part of

Table 23.1 Plant details for the proposed desalination plant

Plant details	Description
Plant capacity	100 MLD (4166.67 m^3/hr)
Plant intake	237 MLD (9875 m^3/hr)
Brine outfall	137 MLD (5708.33 m^3/hr)
Seawater concentration	35 PSU
Brine concentration	48 PSU (13 PSU in excess)
Intake pipe diameter	1.6 m
Outfall pipe diameter	1.4 m

India along the Bay of Bengal in Tamil Nadu with a major port. It has three major seasons, winter, summer and monsoon (SW monsoon and NE monsoon). The average minimum temperature is about 32 ° C and the average maximum temperature is about 39 ° C and the annual average rainfall is about 600–800 mm. Because of the high traffic of the port and other major industries around the port, the most suitable location of the brine outfall is in the southern part of the Tuticorin port behind the bay where the traffic is minimal and free from major industrial activity.

23.1.2 Plant Details

The proposed sea water reverse osmosis desalination plant at Tuticorin is of 100 MLD capacity, where the total water intake is about 237 MLD of seawater and the brine effluent discharged back into the sea is about 137 MLD. The salinity of the seawater is of 35 practical salinity units (PSU) and the discharged brine is of 48 PSU which is about 13 PSU more than the feed seawater. The discharge point of the outfall is taken about 2–2.5 km away from the shore into the sea for better dilution and transportation of the effluent where the depth of the water is about 7–9 m. The outfall point is located at 8.68°N latitude and 78.18°E longitude. The outfall pipe has a diameter of 1.4 m and the intake pipe's diameter is about 1.6 m. The brine effluent is discharged through a diffuser at the rate of 1.58 m^3/s with the flow velocity of 1 m/s at an angle of inclination of about 45° for a better initial near-field physical dispersion as well as a far-field natural dilution (Kumar et al. 2014). The details are given in Table 23.1.

23.2 Material and Methods

The advection-dispersion module solves the advection-dispersion equation, which is a mass conservative equation, for the dissolved or suspended substances in two dimensions. The module is adopted for MIKE-21 Flow Model (DHI Water and Environment 2007):

$$\frac{\partial}{\partial t}(hc) + \frac{\partial}{\partial x}(uhc) + \frac{\partial}{\partial y}(vhc) = \frac{\partial}{\partial x}\left(h.D_x.\frac{\partial c}{\partial x}\right) + \frac{\partial}{\partial y}\left(h.D_y.\frac{\partial c}{\partial y}\right) - F.h.c + S$$

where c is the compound concentration, u and v (m/s) the horizontal velocity components in the x- and y-directions respectively, h (m) the water depth, D_x and D_y (m^2/s) the dispersion coefficient in the x- and y-direction respectively, $F(s^{-1})$ the linear decay coefficient, S may be set as a source/sink discharge Qs ($m^3/s/m^2$) or as a concentration of compound in the source/sink discharge (c_s-c).

The linear decay coefficient $F(s^{-1})$ is zero, because there are no biological suspended particles. The source term S is the brine discharge at 1.58 m^3/s. The densimetric Froude number works out to be 23.43, which is much greater than 1 and accounts for the negatively buoyant flow.

23.3 Results and Discussion

Tides experienced in the Tuticorin coast are of a semidiurnal type with two high tides and two low tides occurring every day. The prediction for the Tuticorin coast for 2012 showed that the average spring tide would be about 0.7 m and the average neap tide about 0.16 m (Anonymous 2011) (Figs. 23.1 and 23.2).

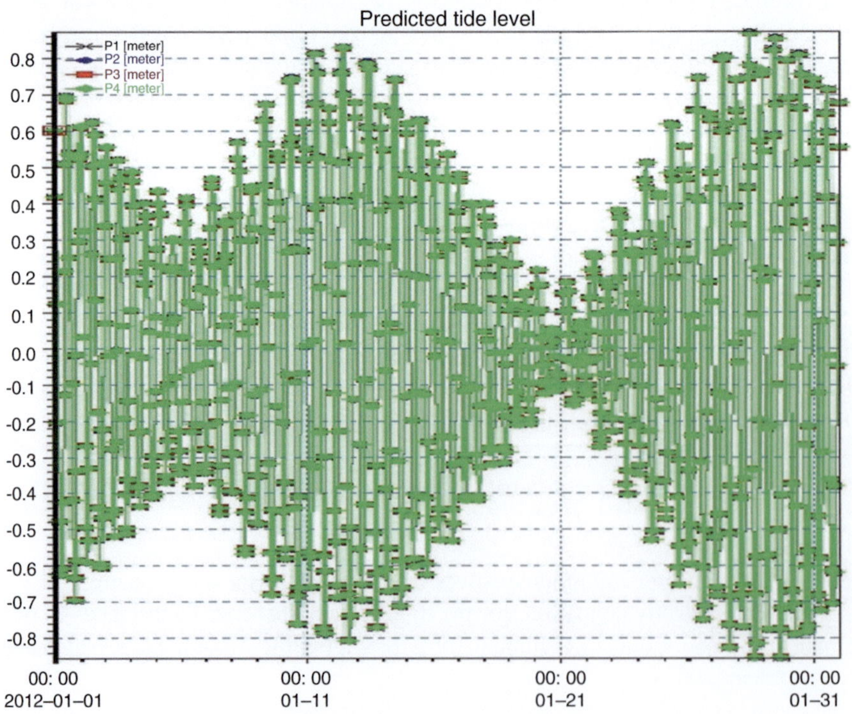

Fig. 23.1 Predicted tide (sample data) for India for the month of January 2012

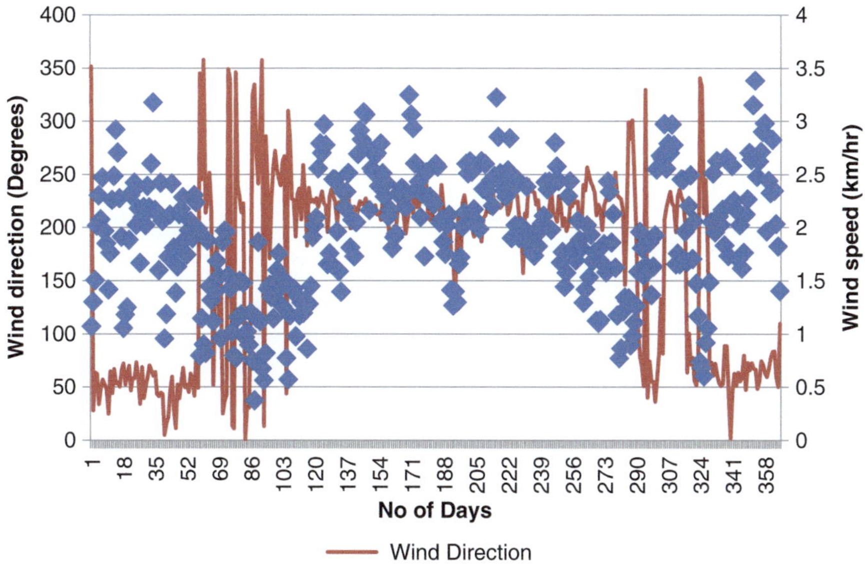

Fig. 23.2 Average speed and wind direction for Tuticorin coast for the year 2012

The daily average for the entire year's wind data is derived from the Centre for Satellite Application and Research (STAR), National Environmental Satellite Data and Information Service (NESDIS) of ASCAT satellite NOAA of 25 m resolution which is predominant during the southwest monsoon from June till September; the average wind speed and average wind direction is about 2.23 km/h and 214° respectively and the average wind speed and direction vary to the seasons (Figs. 23.3, 23.4, and 23.5, Table 23.2).

The current can be predicted knowing the bathymetry and weather. The currents are wind- dominant and the effect of the tide on the currents is negligible. This is incorporated in the MIKE-21 hydrodynamic module during the simulation run. The current pattern varies accordingly with the bathymetry and the wind speed and direction during the different seasons.

The bathymetry is derived from the interpolation of nodes and arcs acquired from the C-Map MIKE module according to the required domain where the grid spacing is about 20 m and there are 10 grids along both the *x* and *y* directions. Tuticorin has a rocky and shallow bathymetry (Figs. 23.6, 23.7, 23.8, 23.9, 23.10, 23.11, and 23.12).

The simulation model which was used for the prediction of the fate of the brine effluent dispersion for the proposed SWRO desalination plant at Tuticorin is the MIKE-21 flow model using the advection-dispersion equation. The model has been

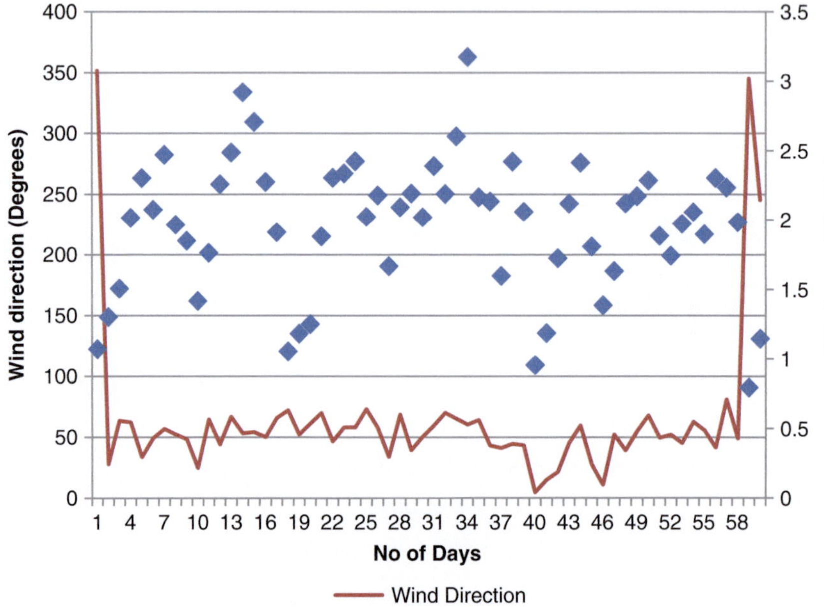

Fig. 23.3 Average wind speed and direction during winter 2012

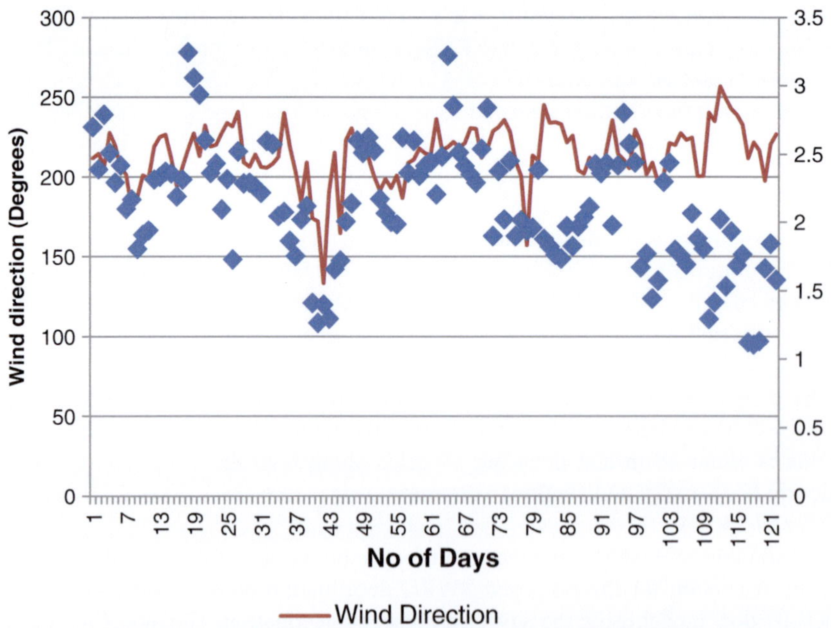

Fig. 23.4 Average wind speed and direction during southwest monsoon 2012

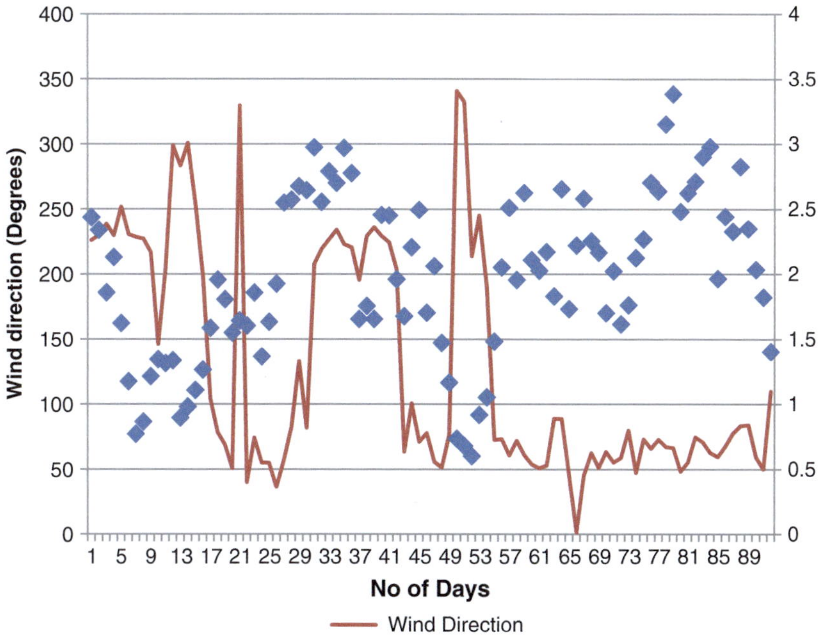

Fig. 23.5 Average wind speed and direction during northeast monsoon 2012

Table 23.2 Average seasonal wind speed and direction for Tuticorin coast for 2012

Season	Month	Wind speed (km/h)	(Degrees)
Winter	Jan–Feb	1.95	64
Summer	Mar–May	1.63	210
South West Monsoon	Jun–Sep	2.23	214
North East Monsoon	Oct–Dec	1.98	130

simulated by using the bathymetry, predicted tides, collected wind data and the predicted current, in addition to variations in temperature and salinity. The temperature is the same as the ambient temperature in SWRO technology. Thus the salinity variation is predominantly taken into consideration while running the simulation. The ambient salinity level is about 35 PSU, which is equivalent to a density of about 1035 kg/m^3, and the brine concentration is about 48 PSU, which is of density 1048 kg/m^3. The point of the source where the outfall is located is about 8.68°N latitude and 78.18°E longitude.

The simulation model has been carried out for winter, southwest (SW) and northeast (NE) monsoon seasons using the average constant wind speed and direction for the year 2012 to predict the fate of the brine effluent dispersion. The model clearly shows that the dispersion of the effluent and distribution of the different salinity concentrations to be about 0.01–0.15 PSU.

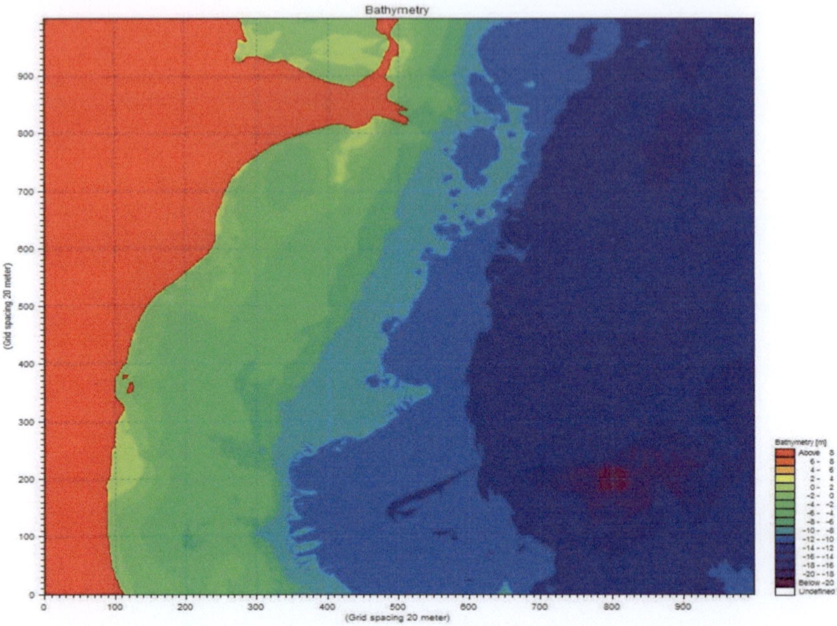

Fig. 23.6 Typical bathymetry of the Tuticorin coast

23.4 Conclusion

The simulation model clearly shows the current and its corresponding dispersion for the winter, southwest and northeast monsoon seasons. There is no brine effluent that has moved towards the shore or accumulated. Moreover, the point of outfall is about 2–2.5 km into the sea at about a depth of 7–9 m. Since the initial momentum of the discharge from the outlet is high and the depth of the outlet is small, the brine surfaces at the point of discharge. Even so, there is no significant effect on the marine environment. The difference in salinity of the brine which dispersed into the seawater is about 0.01–0.15 PSU from the sea water which is not much different from the ambient condition and is acceptable. Thus, all these parameters show that the model predicted for the proposed desalination plant from the point of outfall considered for the brine effluents is optimum and there is not much effect on the surrounding marine environment.

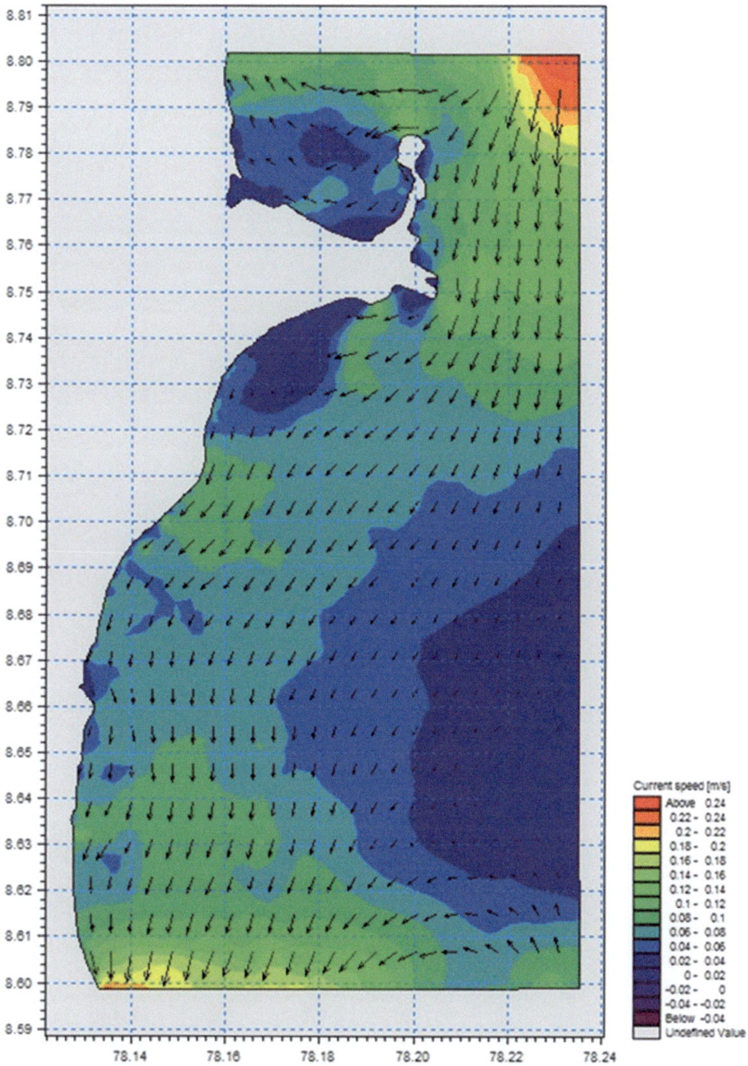

Fig. 23.7 Current during winter

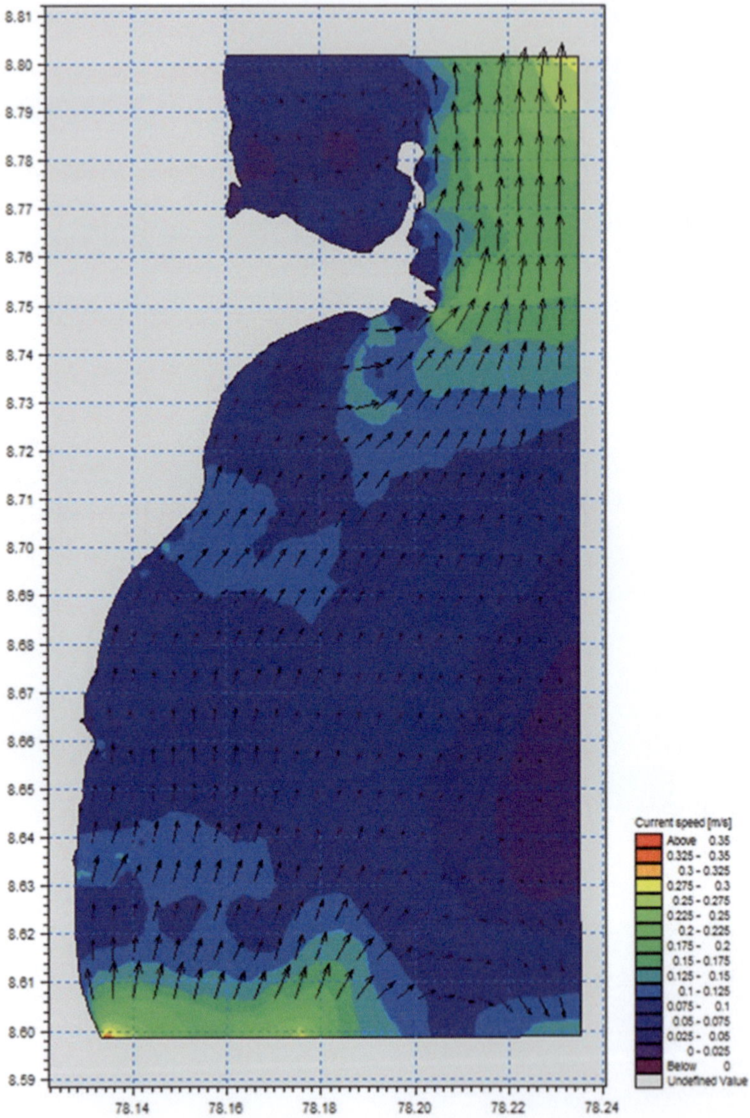

Fig. 23.8 Current during southwest monsoon

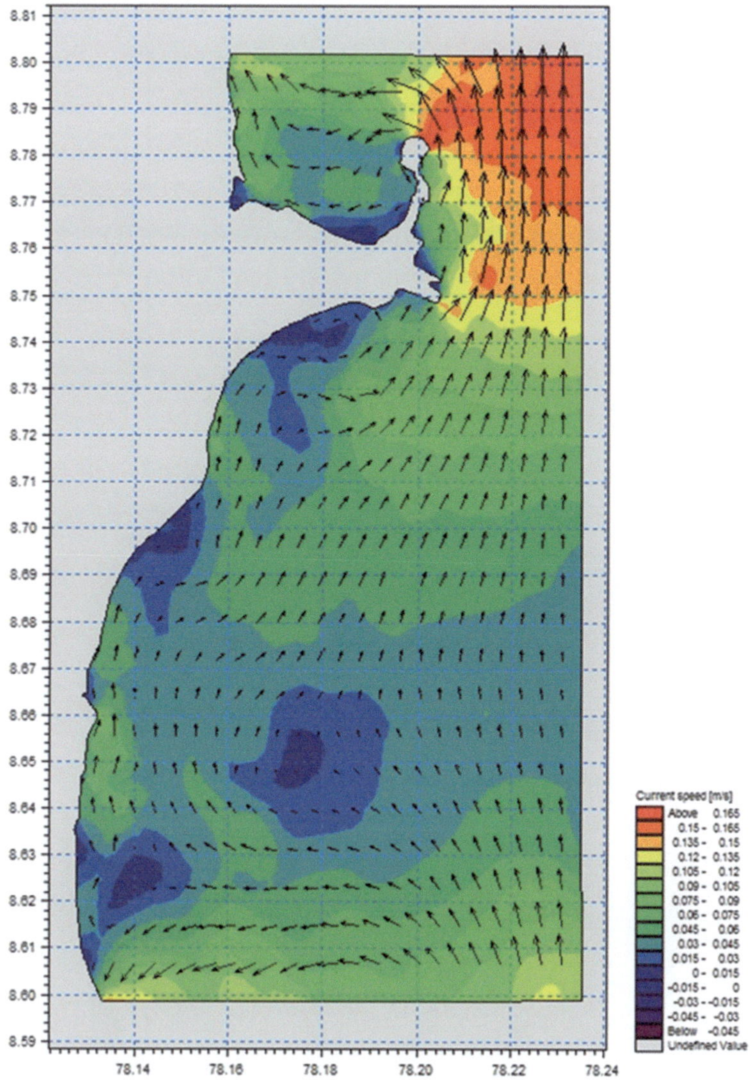

Fig. 23.9 Current during northeast monsoon for the Tuticorin coast

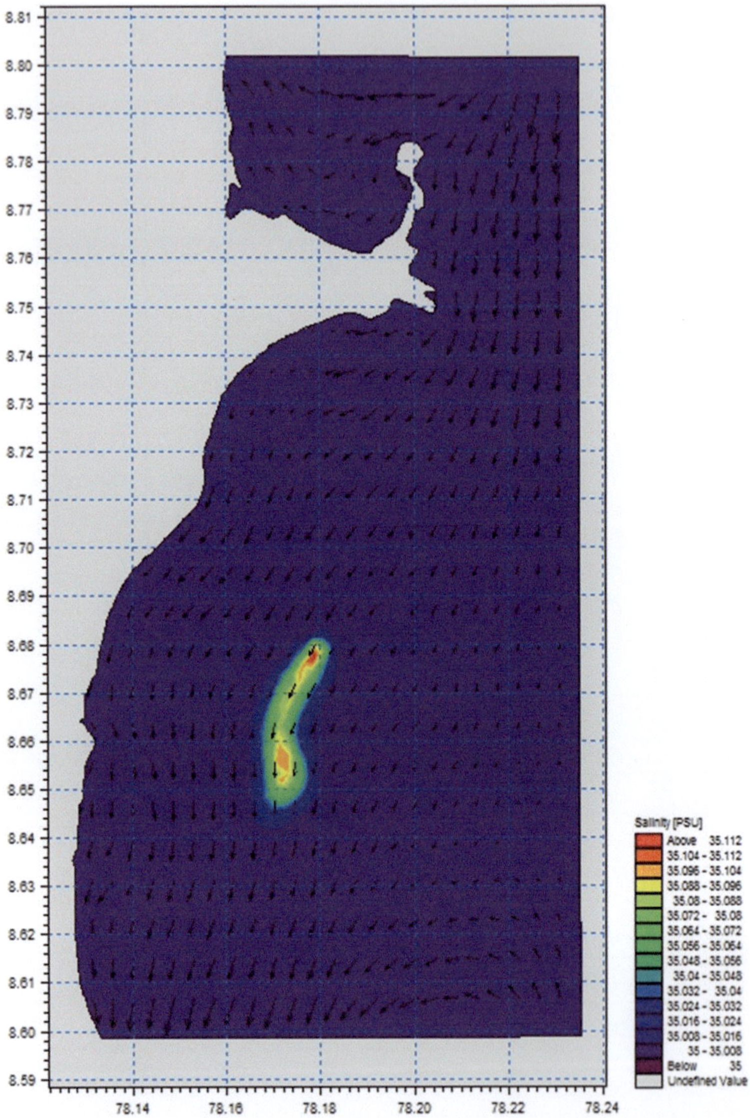

Fig. 23.10 Dispersion flow model during winter

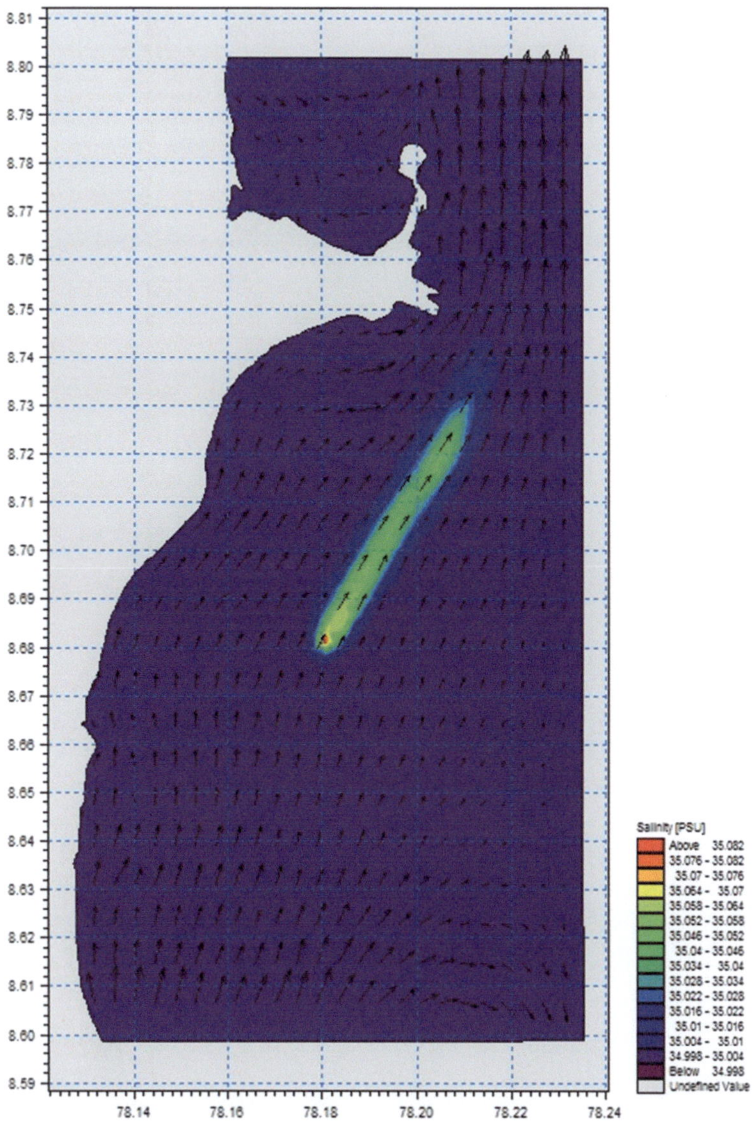

Fig. 23.11 Dispersion flow model during the southwest monsoon

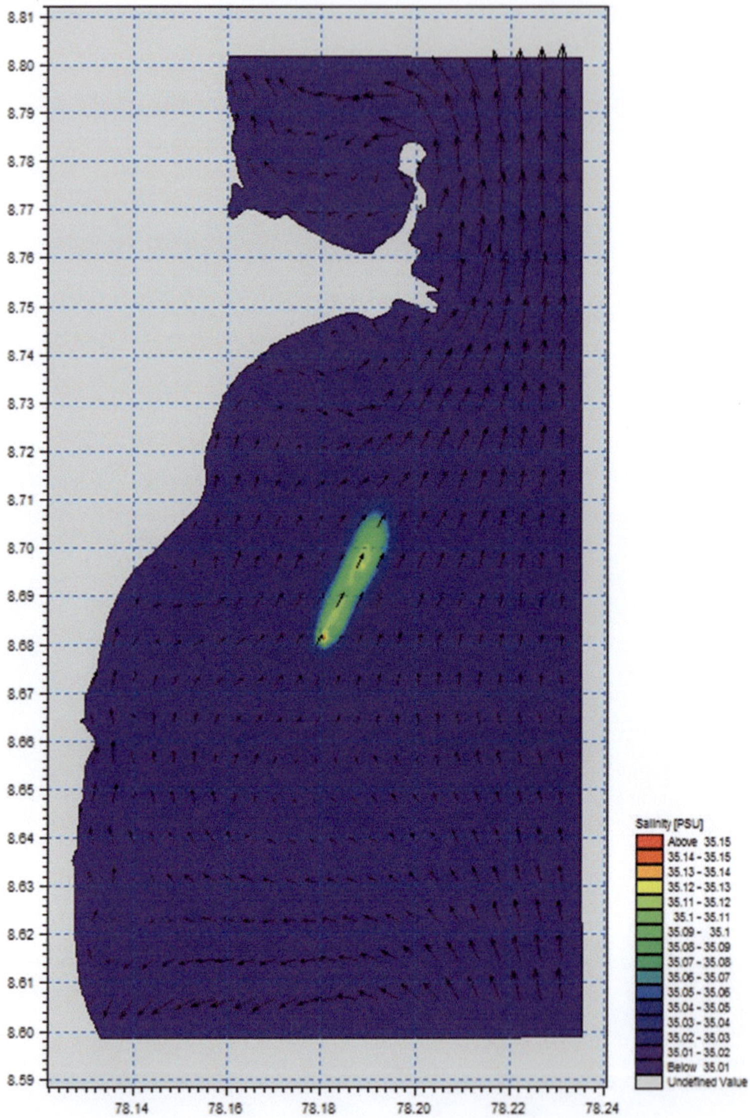

Fig. 23.12 Dispersion flow model during the north east monsoon

List of Figures

References

Anonymous (2011) The Indian Tide Tables – Part I, Office of the Geodetic and Research Branch. Surveyor General of India, Dehradun, 238 pp

Bleninger T, Jirka GH (2010) Environmental planning, prediction and management of brine discharges from desalination plants. Final report of MEDRC Project 07-AS-003, Middle East Desalination Research Centre, Muscat

Centre for Satellite Application and Research (STAR) (2012) National Environmental Satellite Data and Information Service (NESDIS), http://manati.star.nesdis.noaa.gov/datasets/ASCATData.php/

DHI Water and Environment (2007) MIKE-21 Flow Model, Advection-Dispersion Module, Scientific Documentation, DHI Software 2007

Ekebjaerg L, Justesen P (1991) An explicit scheme for advection-diffusion modelling in two dimensions. Comput Methods Appl Mech Eng 88:287–297

El-Gamal HF, Abdrabbo MF (2004) Environmental impact and cost analysis of desalination systems. In: Proceedings of the 8th international water technology conference (IWTC 2004), Alexandria, 26–28 March 2004

Fernandez-Torquemada Y, Gónzalez-Correa JM, Loya A, Ferrero LM, Díaz-Valdés M, Sánchez-Lizaso LJ (2009) Dispersion of brine discharges from sea water reverse osmosis desalination plants. Desalin Water Treat 5(1–3):137–145

Kumar DPK, Kumar NPK, Muraleedharanan KR (2014) Identification of optimum outfall location for desalination plant in the coastal waters off Tuticorin, India. Int J Environ Res 8(1):69–78

Palomar P, Losada IJ (2011) Impacts of brine discharge on the marine environment: modelling as a predictive tool. In: Schorr M (ed) Desalination, trends and technologies. InTech, Rijeka, pp 279–310

Purnama A, Al-Barwani HH, Al-Rawahi MS, Al-Harthi TM (2004) Dispersion models for brine discharges from desalination plants of Oman. Kuwait J Sci Eng 31(1):1–13

Safrai J, Zask A (2006) Environmental regulation for discharging desalination brine to the sea and its possible impacts. Ministry of Environmental Protection, Jerusalem

WHO (2007) Desalination for safe water supply, guidance for the health and environmental aspects applicable to desalination. Public Health and the Environment, World Health Organization, Geneva

MIX
Papier aus verantwortungsvollen Quellen
Paper from responsible sources
FSC® C105338

If you have any concerns about our products,
you can contact us on
ProductSafety@springernature.com

In case Publisher is established outside the EU,
the EU authorized representative is:
Springer Nature Customer Service Center GmbH
Europaplatz 3, 69115 Heidelberg, Germany

Printed by Libri Plureos GmbH
in Hamburg, Germany